Olov Sterner

**Chemistry, Health,
and Environment**

Related Titles

Smart, R. C., Hodgson, E. (eds.)

Molecular and Biochemical Toxicology

2008

ISBN: 978-0-470-10211-4

Dean, J. R.

Bioavailability, Bioaccessibility and Mobility of Environmental Contaminants

2007

ISBN: 978-0-470-02578-9

Frumkin, H. (ed.)

Environmental Health

From Global to Local

2005

ISBN: 978-0-7879-7383-4

Angerer, J., Greim, H. (eds.)

Essential Biomonitoring Methods

from The MAK-Collection for Occupational Health and Safety

2006

ISBN: 978-3-527-31478-2

Külpmann, W. R. (ed.)

Clinical Toxicological Analysis

Procedures, Results, Interpretation

2009

ISBN: 978-3-527-31890-2

Haslberger, A., Greßler, S. (eds.)

Epigenetics and Human Health

Linking Hereditary, Environmental and Nutritional Aspects

2010

ISBN: 978-3-527-32427-9

Knasmüller, S., DeMarini, D. M., Johnson, I., Gerhäuser, C. (eds.)

Chemoprevention of Cancer and DNA Damage by Dietary Factors

2009

ISBN: 978-3-527-32058-5

Olov Sterner

Chemistry, Health, and Environment

Second, Completely Revised Edition

The Author

Prof. Dr. Olov Sterner
University of Lund
Department of Organic Chemistry
POB 124
221 00 Lund
Sweden

Library of Congress Card No.: applied for

British Library Cataloguing-in-Publication Data
A catalogue record for this book is available from the British Library.

Bibliographic information published by the Deutsche Nationalbibliothek
The Deutsche Nationalbibliothek lists this publication in the Deutsche Nationalbibliografie; detailed bibliographic data are available on the Internet at http://dnb.d-nb.de.

© 2010 WILEY-VCH Verlag GmbH & Co. KGaA, Weinheim, Germany
Wiley-Blackwell is an imprint of John Wiley & Sons, formed by the merger of Wiley's global Scientific, Technical, and Medical business with Blackwell Publishing.

Cover Design   Grafik-Design Schulz, Fußgönnhein
Typesetter   Toppan Best-set Premedia Limited, Hong Kong
Printing and Binding   T.J. International Ltd. Padstow

Printed in Great Britain
Printed on acid-free paper

ISBN: 978-3-527-32582-5

Contents

Chemistry, Health, and Environment. Olov Sterner
© 2010 WILEY-VCH Verlag GmbH & Co. KGaA, Weinheim
ISBN: 978-3-527-32582-5

Preface

Modern society makes use of a very large number of chemicals, and although we enjoy and profit from their various benefits, such as the development of safe and effective medicines, most of us worry about the possibility that all these chemicals are somehow hazardous to our health and to the environment. Our worries are well founded. During the last century, many incidents in which the improper use of chemicals has led to severe damage to both people and the environment have occurred. Today, we are well aware that most tumors that afflict humans are caused by chemicals and that the release into the atmosphere of enormous amounts of carbon dioxide and other greenhouse gases produced as a result of human activities poses a serious threat to the climate. The hazards associated with a chemical do not arise by pure chance, but depend on a number of properties of that chemical, for example, its solubility and reactivity. These determine how it will interact with the biochemical components of an organism, such as a human, and therefore how toxic it is. Although it is difficult to predict chemical hazards theoretically, surprisingly many of the properties that determine the behavior of a chemical can be deduced from a simple visual inspection of the chemical structure. In essence, this is what this book is about. By identifying substructures or functional groups that we know are associated with toxicity or environmental effects we can look out for potential hazards of a chemical before it appears as the 'star' of a newspaper headline. We should at the same time be strongly aware of the ways that chemicals are modified by metabolic conversion in organisms and by chemical transformations in the environment, because the products of these changes will, of course, have significantly different chemical properties and therefore present significantly differently hazards.

It is our belief that chemists, chemistry students, and in fact anyone who works with chemicals and who may be responsible for the health and safety of others would benefit from being aware of the links between chemical structure and toxicity, and the intention of the book is to raise this awareness. Although the basis for the discussion is chemical, a further intention is that those with relatively limited knowledge of chemistry should also find the text informative. For that reason the second edition has been extended by the addition of a new chapter that describes the effects of well-known compounds that are typically used as fuels, pesticides, food additives, and pharmaceuticals.

Chemistry, Health, and Environment. Olov Sterner
© 2010 WILEY-VCH Verlag GmbH & Co. KGaA, Weinheim
ISBN: 978-3-527-32582-5

As the author, I sincerely hope that the readers of this book will find it interesting and inspiring, and if anyone is able to make use of the knowledge presented here to prevent a chemical accident I will be a happy man!

Lund, February 2010   *Olov Sterner*

1
Chemicals and Society

Life on planet earth can be scientifically described at a number of different levels. A lake or a forest can be defined as an ecosystem which consists of many different species and many individuals of each species. Some organisms, for example, mammals, are composed of various organ systems, such as the respiratory system, which is made up of organs composed of tissues and cells. A cell is normally considered to be the smallest unit of life, but cells in turn contain various components which take care of important cell functions such as the generation of energy and the construction of proteins. The composition of such cell organelles may appear complex, but today we know in principle how they are made up of biomolecules such as amphipathic compounds, proteins, and nucleic acids. The various processes that take place in a cell can in most cases be described in detail on a molecular level, and chemical structures and chemical reactions are ultimately responsible for all cellular functions. It has been calculated that the number of chemical reactions that keep a relatively simple unicellular organism like a bacterium alive is approximately a few thousand. The cell of a mammal is of course much larger and more complex, and is still not understood in all details, but it is reasonable to assume that our cells owe their status as living things to a certain number (more than a few thousands but still finite) of chemical reactions. Such chemical reactions comprising life can be affected by many things, and this book will look more closely at how various chemicals can disturb them and what consequences that may lead to.

The first chapter will simply introduce the reader to the background, some general aspects of hazardous chemicals, and how society responds to them. The key conclusion of Chapter 1 is that the lack of knowledge concerning hazardous chemicals is alarming, but that fundamental understanding of the relationships between chemical hazards and chemical structure/properties will help anyone handling chemicals to protect himself or herself as well as the environment.

1.1
Basic Problems

There are a number of important basic problems that must be considered when the effects of chemicals on health and the environment are discussed. Some are

Chemistry, Health, and Environment. Olov Sterner
© 2010 WILEY-VCH Verlag GmbH & Co. KGaA, Weinheim
ISBN: 978-3-527-32582-5

BEWARE OF DIHYDRO MONOXIDE (DHMO)

Recent reports have drawn attention to a chemical that should be handled with the outmost care. The name of the chemical is dihydromonoxide, abbreviated DHMO, a deceptively tasteless and odorless chemical that yearly kills thousands of men, women, and especially children. DHMO is used in enormous quantities by the chemical industry, for instance, in nuclear power plants, and its use as an additive during the preparation of so. called junk food has been demonstrated. The presence of high concentrations of DHMO in tumors has recently been reported. In addition, DHMO is a main component of acid rain, and the large amounts of DHMO in our atmosphere make a significant contribution to the greenhouse effect. Military sources have revealed that thousands of tonnes of DHMO are distributed through special underground tube systems to, among others, secret weapon research plants.

Figure 1.1 A newspaper article that so far has not appeared.

self-evident, for example, the fact that the number of people on earth is increasing rapidly and that more and more chemicals are used for various purposes. Others are more difficult to define. Consider, for example, the fictitious newspaper article shown in Figure 1.1.

At first glance one may be concerned by this information, but a chemist would rapidly realize that the chemical DHMO is nothing but ordinary water. However, there are no lies in the article; everything said is in principle true. Thousands of people are killed by water yearly in drowning accidents, it is certainly used in large amounts by nuclear power plants, it is an ingredient in most 'junk foods', and it is a major constituent in tumor cells just as in any other cell. Rain is water and acid rain is mostly water, gaseous water in the atmosphere will reflect heat radiation from the earth just as the more famous greenhouse gas carbon dioxide, and tap water is distributed to most industries, including secret weapon research plants (this fact could have been revealed by any source, including a military source).

However, whereas a chemist reading this article will quickly see the joke and understand what the name DHMO stands for and what properties this chemical has, a person with no background in chemistry may be caused severe anxiety. If instead of water we choose a less well-known chemical that even chemists are unfamiliar with we may imagine a situation where a chemical is described in different and completely contradictory ways, for example, as a relatively nonhazardous chemical that can be handled safely (by a manufacturer or a person that has worked with the chemical for a long period) or as a dangerous and hazardous chemical with which all contact should be avoided (by someone suffering from chemophobia). Few are able to assess information about chemicals critically and react to it in a completely rational way.

The example in Figure 1.1 illustrates several points:

1) that the hazards of a chemical can be described, formally correctly, in several different ways,
2) that chemical hazards cannot be defined in absolute terms, and
3) that different hazards cannot be compared.

People perceive hazards in very different ways. Some have chosen to be smokers and/or to consume alcoholic beverages in spite of the fact that they are well aware that such habits will expose them to chemicals that in the long run will increase the risk of acquiring lung cancer and/or damaging the liver. The reason may be that they underestimate the risk, or they think the benefits of smoking and drinking outweigh the risk of damage. While such decisions perhaps can be left to the individual, others are more complicated and difficult, affect many people, and should consequently be based on very solid knowledge. For example: Should we produce electricity by nuclear power plants or by burning fossil fuels? We know that accidents in nuclear power plants will leak radioactivity to the environment, and that is certainly a severe hazard, but on the other hand we strongly suspect that the carbon dioxide added to the atmosphere by burning fossil fuels will increase global warming. Which hazard is more dangerous? Ask two people and you are likely to get two answers. There is simply not a straight yes or no answer to a question like: Is chemical X dangerous?

We will get back to this question several times in this book, simply because it is so fundamental. Not only do different people perceive hazards differently, but commercial enterprises (manufacturing and selling chemicals) and government authorities (regulating their use and depending on the taxes they generate) will have their views on chemical hazards that are not necessarily the same as yours or mine. One should also be aware that in some situations there may be pressures, for example economical or political, to twist the truth about chemical hazards in order to gain some kind of advantage.

By definition, a chemical hazard is the result of the intrinsic properties of a chemical or a situation involving chemicals that in particular circumstances could harm humans and/or the environment and/or damage property. Highly toxic compounds may be considerably less hazardous than a relatively nontoxic chemical, depending on the conditions under which they are used. A risk differs slightly from a hazard in that it also considers the probability that a hazardous chemical will cause the harm or damage that is has the potential to do. However, the two terms are often used as alternatives.

1.2
Definition of the Sciences Involved

The issues that will be discussed in this book are interdisciplinary, although they will mainly be described from a chemical viewpoint, and we may note that the following major scientific disciplines are involved:

Organic chemistry describes the properties and reactions of organic compounds, which in principle are all compounds that contain carbon, while *inorganic chemistry* deals with chemicals that do not contain carbon (exceptions are, for example, metal carbonates, carbon dioxide, and carbon monoxide, which are considered to be inorganic). As well as carbon, organic compounds can contain virtually any other element, but the vast majority of them are only composed of carbon, hydrogen, oxygen, and nitrogen. Almost all the chemicals that take part in the chemical processes that keep organisms alive are organic, and the study of such processes is carried out within the discipline *biochemistry*. Most compounds that are known to have toxic effects are organic (most of the toxic inorganic compounds considered in this book contain toxic metals), and it is obvious that the carbon atoms of organic compounds play a central role from the point of view of their structure and properties. As an atom with four valence electrons, the carbon atom has a strong tendency to make chemical bonds to four other atoms in order to achieve the stable electronic configuration of the noble gases. Carbon readily makes chemical bonds to a range of other atoms, including carbon itself, and compounds containing only a few carbon atoms can form a large number of different molecules. Figure 1.2 indicates how the structures of organic compounds composed of a certain number of atoms can be varied almost infinitely, simply by changing the positions of the bonds between the atoms. Even without considering the stereochemistry of the substituents on the rings and the double bonds, there are 44 compounds in Figure 1.2 all having the elemental composition $C_5H_{10}O$ and all having different chemical properties.

The carbon atoms will normally provide the backbone of an organic molecule, while the heteroatoms will constitute functional groups that give the compound its characteristic properties (discussed in Chapters 2 and 3). Adding to the usefulness of organic compounds, the energy of carbon–carbon and carbon–hydrogen bonds is high, which makes organic compounds suitable as energy sources for, for example, the biochemical reactions (discussed in Chapters 3, 4, and 7).

Although we will not get involved in any intricate biological problems, we will discuss some aspects of *toxicology* and *ecotoxicology*. Toxicology in general deals with the study of poisons and their effects on single organisms, especially on humans, while ecotoxicology is concerned with the effects on ecosystems. The difference should be noted; while an effect of one chemical on one or more organisms may be dramatic (e.g., death) the effect on the ecosystem could well be negligible, and vice versa. The term *environmental toxicology* usually indicates that the interest is focused on what effects chemicals have on the environment and how these effects affect humans, for example, the contamination by pesticides of species used by humans as food. Environmental toxicologists are not primarily

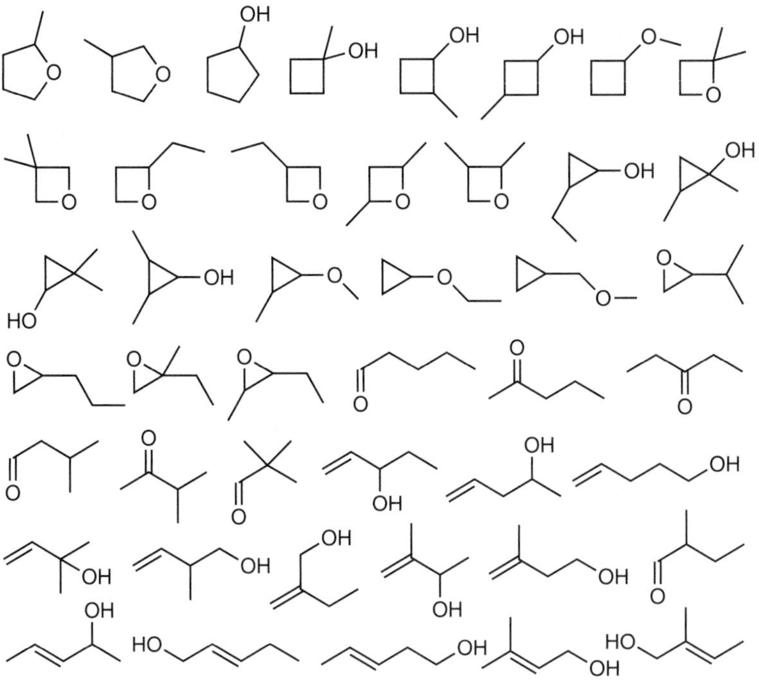

Figure 1.2 Examples of organic compounds having the composition $C_5H_{10}O$.

interested in the effects on ecosystems. The main focus in this text will be on the chemical properties that are associated with toxic effects (directly or indirectly) on humans, for which the term *chemical toxicology* may be the most appropriate.

1.3
Trends and Developments over the Years

Anyone who has been working for a long time where chemicals are handled will have noticed that chemicals are treated differently today compared with, say, 10 or 20 years ago. Attitudes have changed, procedures have been improved, and regulations are more strict. Some decades ago, chemicals were handled in ways that could lead to health problems, but this is no longer considered acceptable. While chemical waste in those days was simply buried or discarded into the nearest river, it is today taken care of and destroyed by means of specially developed processes. Also, new bioactive chemicals were invented that, for example, could be used to treat illnesses and control pests, and that initially were considered to be miraculously suitable for their task but later turned out to be hazardous. Examples are the insecticide DDT (DichloroDiphenylTrichloroethane) and the biocide 2,4-dinitrophenol, which was later used in slimming cures (see Figure 1.3).

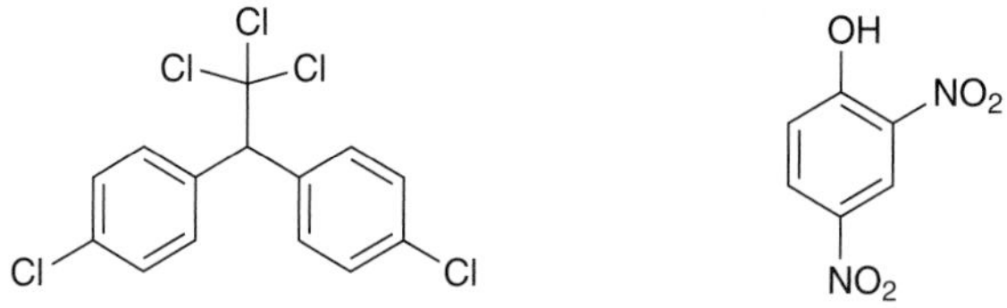

Figure 1.3 The structures of DDT and 2,4-dinitrophenol.

DDT, an old compound that was rediscovered in the 1930s when it was found to be a very efficient insecticide, was used extensively during World War II and the following decade. It was, in those days, important for the successful control of illnesses (typhus and malaria in particular) that are spread by insects, and saved the lives of a great number of people. The effects of DDT were truly remarkable, as the incidence of malaria in a small country like Sri Lanka shows. Here, 2 800 000 cases of malaria were reported in 1946, before the days of DDT, but only 110 cases in 1961 after DDT had been used for several years, and then again 2 500 000 in 1969 after DDT had been banned for 4 years. However, something that was not considered was that DDT is an extremely lipophilic and stable compound which is not degraded in our environment but is instead efficiently extracted by organisms. Some species (e.g., birds of prey) at the top of the food chain eventually accumulated so much DDT that their reproduction was threatened, and had the use of DDT continued on the same scale other species including man would have been in danger. The 'magic bullet' against insects became an environmental nightmare, and we shall return to DDT and similar compounds later in the book.

2,4-Dinitrophenol, an old pesticide, was launched in the 1930s as an efficient slimming agent for people who wanted to lose weight. It is indeed efficient, blocking the production of energy which is normally associated with the degradation of nutrients and stimulating the body to consume more nutrient (e.g., fat) than it needs. The only problem was that a slight overdosing could be fatal, and many accidents took place. It was therefore eventually banned as a slimming agent, and in the 1970s it was even considered to be too toxic even for use as a pesticide and banned for this purpose in many countries.

Thus, knowledge about the hazards associated with the use of chemicals has increased over the years, often in what we may call 'the hard way' – by making mistakes and learning from them. This has led to extensive regulation of the use of chemicals, and is the subject of the next section. The purpose of such regulations can be said to be to protect man and the environment from the hazards of chemicals. Today we may regard the misjudgments made in the 20th century as somewhat foolish and due to lack of knowledge, but it would be very stupid indeed to consider the level of knowledge we have today to be complete. Nor should we count on legislation to protect us from chemical hazards completely, even if everyone followed the rules. Instead, we may be sure that 20–30 years from now experts will look back and comment on the senseless use of chemicals X and Y and how this led to serious problems that could easily have been avoided 'if only they had known what we know now'. And they in turn will be the victims of the same hindsight after a further 20–30 years ...

The conclusion, which is valid most of the time, is: Things have improved, but they are not perfect.

1.4
Legislation

The legislation that regulates the use of chemicals aims to avoid all avoidable risks to both man and environment associated with chemicals, just as traffic rules are intended to protect people from traffic accidents. However, as we are all strongly aware, society as we know it will not function without chemicals or without traffic, and even if the intentions of the legislators are the best it is simply not possible to reduce the risks to zero, whether they are due to traffic or chemicals. So, at the same time as the legislation protects against chemical hazards it must also provide the means to use economically important chemicals, some of which are known to be carcinogenic, for example, but in spite of their toxicity are allowed to be used in workplaces. Nevertheless, modern legislation has improved the working conditions in industry immensely, and has also managed to decrease the pollution of the environment considerably. Any activity involving the handling of chemicals, for example, a business, will be regulated by a number of laws, rules, and regulations, and anyone who intends to take part in such an activity must get acquainted with quite a lot of legislative text. Thus, knowledge of the law is almost as important as knowledge of the hazards, and ideally would be a major topic in a text like this, but for two reasons this cannot be the case. Firstly, the legislation differs from country to country, even if the chemical hazards and the people that should be protected are identical, and it is not possible to discuss 'average' laws. Secondly, in contrast to the chemical hazards that the legislation should protect us against, the laws change frequently following new discoveries, improved or deteriorating financial situations and so forth, and any discussions of laws and regulations therefore become out of date quite quickly.

The legislation may regulate in detail which chemicals are permitted, making sure that only chemicals that have been proven to be sufficiently safe are approved. However, that kind of legislation is unusual and normally only applies to certain compound classes (e.g., pharmaceuticals, pesticides, and food additives). In most countries it regulates the concentrations of the most frequently used chemicals that are allowed in the respiration air in workplaces and how chemical waste from different kinds of activities should be treated. In most countries there are also rules that ensure that the labels on containers for chemicals has information about any associated hazard, that more comprehensive information is available to those who purchase and use a chemical, and that workplaces where chemicals are handled are equipped with safety devices such as gas masks and fire extinguishers. The competence (education and training) of persons handling chemicals is not regulated in any detail, but this is of course a critical point, as anyone with a good

CAUTION
Chemical X may be harmful if inhaled. Chemical X may cause
irritation. Chemical X may cause a rapidly-developing
pulmonary insufficiency, labored breathing, and cyanosis
followed by cor pulmonale and short survival time. Death may
result from cardiac failure or destruction of lung tissue with
resulting anoxia. Skin contact may cause irritation and
dermatitis. Eye contact may cause redness, irritation, and
conjunctivis.

TARGET ORGANS AFFECTED
Eyes, skin, and mucous membranes.

FIRST AID — INHALATION
Remove from exposure area to fresh air immediately. Keep
person warm and at rest. Get medical attention immediately.

FIRST AID — SKIN
Remove contaminated clothing and shoes immediately. Wash
with soap and large amounts of water. Get medical attention.

FIRST AID — EYES
Wash eyes immediately with large amounts of water for 15–20
minutes. Get medical attention.

Figure 1.4 Information about health hazards on the container of chemical X (*cor pulmonale*
is a heart disease resulting from disease of the lungs or pulmonary circulation).

understanding and knowledge about a hazard will be able react adequately. It may
be argued that competence is even better than regulations in some instances,
although a combination of the two may well prove best in the long run. The
example shown in Figure 1.4 illustrates this.

Unlike the example in Figure 1.1, this has been taken from real life and is not
made up. Anyone with a bottle of chemical X in his or her hand will be concerned
on reading the warnings: X is obviously a chemical to treat with caution and
respect. However, chemical X is in fact sea sand, a product that is used in the
chemical laboratory for various purposes but is not associated with any dramatic
hazards. Indeed, many of us dream about spending time on a sunny beach consisting of sea sand, and the idea that this should pose a hazard is for most people
far fetched. It is true that inhaling sea sand would give problems with respiration
and that sand in the eyes is painful, but this really comes as no surprise. This is
in fact an example of a reaction of the producer, the chemical company, to a situation where they may be accused of not warning the consumer of any imaginable
hazard, never mind how unlikely it is. For many, the initial feeling of respect for
chemical X will be replaced by confusion, quickly followed by a feeling that information about health hazards exaggerates the risks. In such a case, the law is of
little value, and users of chemical X may instead choose to rely on their own judgment, based on knowledge about chemical hazards and experience of handling
chemicals.

1.5
Knowledge about Chemical Hazards

Knowledge about the effects of chemicals on man and the environment is crucial, as is the good sense to use that knowledge in a sensible way. Knowledge can be acquired in many ways, but if the effects of a chemical are described in the literature this is the first place to search. Today, huge databases where information about chemicals can be sought are available not only in chemical libraries but also via the internet. However, surprisingly few of the approximately 50 000–100 000 chemicals used today have been studied from a toxicological and ecotoxicological viewpoint, and for only a few percent of these can we can say that we have complete knowledge (which includes long-term toxicity, e.g., carcinogenicity) about their effects on man. In general, the chemicals that have been the most thoroughly studied are those that the legislation requires to be tested before they are approved. A pharmaceutical compound, for example, must be shown to be relatively safe before it may be used. Bulk chemicals, on the other hand, although they are used in large quantities in the chemical industry, are not subject to such regulations, and little is known about the toxicity of most of them. In addition, they may contain considerable amounts of impurities (several %) which are not even declared on the label of contents. Many bulk chemicals have been used for decades, and over the years we have had some experiences with them that indicate to us in what situations they may be hazardous, but even if a chemical appears from experience to be safe this is no proof that it really is so. Taking carcinogenic chemicals as an example, all chemicals that have been proven to be carcinogenic to man have been studied in epidemiological investigations where the exposure to a chemical (at a workplace for example) is correlated with the incidence of cancer 20–30 years later, and this is compared to a nonexposed control group. However, it is quite difficult to demonstrate such correlations (it has only been done with 30–40 chemicals), because it requires either (a) that the chemical is a potent carcinogen (e.g., bis-chloromethyl ether, formerly used for the manufacture of ion exchange polymers), (b) that the group exposed is huge (e.g., cigarette smokers vs nonsmokers) or (c) that the tumor formed is extremely unusual (e.g., vinyl chloride, which is a weak carcinogen, causes a very unusual form of liver cancer). Carcinogenic chemicals that are not potent, are not used by very large or isolated groups, and do not cause unusual cancers will not be detected by epidemiological investigations because their effects will not be noticeable from the number of tumors formed spontaneously.

An interesting example of a chemical that has been used for a long time (more than 100 years) and that we have a lot of experience of is acetylsalicylic acid (see Figure 1.5). This is used as an anti-inflammatory agent, as an antipyretic, and as an analgesic, and recently its ability to inhibit the ability of the blood to coagulate has been taken advantage of for the treatment of thrombosis. Acetylsalicylic acid is a derivative of salicylic acid, as is salicin, the active constituent of the bark of the willow tree (*Salix alba*). In the latter part of the 18th century, Reverend Edward

salicylic acid   acetylsalicylic acid   salicin

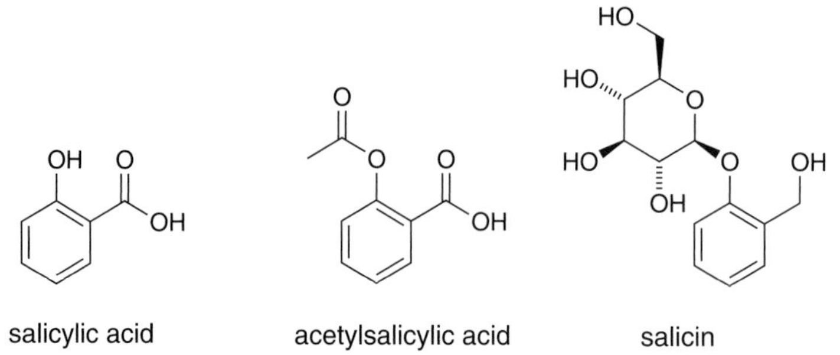

Figure 1.5 The structures of salicylic acid and two of its derivatives.

Stone, inspired by a local woman's tale and by the bitterness of the bark, believed that it would have the same effects as Peruvian bark (containing quinine) and treated a number of people suffering from fevers with the bark. The treatment was successful, and the salicylic acid derivatives obtained from willow bark as well as from the plants meadowsweet and wintergreen became important agents to reduce fever and pain and to relieve gout and arthritis. The acetylated derivative was prepared in the end of the 19th century by the German company Bayer AG and sold under the name Aspirin.

It is not a nontoxic compound (10–15 g may damage the kidneys of an adult), but the long-term effects (e.g., carcinogenicity and teratogenicity) of low doses of acetylsalicylic acid are based on our long experience with human exposure and known to be negligible. However, in animal experiments it has been noted that acetylsalicylic acid has a weak teratogenic effect, and it is doubtful whether it would be approved if it was invented today.

The fact that we use so many chemicals without really knowing much about their effects on man and environment may come as a shock. The reason is of course economic; new knowledge is unfortunately not free, and the costs of assaying the toxicity and ecotoxicity of one chemical can climb to 1 million €. Even though the ambition today is to raise standards and demand more knowledge about all new chemicals that are added to the market, there are no possibilities to thoroughly investigate the toxicity of all the old chemicals already in use. It may seem logical to ban all chemicals until we know about their effects and can regulate their use, but such an action would overturn society. Instead, we will have to learn to live with this situation for many years to come. If knowledge about the hazards of a chemical cannot be found in the literature, and if it is considered to be too expensive to carry out the necessary biological tests by oneself, the only alternative is to make intelligent guesses based on known relationships between chemical structure/chemical properties and toxicity/ecotoxicity for similar compounds. This is essentially what this text aims to do, to show the reader that the effects on man and environment that we want to avoid are caused by chemical properties that we to a large extent can understand by analyzing the chemical structure. In the future it is believed that such intelligent guesses, made by computers that have been fed all available data and taught to analyze the data in a relevant way, will be used routinely to sort out old chemicals that we know too little about but which have chemical properties that the computer has associated with toxicity, as well as chemicals in the pipelines of development but classified as potentially hazardous. However, until the day when computers take over the decision-making, it is good advice try to maintain the ability to make judgments for oneself.

1.6
Acceptable Chemical Risks

At the bottom line we will find that all chemicals have some hazards associated with them, although the differences between those regarded as 'safe' and 'hazardous' are enormous. Toxicity and ecotoxicity do not come in black and white but as a gray scale, and apart from deciding if and how a chemical is hazardous we need to know something about its potency. This can be done by relating the toxic response of a chemical to the dose, and an example is shown in Figure 1.6, where the carcinogenicity of two chemicals is plotted against the dose. There is an obvious relationship between dose and response, which can be approximated with a straight line (obtained by statistical analysis of the data or simply by putting a ruler along the crosses and circles). The slope of that line would be a measure of the potency of the compound. The slope of the line corresponding to the crosses is greater than that corresponding to the circles, and it is thus possible to compare the carcinogenicity of the two chemicals.

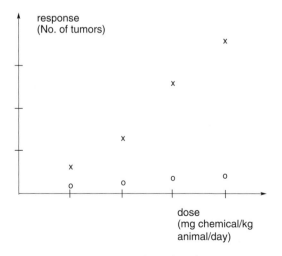

Figure 1.6 Dose-response relationships for two carcinogenic chemicals.

Before we start to use them, let us define some common terms:

in vitro   This means 'in the glass' and indicates experiments carried out with cells or tissues in test tubes or equivalents.

in vivo   This means 'in life' and indicates experiments carried out with living organisms.

The next difficulty is to translate our knowledge about the toxicity of a chemical, which in almost all cases refers to *in vitro* assays and animal experiments, to the species that we are interested in, namely *Homo sapiens*. It is also not obvious whether it is relevant to use data from animal experiments where the doses used are much higher than those to which humans would be exposed. These problems will be discussed in more detail in Chapter 7, but in general the knowledge that we have acquired is not enough to make an absolutely certain assessment of the toxicity to man, and approximations have to be made. Finally, knowing about the ifs, the hows and the how muches, we have to decide which risks are acceptable in different activities. Obviously we have to accept all kinds of chemical hazards, even genotoxicity, but we can still regulate, for example, what concentrations of a carcinogenic chemical we consider acceptable in the respiration air at a place of work. Some chemicals, for example, pesticides and preservatives, are considered by many to be unacceptable as they in general are toxic to at least some organisms, but they also save human lives by improving crops, killing disease carriers (e.g., DDT, *vide supra*), and avoiding food poisoning by bacteria. Decisions on what risks we are ready to accept in order to benefit from the various uses of chemicals are extremely difficult, as we have already noted, as individuals as well as authorities and companies value risks and benefits associated with chemicals very differently. In addition to the scientific and humanistic dimensions, there is also a political aspect, as the safety of people and the protection of the environment have to be considered together with economic factors. Looking back, there are examples of chemicals that have received exaggerated attention (because of lack of knowledge, suspicion, fear, and/or political opportunism) while others have not received the attention they merit. One of the major threats today is that the rapidly increasing concentrations of carbon dioxide and methane in the atmosphere, formed by combustion of mineral fuels and by anaerobic fermentation, will affect the climate and thereby the weather systems of the earth. Little is today known about this so-called greenhouse effect (it will be further discussed in Chapter 10), but if it exists and the additional carbon dioxide and methane do change the climate in a dramatic way, we will be very sorry that we did not take action sooner.

2
The Chemicals of Life and Nature

An ecosystem can be defined as a functional unit of life on earth, with animals, plants, microorganisms, and minerals that together provide everything necessary for their co-existence. In principle it is not possible to identify isolated ecosystems as the whole planet is one gigantic ecosystem. However, for practical reasons it can still be useful to refer to, for example, a forest or a lake, with natural boundaries between it and its surroundings, as an ecosystem. In an ecosystem, the species depend on each other in sensitive balances that have evolved with time. If the conditions in an ecosystem change, for example, the climate, individual species will be favored or disfavored, and eventually the change will lead to the establishment of a different ecosystem with new balances. Small changes may have dramatic consequences and may even lead to the extinction of species, because the species are in most cases highly specialized and depend heavily on the ecological equilibria. This has led to big differences in color, size, and function of the organisms of the world, but however different they may appear they all share the basic chemical features of life. They are all composed of cells, one cell in unicellular organisms and many in multicellular organisms (approximately 10^{15} in humans), and all living cells have the same basic chemical components and use the same reactions for the primary life processes. As the toxicity of chemicals is due to their interference with the chemical components and reactions of a cell, the basic principles of chemical toxicology are the same for all organisms and can to some extent be generalized. Besides the primary life processes, there are of course a number of secondary processes responsible for various specialized functions that may be completely different if we compare one species with another. As the effects of many toxic chemicals depend on their interference with such secondary processes, one species can be much more sensitive to a certain chemical than another. In later chapters, several examples of such differences will be encountered and the reasons for the differences in sensitivity will be discussed, but in this chapter we focus on the chemical constituents of nature. As mentioned in Chapter 1, the species *Homo sapiens* is central to our interest and will get more attention than other species.

Chemistry, Health, and Environment. Olov Sterner
© 2010 WILEY-VCH Verlag GmbH & Co. KGaA, Weinheim
ISBN: 978-3-527-32582-5

Table 2.1 The 10 organ systems of humans.

Organ system	Major organs or tissues	Major functions
Musculo-skeletal	Bone, ligament, joints, muscle	Support, protection, motion
Integumentary	Skin	Protection, defense, regulation
Circulatory	Heart, vessels, blood	Transportation
Respiratory	Throat, trachea, bronchi, lungs	Exchange of gases, regulation
Digestive	Stomach, intestines, pancreas, liver	Digestion, absorption, excretion
Urinary	Kidneys, bladder	Regulation, excretion
Nervous	Brain, spinal cord, peripheral nerves	Detection, coordination, control
Immune	Lymphocytes, thymus, bone marrow	Regulation, surveillance, defense
Endocrine	Hormone secreting glands	Regulation, coordination
Reproductive	Testes, ovaries	Reproduction

2.1
The Cell

A human being consists of 10 organ systems that have special tasks and functions to perform but collaborate closely with each other. A brief description is given in Table 2.1.

The organ systems are composed of organs and organ subunits, which in turn are made up of tissues (epithelial tissue, connective tissue, muscle tissue, and nerve tissue), and cells. A human body contains approximately 200 significantly different cell types, although they can be classified according to their basic functions as either epithelial cells, connective cells, muscle cells, or nerve cells. The cell is the lowest common denominator of life, and although cells of simpler organisms (e.g., bacteria) are quite different from human cells it is nevertheless natural to have the cell as a starting point for our discussions about the effects of harmful chemicals.

2.1.1
Human Cell Types

Epithelial Cells. The epithelial cells are found on all surfaces that cover organs and organ subunits, and form the border between different macroscopic components and functions. They are responsible for the exchange of compounds between the body and the environment, for example, in the gastrointestinal tract (between gut contents and the blood) and the lung (between the inhaled air and the blood), as well as between different compartments of the body. The skin, which forms the boundary between the body and the rest of the environment and protects us, is also composed of epithelial cells. These are characterized by a relatively rapid division rate, something that makes them sensitive to agents that affect their genetic material and more prone than other cells to be transformed into cancer cells (see Chapter 8).

Connective Cells. A major function of connective cells and connective tissue is to give support to various structures in the body, giving the organs and the organ subunits their form and shape. In most organs the connective cells make up the bulk of the tissue, forming a cellular network to which other cells may attach themselves and be anchored. This is done by secreting a matrix consisting of a polymer in a polysaccharide gel into the extracellular space. Bone cells secrete calcium phosphate which is crystallized to form bone, while ligaments connecting muscles are produced by connective cells secreting the proteins collagen and elastin. As well as this, connective cells perform a variety of functions; examples of cell types that are included in this category are the adipose cells, storing fat, and the red and white blood cells.

Muscle Cells. The muscle cells have the ability to contract, and if a muscle cell is attached at both ends the contraction may generate motion. Aside from the most obvious function of muscle tissue, to enable the body to move around, it should be remembered that the heart consists of muscle tissue, and that the tubes through which body fluids like the blood flow are surrounded by muscle cells that can regulate the diameter of, for example, a blood capillary and thereby the flow rate.

Nerve Cells. Nerve cells can receive and transmit signals from and to other cells as well as from the environment. For example, muscle cells that contract do so when they receive a signal from a nerve cell. One of the central functions for nerve tissue is therefore to control and coordinate the actions of other cells, and indirectly of organs (via, for example, the stimulation of a gland cell producing a hormone). The signals are electrical within nerve cells and chemical between the cells: a more detailed discussion about the normal function of nerve cells and how this is affected by neurotoxic chemicals can be found in Chapter 7.

The fact that all cells in a human body originate from one single cell, namely the ovum after conception, deserves a moment of reflection. All the different shapes and functions must be preprogrammed in this cell and this information must be made available at exactly the right time. This is one of the many wonders of life, and we will return to it in Chapter 8.

2.1.2
The Basic Components and Functions of a Cell

The cell is the basic unit of life, and all organisms consist of one or more cells. By definition, a cell has the ability to absorb nutrients from its surroundings, to use them for the production of energy and the molecules it needs for its various activities, and to multiply. As has been discussed above, cells differ tremendously in size, appearance, and function. Bacteria, for example, are differently organized, and are between 1000 and 1 000 000 times smaller than the average human cell, although the basic chemical machinery that sustains life is roughly the same. Several bacteria have been so thoroughly investigated that one knows more or less in detail how they work. In bacteria, life can be said to consist of a few

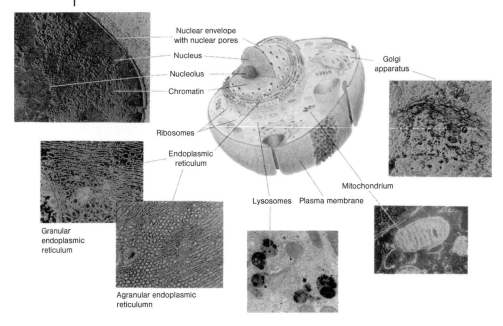

Figure 2.1 An average mammalian cell.

thousand, say 3000, chemical reactions. These will together perform all the functions discussed below and provide all things necessary for the normal life of the bacterium. Human cells are not only bigger but of course also considerably more complicated than bacteria, and not all cellular functions are known on the molecular level. The number of chemical reactions in our cells is therefore much greater than 3000, but it is still finite and will in the near future be determinable.

The most important parts of a mammalian cell are indicated in Figure 2.1. The plasma membrane surrounds the cell and provides a boundary across which the flow of at least some chemicals can be regulated by the cell. This is also where the receptors that convey chemical signals from other cells are situated. Inside the plasma membrane is the cytoplasm, a thick aqueous solution containing several cellular structures called organelles and the nucleus (in eukaryotic cells, the cells of animals, fungi, and plants). The nucleus is the largest structure of a mammalian cell: it is surrounded by the nuclear envelope and it contains the genetic material. The genetic material (chromatin) is organized in 46 strands (in a human cell nucleus), which is condensed to chromosomes that can be observed in a light microscope at the time of cell division (see also Chapter 8). In simpler organisms (bacteria) the cell nucleus is missing (prokaryotic cells) and the genetic material is present in the cytoplasm To facilitate the movement of molecules and molecular complexes between the cytoplasm and the nucleus, the nuclear envelope has a number of openings called nuclear pores. The nuclear envelope is made up of two membranes separated by a small space. The endoplasmic reticulum is a large organelle also made up of two membranes (with a space between them) that traverse the cell, and the intermembrane space of the nuclear envelope and the endoplasmic reticulum are interconnected. Parts of the endoplasmic reticulum may be granular (rough), containing ribosomes that synthesize proteins, or agranular (smooth), in which synthesis of steroids and fatty acids as well as the metabolic conversions of exogenous chemicals take place. All cells contain many mitochondria, small oval-shaped organelles in which energy is produced, and lysosomes, which are responsible for the recycling of nonfunctional cellular components by the action of their degrading enzymes. The Golgi apparatus is another membrane structure, whose function is to concentrate and modify proteins to be secreted from the cell.

The size of an average mammalian cell is 10–20 µm, which make them invisible for the naked eye but only just (the limit for our eyes is approximately 100 µm). The three-dimensional form of cells, which can be anything from a sphere to an extended structure, is largely maintained by filaments and microtubules, which (especially the latter) provide a cytoskeleton to which organelles can be anchored. Transportation inside a cell is normally no problem for molecules that are dissolved, as molecules rapidly diffuse around the cell. However, for larger structures there are active cellular transport systems that, for example, make use of the cytoskeleton, and some cells (nerve cells which can have a length of 1 m) depend on such systems (which can transport vesicles up to 40 cm per day) for their function.

Before we take a closer look at the most important molecular materials of the cell, the membrane, the proteins, and the genetic material, we need to briefly discuss chemicals in general and biochemicals in particular.

2.2
Chemicals

We have already defined the difference between organic and inorganic chemical compounds, or chemicals, and have been impressed by the diversity of organic compounds. Basically, the organic compounds are either hydrocarbons, being made up of only the elements hydrogen (H) and carbon (C), or are based on a hydrocarbon skeleton. An organic compound may be more or less oxidized (more if there are unsaturations, i.e. double bonds, triple bonds, and/or rings), and may also contain heteroatom substituents (a heteroatom is any atom except carbon and hydrogen). The major classes of organic compounds, essentially in order of increasing oxidation level, are the following (see also Figure 2.2):

Hydrocarbons
Saturated hydrocarbons (alkanes) which contain only single bonds between the carbon atoms and no functional groups. As substituents they are called alkyl groups.

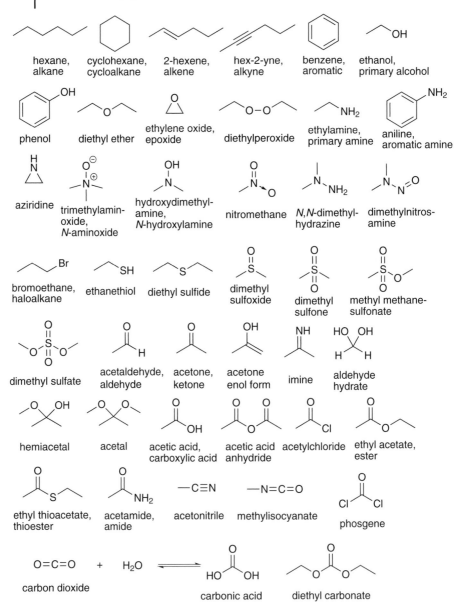

Figure 2.2 Examples of compounds at the oxidation level of carbonic acid.

Unsaturated hydrocarbons with one or several carbon–carbon double (alkenes) or triple (alkynes) bonds, still lacking functional groups.

Aromatic hydrocarbons with one or more aromatic rings and thereby possessing unique properties. As substituents they are called aryl groups.

Alcohols and ethers

The alcohols are oxygenated organic compounds that contain a hydroxyl group and consequently have -*ol*, as in ethanol, as a suffix. If the saturated carbon to which the hydroxyl group is attached has one additional carbon atom bound to it, the alcohol is called primary, if it is two or three carbons the alcohol is classified as a secondary or a tertiary alcohol, and if the hydroxyl group is attached directly to an aromatic ring the compound is called a phenol. Ethers have two alkyl/aryl groups bound to the oxygen, and consequently no hydrogen. They may be cyclic, and the epoxides (3-membered rings) are of special interest. The alcohol/ether function may be oxidized to a peroxide.

Amines

Amines contain one or more nitrogen atoms, which may be substituted with one (primary amines with an $-NH_2$ group), two (secondary amines with an $-NH-$ group) or three (tertiary amines with no hydrogens bound to the nitrogen) alkyl/aryl groups. The nitrogen may even bind four alkyl groups in quaternary ammonium ions (e.g., the tetramethylammonium ion), when the unshared electron pair of nitrogen has been used for the fourth bond and the nitrogen has become electron-deficient (and consequently has a positive charge). The nitrogen may be oxidized to form *N*-aminoxides, *N*-hydroxylamines, hydrazines, nitroso, and nitro compounds, and *N*-nitrosamines. Also, amines may be cyclic.

Halogenated hydrocarbons

Any of the four halogens fluorine, chlorine, bromine, and iodine can be a substituent in a halogenated hydrocarbon, chlorinated compounds being the most common.

Functionalities containing sulfur

Sulfur can replace oxygen in alcohols and ethers resulting in thiols and sulfides. As the sulfur atom is bigger than oxygen it can form additional bonds, and is frequently oxidized to sulfoxides, sulfones, sulfonates, and sulfates.

Aldehydes, ketones, and imines

Aldehydes and ketones have a carbonyl group with an oxygen connected to a carbon by a double bond. Aldehydes, with the suffix -*al*, have at least one hydrogen bound to the carbonyl group, while ketones, with the suffix -*one*, have alkyl/aryl groups on both sides. Imines have a nitrogen connected to a carbon by a double bond, and are often easily hydrolyzed (react with water) to form the corresponding carbonyl derivative and the amine. Aldehydes, ketones, and imines with at least one hydrogen on an α-carbon are in equilibrium with their corresponding enol and enamine.

Hemiacetals and acetals

The carbonyl group of an aldehyde or a ketone may be transformed to a hydrate by the addition of water, or to a hemiacetal or an acetal by the addition of an alcohol to the carbonyl group. These reactions are in principle reversible, and in the presence of water hemiacetals and acetals are normally transformed to the aldehyde/ketone. The corresponding products may form if amines, thiols, or halide ions add to the carbonyl.

Carboxylic acids, anhydrides, and halides
The carboxylic acids are characterized by their acidity, although few are strong acids. Carboxylic acid anhydrides and carboxylic acid halides are derivatives of carboxylic acids.

Carboxylic acid esters, thioesters, and amides
In a carboxylic acid ester, the hydrogen in the carboxylic acid group has been exchanged for an alkyl/aryl group, while in a carboxylic acid thioester the adjacent oxygen is a sulfur. A carboxylic acid amide has instead a nitrogen in this position. Esters, thioesters, and amides may be cyclic and are then called lactones, thiolactones, and lactams. All these compounds can be hydrolyzed to form the corresponding carboxylic acid and the alcohol, thiol, or amine, respectively.

Nitriles
Nitriles contain a carbon with a triple bond to nitrogen–a cyanide group. In the cyanide group, the carbon is on the same oxidation level as that of the carbon in a carboxylic acid group.

Fully oxidized organic compounds
Carbon in organic compounds may be fully oxidized, that is, on the same oxidation level as that of carbon dioxide and carbonic acid. Examples are the isocyanates, derivatives of carbonic acid such as diethyl carbonate, and phosgene.

Note that by going from the hydrocarbon methane through the list to the carbon dioxide derivatives, we pass through all oxidations states of carbon. This is summarized in Figure 2.3.

Aside from their chemical functionalities, chemicals may also be classified in other ways, and in this context it is convenient to differentiate the chemicals that are endogenous to organisms from those that are exogenous, as well as those that are natural from those that are synthetic.

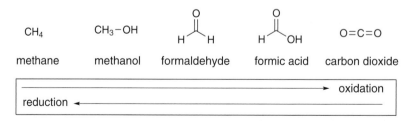

Figure 2.3 The various oxidation states of carbon.

2.2.1
Endogenous vs Exogenous Chemicals

Endogenous chemicals have a natural place in the normal biochemical processes of the organism, being either essential or nonessential compounds. Compounds or elements without which an organism cannot survive and which the organism cannot make itself (at least in sufficient amounts) and that consequently have to be provided from the environment, are called essential. For humans we know of approximately 50 essential compounds and elements, for example, water, molecular oxygen, 9 amino acids, some unsaturated fatty acids (e.g., linoleic acid, linolenic acid and arachidonic acid), approximately 20 vitamins, and minerals. Approximately 25 elements are assumed to be essential to life. Some of these are present in large amounts (>1% of the body weight, for example, carbon, hydrogen, nitrogen, oxygen, and phosphorus) and form part of the molecules which compose the main building blocks used in the macromolecules of cells. Others (e.g., the trace metals) are only present in minute amounts and are important for the function of enzymes. Nonessential but still endogenous chemicals are the remaining 11 amino acids, many carbohydrates, and all other compounds that take part in the biochemical processes. An example of a nonessential but endogenous chemical is glucose, which we use to generate energy (our brains burn more than 100 g glucose per day). Although we depend on glucose for the generation of energy, we can survive on a glucose-free diet because our metabolism can produce glucose in adequate amounts from other chemicals present in our bodies, although if it is present in the food we will of course use it as it is.

Shortage of essential compounds and elements will result in a less than optimal capacity of that organism, although the effect does not have to be dramatic, while too rich a supply of essential compounds and elements will also be harmful and may give toxic effects. An example is molecular oxygen, which we absolutely need for our survival but which is harmful in high concentrations for prolonged periods of time. Our bodies are designed to live in an environment containing approximately 20% oxygen in the air we inhale, which will enable our blood to transport oxygen safely and only leave a low concentration of free oxygen in our tissues. However, if we breathe air containing more oxygen for a longer time, the blood will not be able to bind all of it and the amount of free oxygen that can participate in potentially harmful reactions (Section 5.2.3) increases.

Exogenous chemicals. Exogenous chemicals, or xenobiotics (from the Greek: 'xenos' and 'bios' make 'stranger to life'), are chemicals that do not occur naturally in an organism. Exogenous compounds may in rare cases be inert, like molecular nitrogen, but should in principle be considered to be potentially toxic.

The differences between essential, other endogenous, and exogenous chemicals are indicated in Figure 2.4. For essential chemicals there is always an optimal concentration for the organism, above which the favorable effect will diminish and eventually become a toxic effect. In some cases the maximal favorable effect is obtained within a broad concentration range while in other cases the concentration of the essential compound is more critical. Non-essential endogenous

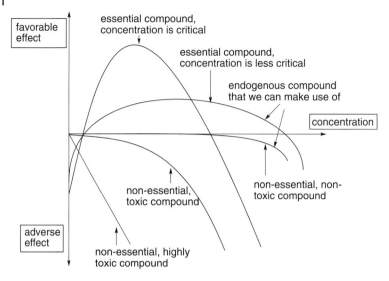

Figure 2.4 The effects of essential and nonessential chemicals on an organism.

compounds such as glucose can be imagined either to give a favorable effect, because they relieve the metabolism, or to give no effect at all (in reasonable concentrations), because the organism will cope with the situation anyway. Exogenous compounds will never give a favorable effect. If they are non-toxic, as molecular nitrogen, they will give an adverse effect in high concentrations (>90% molecular nitrogen in the respiration air is dangerous because the concentration of oxygen is too low).

2.2.2
Natural vs Synthetic Compounds

Quite often one encounters the terms 'natural' and 'synthetic' compounds, these terms being used to distinguish between compounds that are formed by organisms (natural) and by man in artificial ways (synthetic). Some believe that natural compounds are less toxic that unnatural, since the human species has evolved on earth in the presence of natural compounds and has had the possibility to adapt to their presence. Synthetic compounds have only existed for a short time, and it is obvious that organisms have not had the same chance to learn how to deal with them (although microorganisms have shown a remarkable ability to become resistant to newly developed antibiotics). To some extent this is of course true, and natural compounds are for instance more easily biodegradable compared to many man-made compounds (e.g., DDT and PCB) because of the co-evolution of organisms and natural products. However, the most toxic compounds known are natural, being produced by organisms that possibly use them to resist or control

fungal metabolites:

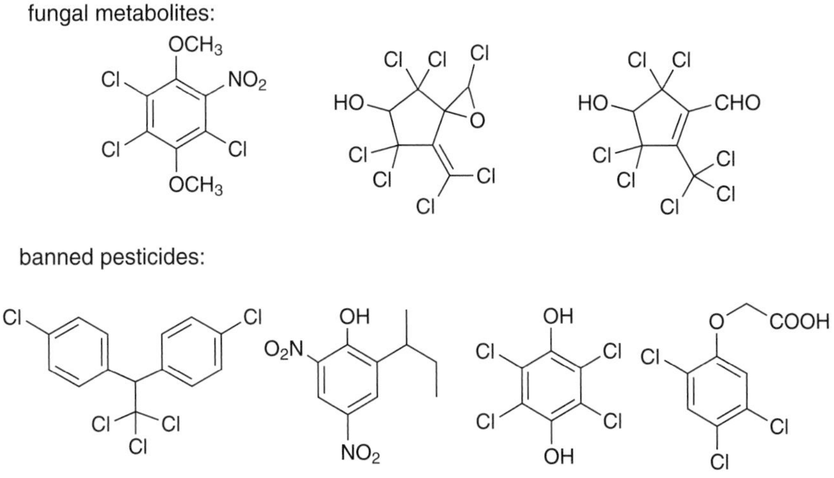

banned pesticides:

Figure 2.5 Natural compounds can be as toxic as synthetic.

competitors for food and space. The production of biologically active and potentially toxic compounds in nature is enormous, and as can be seen in Figure 2.5 where the structure of some fungal metabolites are compared with those of a few notorious pesticides. It is not always obvious from the structure which compounds are natural and which are synthetic!

Many of the hazardous synthetic compounds that cause problems today are halogenated, and traditionally the presence of halogen atoms in an organic compound has been associated with a synthetic origin. However, the natural production of halogenated compounds far exceeds the synthetic production, although the latter compounds are in some cases poorly biodegradable.

We also have to be aware that the borders between endogenous and exogenous as well as natural and synthetic compounds are not permanent, and that new scientific discoveries may change the status of a compound. A recent example is nitric oxide (NO), which for decades has been known as a noxious gas formed, for example, when metallic copper is dissolved in nitric acid or in a petrol engine when N_2 is oxidized by O_2 at high temperature. Besides damaging the lungs (causing pulmonary edema, Section 7.3.5), it has been shown to affect the ozone in the stratosphere (Section 10.4) and to be a component of acid rain. However, recently nitric oxide has been found to have several important physiological functions, for example, as a transmittor of nerve signals and as an inhibitor of platelet aggregation. Nitric oxide also dilates the capillaries delivering blood to, for example, the heart muscle, and has been used in the form of nitroglycerin (which is converted in the body to nitric oxide) to treat vascular spasms (e.g., angina pectoris). In addition, nitric oxide is one of the components produced by the immune system (the macrophages) and used to kill infecting microorganisms.

2.3
Biochemicals

2.3.1
Water

The earth contains approximately 1.4 billion km^3 water, making the water molecule (H_2O) the most abundant molecule on earth. Water is the medium in which life originally formed and evolution started, and is the solvent in which biomolecular processes take place. As far as we know, the evolution of any form of life anywhere in the universe would depend on the presence of water. Water has a number of qualities that makes it exceptional in this respect. For example, the density of water is highest at $+4\,°C$, something that not only protects water-living organisms in cold climates but also is important for the transformation of stone to dust (shattering by freezing water in cavities). The intermolecular forces between water molecules and between water molecules and polar compounds as well as ions dissolved in water are exceptionally strong, giving water an unexpectedly high boiling point and enabling it to transport a variety of chemicals in biological systems. Because of the high dielectric constant of water, ions dissolved in it will be efficiently shielded from each other and can stay in solution. Water absorbs IR radiation and thus contributes to the natural greenhouse effect that keeps the temperature on earth at a favorable level. The heat capacity of water is also exceptionally high, and seas and oceans therefore function as buffers against variations in temperature. So is its heat of evaporation, which makes it suitable for cooling organisms by transpiration. The surface tension of water is the highest known for liquids (except liquid mercury), important for the interaction of water with other chemicals as well as for the formation of raindrops in the atmosphere. (Several of the chemical properties of water that arise from these qualities will be discussed in Chapter 3.)

Water is considered to be the solvent of life, in which the molecular reactions on which life is based take place. However, a cell contains approximately 80% water and 20% other material, and the chemical properties of the thick water solution present inside a cell cannot be compared with those of pure water. Water is an essential chemical, and the only hazard that most people associate with it is its potential to cause drowning. However, the consumption of large quantities of pure water (distilled or desalted) may also be dangerous, as it may bleed the body of minerals.

2.3.2
Carbohydrates

This class of organic compounds received its name because the elemental composition of many (but not all) of its members corresponds to $C_x(H_2O)_y$. Take for example glucose (blood sugar) and sucrose (table sugar) with the molecular formulas $C_6H_{12}O_6$ or $C_6(H_2O)_6$, and $C_{12}H_{22}O_{11}$ or $C_{12}(H_2O)_{11}$, respectively. The simpler

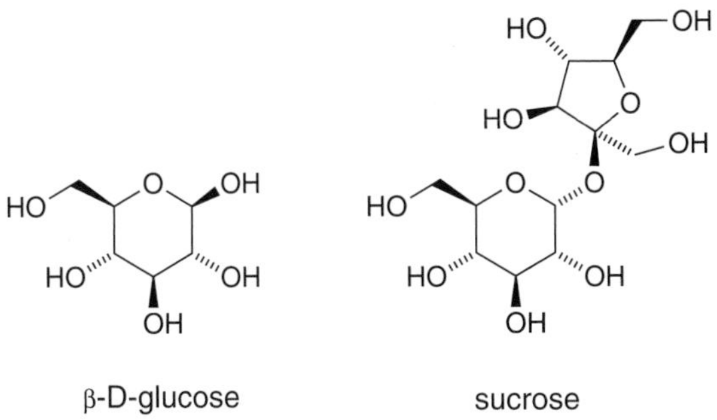

Figure 2.6 The structures of glucose and sucrose.

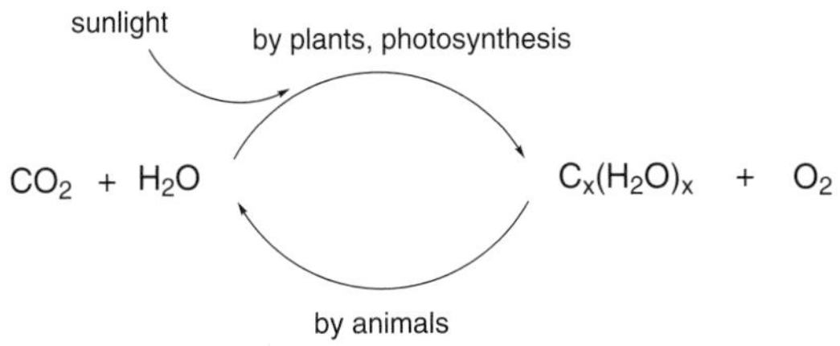

Figure 2.7 The energy consumed by organisms on earth originally comes from the sun.

members of the carbohydrates are also called saccharides, and the monosaccharides, with the general formula $C_xH_{2x}O_x$ (x = 3, 4, 5, 6, 7 or 8) (for example glucose) are the monomers of more complicated carbohydrates. Disaccharides are made from two monosaccharides, for example sucrose (made from one glucose and one fructose), maltose, and lactose, while trisaccharides are made from three monosaccharides. Oligosaccharides contain between four and ten monosaccharides, while polysaccharides (e.g., starch, cellulose, and glycogen) contain a large number. (As can be seen in Figure 2.6, we can indicate the direction in space of chemical bonds in a molecule. In this text, a bold wedged bond indicates that a bond rises above the plane of the paper while a hashed bond indicates that it is directed below the plane of the paper.) In addition, like most natural compounds, monosaccharides are chiral (the mirror image of the molecule is not superposable on the molecule itself) and are essentially produced in the biological world as only one enantiomer (mirror image). The vast majority of carbohydrates are members of the dextro series, indicated by the prefix D-, and illustrated by the structures of glucose and sucrose shown in Figure 2.6.

The carbohydrates are formed by plants and green algae, which convert water and carbon dioxide into carbohydrates and molecular oxygen, using the energy of sunlight to make this energetically unfavorable reaction proceed. This is called photosynthesis, and generates the fuel that other organisms (including ourselves) depend on (see Figure 2.7).

Besides storing chemical energy in, for example, starch and glycogen, carbohydrates are also used by organisms as construction materials (e.g., cellulose in plants, polysaccharides in the cell walls of bacteria, and chitin in the exoskeletons of insects and crustaceans). Cellulose is actually the most abundant organic compound in the world, containing more than half of all carbon bound in organic molecules. Carbohydrates are also used by our cells as 'flags' linked to proteins and lipids and positioned in cell membranes. Such signals can be recognized by the immune system and used for identification, and our different blood types (A, B, O, etc.), for example, are determined by different carbohydrates bound to the membrane of the red blood cells. A person having type A blood cannot be given type B blood because his immune system will identify the type B blood cells as foreign and destroy them. Additional important functions of carbohydrates are as components of DNA and various metabolic cofactors (see below).

2.3.3
Fatty Acids

A fatty acid is a long unbranched carboxylic acid with 12–20 carbon atoms (see Figure 2.8 for examples), essentially derived from the hydrolysis of natural fats, oils, and waxes. Fats (solid at room temperature) and oils (liquid) are triesters of fatty acids with glycerol, while waxes are esters of a fatty acid with a long-chained alcohol. Fatty acids can be saturated, like palmitic acid and stearic acid, or unsaturated, like oleic acid. Polyunsaturated fatty acids, for example, linoleic acid and linolenic acid, contain two or more carbon–carbon double bonds. Unsaturated fatty acids are less stable than saturated ones, and take part in oxidative reactions (lipid peroxidation) that will be discussed in Chapters 5 and 7. Linseed oil and other drying oils contain the triglyceride of linolenic acid, and linseed oil paint will dry not because a solvent evaporates as in normal paints but because the linolenic acid reacts with molecular oxygen in the air and polymerizes.

Just like carbohydrates, fatty acids (in the form of triglycerides) are used to store chemical energy, and because fatty acids are less oxidized than carbohydrates

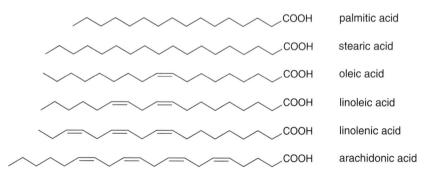

Figure 2.8 Common and important fatty acids.

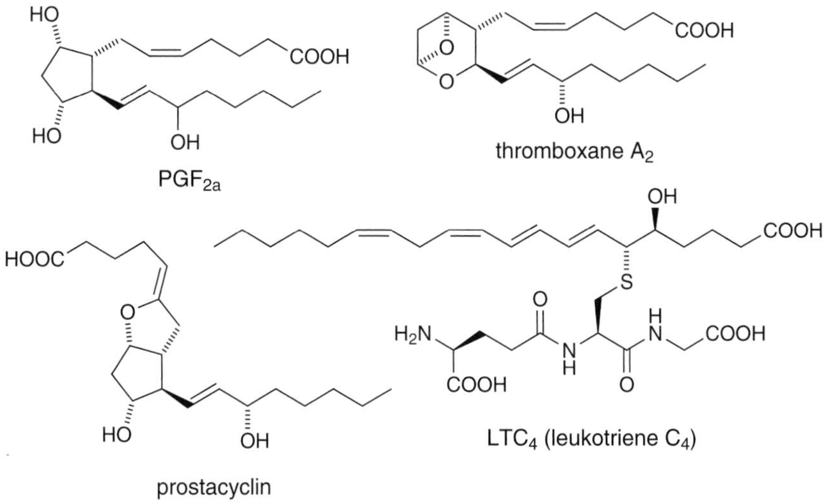

Figure 2.9 Some physiologically important compounds derived from arachidonic acid.

their energy content is larger (complete oxidation of fatty acids yields approximately 9 kcal/g while carbohydrates yield 4 kcal/g). Fatty acids are also important components (as phospholipids and glycolipids) of biological membranes (see below), and some are essential compounds with important functions in our biochemistry. Arachidonic acid, for example, is used by our bodies as a starting material for the biosynthesis of the prostaglandins (e.g., $PGF_{2\alpha}$, which stimulates contraction especially of uterine smooth muscle, and prostacyclin, which dilates blood vessels), the leukotrienes (e.g., LTC_4, which stimulates the contraction of especially lung smooth muscle), and the thromboxanes (e.g., thromboxane A_2, which triggers the constriction of injured blood vessels and the aggregation of blood platelets) (see Figure 2.9).

2.3.4
Terpenoids

The terpenoids are in general not directly involved in the basic metabolism that provides us with energy, but are instead products of what is called secondary metabolism (see also Chapter 5). They are found in all organisms, and are, for example, responsible for the smell of flowers. In humans, the major class of terpenoids is the steroids, represented in Figure 2.10 by cholesterol. Most of the cholesterol produced in a human (an adult contains approximately 150 g) is present as a component of the cell membrane, and the membranes of nerve tissue are particularly rich in cholesterol (containing 10%). Cholesterol is also used as a precursor in the synthesis of the sex hormones (e.g., testosterone in men and progesterone in women), the glucocorticoid hormones (e.g., cortisone), which, for

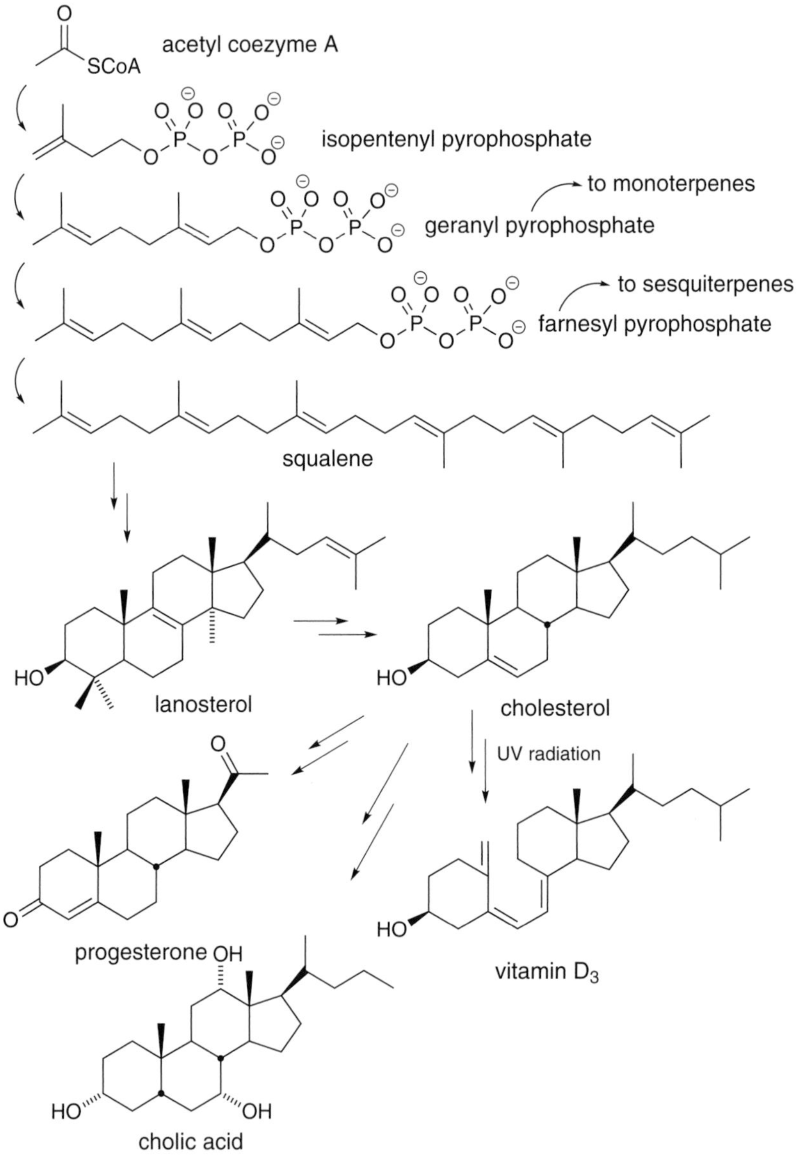

Figure 2.10 Some important steroids and their formation.

example, take part in the inflammatory process and regulate the metabolism of carbohydrates, cholic acid, which acts as an emulsifying agent in the intestines, and vitamin D, formed in a photochemical reaction (see Figure 2.10). The amounts of vitamin D (actually a group of structurally related compounds) that are made from cholesterol are not sufficient, and it is still an essential compound. Cholesterol is formed by the terpene biosynthetic pathway via lanosterol, an extremely elegant procedure that synthetic organic chemists so far can only dream of mimicking.

In addition, vitamin A is a terpenoid (formally a diterpene with 20 carbon atoms), while in vitamin E and vitamin K a major part of the molecule is a terpene. The four vitamins (A, D, E, and K) discussed in this subsection are relatively nonpolar compared to the other vitamins and are fat-soluble.

2.3.5
Amino Acids

Amino acids, or more correctly α-amino acids (as the amino group is positioned on the carbon α to the carboxylic acid), are the building blocks of peptides (up to approximately 20 amino acids), polypeptides (>20 amino acids) and proteins (one or several polypeptides and a molecular weight exceeding 5000 g/mol). The general structural formula of an α-amino acid is shown in Figure 2.11, and we should note a couple of important features. As discussed in Section 2.2, amines are basic and the nitrogen will be protonated by an acid, while carboxylic acids are acidic and will be deprotonated to form a carboxylate anion in the presence of a base, and the α-amino acid moiety is at physiological pH actually present as a zwitterion, with one positive and one negative charge (i.e., no net charge). The solvent water is acidic and basic enough to both protonate amines and deprotonate carboxylic acids. As zwitterions, amino acids are more polar and water soluble than one would expect from their normal structure. The α-amino acid moiety is also the link that connects the amino acids in peptides and proteins by forming an amide bond (called peptide bond) between two amino acids. The α-carbon is chiral in most α-amino acids, and these are mainly found in the laevo (i.e. not dextro) form, denoted by the L- prefix, in biological systems (Figure 2.12).

The structures of the 20 amino acids common in proteins are shown in Figure 2.12. The α-substituents differ in size, polarity, and charge, and the combination

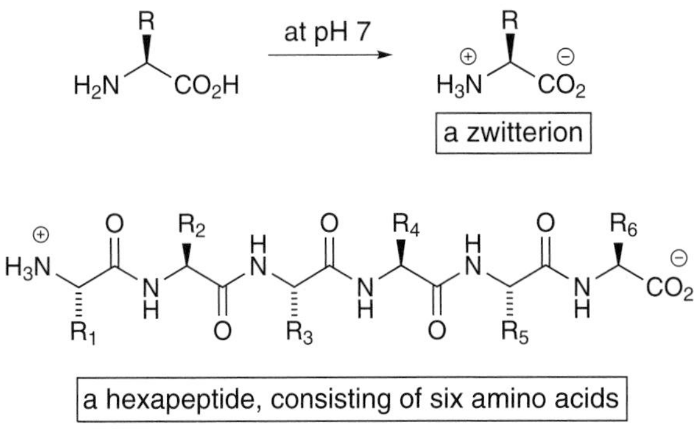

a zwitterion

a hexapeptide, consisting of six amino acids

Figure 2.11 The amino acid as a zwitterion, and in peptide bonding.

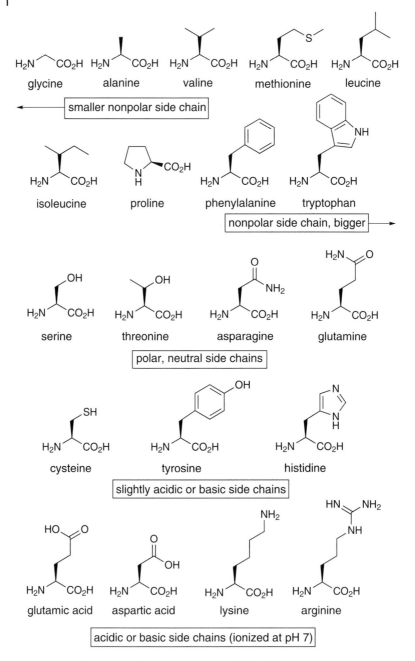

Figure 2.12 The amino acids (L-isomers) commonly found in proteins.

of amino acids, resulting in a chain with a combination of different α-substituents, is what gives the peptide or protein its characteristics. In Figure 2.12 they are presented according to their chemical features, first those with nonpolar side chains of different sizes, then those with polar but neutral side chains that are able to form strong hydrogen bonds, followed by weakly acidic (cysteine and tyrosine) and weakly basic (histidine) amino acids. The latter are only deprotonated/protonated to a minor extent at pH 7, but this will nevertheless influence their behavior. The last group of amino acids are more or less completely deprotonated (glutamic acid and aspartic acid) or protonated (lysine and arginine) at pH 7, and consequently have a net charge. In addition to size and polarity, we can note that some amino acids have chemical moieties that should readily participate in certain chemical reactions, something that we shall return to in the next chapter.

2.3.6
Nucleosides

A nucleoside a molecule composed of one of five heterocyclic aromatic amine bases (adenine, cytosine, guanine, thymine or uracil, derived from purine or pyrimidine) and one of two monosaccharides (ribose or 2-deoxyribose), linked together with a β-*N*-glycoside bond (see Figure 2.13). Nucleoside monophosphates (called nucleotides) have important functions as monomers in DNA and RNA (see below) and as secondary messengers that relay the message delivered to the cell membrane by hormones into the cell (e.g., cyclic AMP [AMP = adenosine monophosphate]), and are part of coenzymes such as coenzyme A and ATP (ATP = adenosine triphosphate).

2.4
Biomacromolecules and Cellular Constituents

2.4.1
Cell Membranes

The major functions of the membranes in a cell have already been mentioned, and in Chapter 4 we will discuss the transport of chemicals through membranes in more detail. Cell membranes are basically made up of a double layer of amphipathic phospholipids, ordered in the sense that the lipophilic parts of the molecules are inside the double layer while the hydrophilic parts are on the outside in contact with the surrounding water solution. This is also called a lipid bilayer (see Figure 2.14), and enables amphipathic compounds to spontaneously arrange themselves in water, as it minimizes the interactions between lipophilic (parts of) molecules and water and maximizes the number of stabilizing hydrogen bonds that can be formed. The phospholipids are diesters of phosphoric acid with diacylglycerol (the fatty acid of the diacylglycerol is normally palmitic, stearic, or oleic acid) and, for

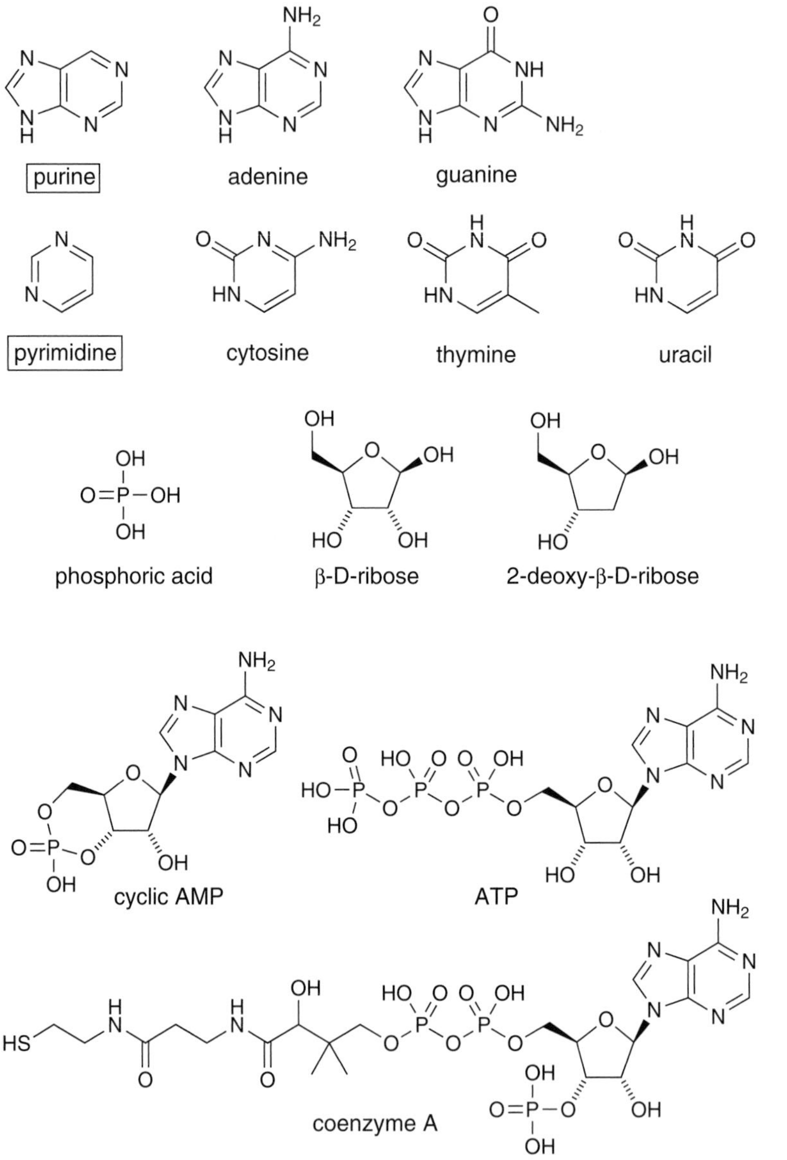

purine adenine guanine

pyrimidine cytosine thymine uracil

phosphoric acid β-D-ribose 2-deoxy-β-D-ribose

cyclic AMP ATP

coenzyme A

Figure 2.13 The components of nucleosides, and some important nucleoside derivatives.

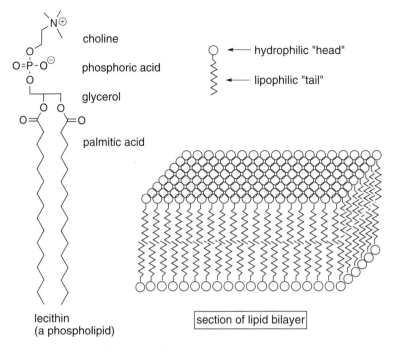

Figure 2.14 The lipid bilayer of cell membranes.

example, the alcohol choline (resulting in lecithin) or ethanolamine (resulting in cephalin). The fatty acid esters will give that part of the molecule lipophilic character, while the phosphate group is charged and hydrophilic. Amphipathic compounds in a membrane can move relatively freely in two dimensions, but will not be able to change sides.

Besides phospholipids, membranes also contain approximately equal amounts of proteins, although the proportions are different in different membranes (varying from 1:4 to 4:1). The proteins may be associated with the inside or the outside of the membrane or they may penetrate it, and their various functions will be discussed in the next section. In addition, the membranes contain small amounts of cholesterol, which, together with the nature of the fatty acid in the phospholipid, determine the fluidity of the membrane, and glycolipids, which reach out signaling carbohydrate structures into the extracellular space. A cell membrane is consequently very complicated and not completely understood. Figure 2.15 shows how one can imagine a cell membrane with all its different constituents, although it should be emphasized that this is only a relatively simple picture and that large variations between different membranes exist.

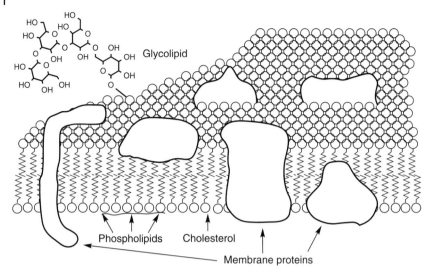

Figure 2.15 A 'true' membrane.

2.4.2
Proteins

The proteins may be described as the workhorses of cells, with a number of different functions. For example, they catalyze most biochemical reactions (enzymes), transform exogenous chemicals so that we can get rid of them (described in Chapter 5), make cells respond to signals (receptors), regulate the flux of chemical through membranes (described in Chapter 4), transport oxygen (hemoglobin), make cells mobile (actin and myosin), recognize foreign structures (antibodies), give cells and tissues their structure and form (collagen, elastin, keratin), etc. Peptides have functions as transmitters of signals between nerve cells and between cells of the immune system, and as hormones. Many peptides and proteins are also highly toxic, for example, bee and snake venoms, mushroom poisons, as well as diphtheria and botulin toxins.

Proteins are composed of a large number of amino acids, normally more than 100, often as a combination of several polypeptides. The chain of amino acids arranges itself in various regions of the polypeptide as certain conformations are more stable, for example as an α-helix, and will fold in space to form a three-dimensional structure whose form ultimately depends on the number and order of the various amino acids in the chain (see Figure 2.16). In addition, the form of a polypeptide is often stabilized by cross links between cysteine side chains, which form disulfide bonds. In principle, if a polypeptide is stretched out it will spontaneously fold back into its original form because this represents the most stable conformation. However, many proteins are only stable under special

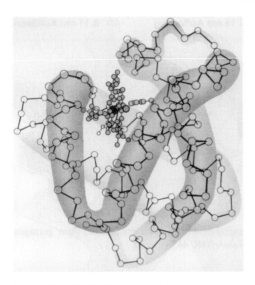

Figure 2.16 The folding of a protein.

circumstances, for example, when they are situated in a membrane, and will not fold to a functional three-dimensional form in a water solution. As well as the polypeptide(s), many proteins (mostly enzymes) also need a cofactor. This can be a simple metal ion or more complicated organic molecules (called coenzymes).

In enzymes and receptors only a small part of the protein is directly involved in the chemical conversion, or recognition of a messenger, and this is called the active site. In the active site, a few critical amino acids positioned at different locations in the polypeptide are brought together by the folding of the polypeptide. Evidently, chemicals may interfere with the function of such proteins by binding to or reacting with components of the active site, but it should be remembered that the interaction of a chemical with another part of the protein may affect its overall structure and thereby indirectly its active site.

2.4.3
Nucleic Acids

The nucleic acids DNA (deoxyribonucleic acid) and RNA (ribonucleic acid) are biopolymers of nucleotides joined together as a chain of phosphodiesters, as indicated for DNA in Figure 2.17. While the function of DNA is to be the data base of the cell and provide the information about how and when proteins should be prepared as well as enabling the cell to pass on the information to future cell generations, RNA is used to transcribe and translate the information provided by DNA. DNA makes use of the four bases adenine (A), cytosine (C), guanine (G) and thymine (T) (in RNA thymine has been exchanged for uracil [U]). The

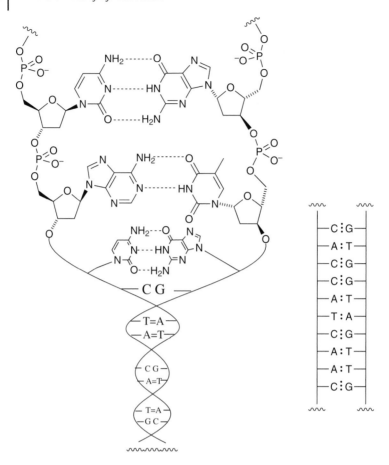

Figure 2.17 A schematic picture of DNA.

monosaccharide is deoxyribose (ribose in RNA) and occurs as two complementary and paired strands twisted into a double helix (as indicated in Figure 2.17) (RNA is single-stranded). In DNA, adenine will only form a base pair with thymine, while guanine only will form a base pair with cytosine, making A–T, T–A, G–C and C–G the only base pairs observed. Each one of our cells contains approximately 6 billion base pairs on 46 pairs of strands. The double-stranded nature of DNA makes it a very stable molecule and also facilitates the replication of DNA that precedes a cell division as each strand can serve as a unique template for the cell to make two identical copies of its DNA.

The information stored in DNA is called the genetic information, and the DNA of a cell or an organism is called the genetic material. The genetic information is encoded by the sequence of the bases in DNA, and, being a chemical memory, it is sensitive to chemical interference. Genotoxic chemicals (toxic to the genetic material) will affect the genetic information, and this can for example lead to the loss of cellular control systems that facilitate the development of cancer cells. An obvious target for genotoxic chemicals is the DNA bases, which have to fit exactly into the double helix and bind the complementary bases with the correct set of hydrogen bonds. As the bases are relatively good nucleophiles they may react with electrophiles to forms that cannot function in DNA. In addition, the double helix is sensitive to agents that cause breaks in the phosphodiester-deoxyribose polymer, and we shall return to this in Chapter 8.

2.5
Basic Environmental Chemistry

The chemistry of the earth can be described in one word – complex. So far we have discussed the elements of biochemistry, or the chemistry of organisms, which constitutes what is called the biosphere of the earth. In addition, we can identify several other distinct 'spheres', the atmosphere (the air stratum surrounding earth), the hydrosphere (oceans, seas, lakes, rivers, and so forth), the lithosphere (stone and rock material of the earth's crust), and the pedosphere (the soil, formed from the weathering of the lithosphere and the site of much of the life on earth). The biosphere is the part of the world that is inhabited by living organisms, approximately 3 km of the earth crust, up to a depth of some 10 km in the oceans, and 25 km up in the atmosphere. The elemental composition of the earth is completely different from that of organisms and of the cosmos, as can be seen in Table 2.2 where the 12 most common elements in various materials are shown.

Table 2.2 The amounts (wt %) of the 12 most common elements of (a) the world (the explorable parts, including the earth's crust, the atmosphere and the hydrosphere), (b) the earth's crust (excluding atmosphere and hydrosphere), (c) man, and (d) the cosmos. The values given are estimates.

World		Earth's crust		Man		Cosmos	
O	49.5	Fe	36.9	O	63	H	74.6
Si	25.8	O	29.3	C	19	He	23.7
Al	7.6	Si	14.9	H	9	O	0.81
Fe	4.7	Mg	6.7	N	5	C	0.28
Ca	3.4	Al	3	Ca	1	Fe	0.14
Na	2.6	Ca	3	P	1	N	0.09
K	2.4	Ni	2.9	S	0.6	Ne	0.08
Mg	2	Na	0.9	K	0.2	Si	0.08
H	0.9	S	0.7	Cl	0.2	Mg	0.07
Ti	0.4	Ti	0.5	Na	0.1	S	0.04
Cl	0.2	K	0.3	Mg	0.04	Ar	0.02
P	0.1	Co	0.2	Fe	0.01	Al	0.01

The 12 most common elements of our world together comprise more than 99.5% of the total mass, and carbon, for example, a basic element in biochemistry, can be regarded as a trace element. Oxygen is the most common, comprising approximately 50% of the mass. Most of it is present in the lithosphere and hydrosphere, while the atmosphere contributes surprisingly small amounts of this element.

2.5.1
The Atmosphere

There is no sharp limit between the atmosphere surrounding the earth and the rest of the cosmos, but for practical reasons the limit is set to 1000 km. The first 10 km above the sea level is called the troposphere, the stratosphere is the space between 10 and 50 km, the mesosphere between 50 and 100 km, while the thermosphere constitutes the space between 100 and 1000 km. The pressure at the upper end of the troposphere (10 km altitude), the absolute limit for humans without technical equipment, is 25% of that at the sea level, while at the end of the atmosphere (1000 km altitude) it is only 0.000 000 000 01%. Nevertheless, the N_2 and O_2 present in the thermosphere absorb the most energetic radiation (with wavelengths between 120–220 nm) from the sun. The energy release is enormous, and the 'temperature' in the thermosphere is above 1000 °C, but because of the low pressure this value cannot be compared with 1000 °C at sea level. Most of the ultraviolet radiation, with a wavelength above 220 nm, still very dangerous to man, passes through both the thermosphere and the mesophere but is absorbed in the stratosphere by O_2 and ozone (O_3). Some chemicals, for example, the freons, hamper this process and increase the influx of UV light, and this will be discussed in Section 10.2.1.

In view of the central functions of the atmosphere in regulating the influx of radiation and the outflux of heat, combined with the facts that it contains only a small part of the total mass of the world and that the transport of chemicals in the atmosphere is very efficient (see below), it is not surprising that several of the environmental threats discussed in Chapter 10 are linked to the atmosphere. An example of how sensitive the atmosphere is to chemical pollution is a massive volcanic eruption, a natural event that occurs regularly, and which in a short time (within a year) may have a significant impact on the whole world, mainly because of the release of particles. The composition of the atmosphere appears simple (all figures, including those of Table 2.3, are given for dry air at ground level in 1992 [the average concentration of water in air close to the earth is approximately 0.05%]), more than 99.96% consisting of the three gases N_2 (78.08%), O_2 (20.95%) and argon (Ar) (0.93%). However, the remaining 0.04% is a complex mixture (called the trace gases) and contains several compounds (natural as well as anthropogenic, i.e. released by human activities) that are causing some of the negative effects discussed primarily in Chapter 10. In Table 2.3, the concentration of the various trace gases in the atmosphere is shown.

Table 2.3 Approximate concentrations (in ppb, litres per 1 000 000 m3 air) of the trace gases in the atmosphere (dry air at earth level in 1992). 'Stable' signifies a half-life of at least 10 years, 'unstable' a half-life of days while 'very unstable' compounds have a half-life of minutes or shorter. The concentration of carbon dioxide has from 1992 to 2009 increased to approximately 385 000 ppb.

Compound	Structure	Conc. in ppb	Notes
Carbon dioxide	CO_2	360 000	Increasing
Neon	Ne	18 000	Noble gas
Helium	He	5 200	Noble gas
Methane	CH_4	1 800	Increasing
Krypton	Kr	1 100	Noble gas
Hydrogen	H_2	500	
Dinitrogenoxide	N_2O	300	Incr., stable
Carbon monoxide	CO	100	Fluctuates
Xenon	Xe	87	Noble gas
Ozone	O_3	50	Fluctuates
Nitrogen dioxide	NO_2	50	Fluct., very unstable
Nitrogen monoxide	NO	50	Fluctuates
Sulfur dioxide	SO_2	10	Fluct., unstable
Chloromethane	CH_3Cl	0.6	
Dichlorodifluoromethane	CCl_2F_2	0.5	Incr., stable
Carbon oxidesulphide	COS	0.5	
Trichlorofluoromethane	CCl_3F	0.3	Incr., stable
Ammonia	NH_3	0.3	Fluctuates
Formaldehyde	HCHO	0.3	Fluct., unstable
1,1,1-Trichloroethane	CH_3CCl_3	0.1	
Tetrachloromethane	CCl_4	0.1	
Nitric acid	HNO_3	0.1	Fluct., unstable
Tetrafluoromethane	CF_4	0.07	Stable
Carbon disulfide	CS_2	0.05	Fluct., unstable
Dimethylsulfide	CH_3SCH_3	0.05	Fluctuates
Peroxyacetylnitrate	$CH_3CO_3NO_2$	0.02	Fluctuates
Methanethiol	CH_3SH	0.02	Fluctuates
Hydroperoxyl radical	HO_2.	0.004	Very unstable
Bromtrifluoromethane	$CBrF_3$	0.002	
Hydrogen peroxide	H_2O_2	0.001	Unstable
Carbon hexafluoride	SF_6	0.0005	
Hydrogen sulfide	H_2S	0.0001	
Hydroxyl radical	OH.	0.00004	Very unstable

2.5.2
The Hydrosphere

Water as a chemical has been discussed in Section 2.3.1. It is not only the solvent in which the life processes take place, but also participates in many biochemical reactions of which perhaps the photosynthetic reactions producing oxygen and carbohydrates are the most important. The total amount of water in the world is

Table 2.4 The distribution of water in the world.

	% of total
Oceans and seas	97.2
Snow and ice	2.1
Subsoil water	0.7
Lakes and rivers	0.009
(Atmosphere	0.0009)
(Organisms	0.00004)

$1.4 \times 10^9\,km^3$. Measured in liters, the amount is 1.4×10^{21}, an almost inconceivably large number that still only represents the approximate number of water molecules in one drop of water. Its distribution in the various parts of the hydrosphere (and other spheres) is shown in Table 2.4.

The mixing of various natural waters takes much longer than the mixing of the air masses of the atmosphere (see below), and the detailed chemical composition of different waters may vary considerably. Besides having a longer mixing time, the hydrosphere has 300 times the mass of the atmosphere. While pollutants released into the atmosphere may result in global effects, those released into the hydrosphere are more likely to result in local (within a range of 100 km) or regional (within a range of 1000 km) damage.

2.5.3
The Lithosphere and the Pedosphere

The lithosphere extends approximately 100 km below the surface of the earth and mainly consists of rock, but more interesting is the pedosphere – the top layer (on average a few decimeters or meters deep) of weathered stone that we call soil – because it is here that we find most life forms. Soil is composed of complex mixtures of organic and inorganic chemicals in varying amounts, but the components minerals, decayed or decaying dead organisms (organic material), living organisms, water, and air are present in most soil. Water and air fill the pores and may together constitute 50% of the volume of soil. The minerals (as clay particles) are formed by erosion (flowing water and winds) and by the action of freezing water on stones and rocks, and these provide many essential elements as inorganic salts that can be dissolved in the water and absorbed by organisms. The organic remains of dead organisms provide the basis for the organisms living in soil. As indicated in Table 2.5, the number of organisms living (on average) in the top 30-cm deep layer of a square meter of soil is enormous, and the amounts of organic material continuously added to the soil are large.

The various organisms have adapted to each other's presence, and to a large extent they live separately from each other, although parasitic and symbiotic relationships have also emerged. An example of an extremely important symbiosis is

Table 2.5 Typical numbers of various organisms found in the uppermost 30 cm layer of 1 m² soil.

Organism (group)	No.
Bacteria	60 000 000 000 000
Fungi	1 000 000 000
Other microorganisms	500 000 000
Nematodes	10 000 000
Algae	1 000 000
Mites	150 000
Springtails	100 000
White earthworms	25 000
Earthworms and fly larvae	200
Millipedes	150
Beetles	100
Snails, spiders, wood-lice, centipedes	50
Vertebrates	0.001

the one between plants and fungi, which exchange nutrients to the benefit of both organisms. Any negative influence on the conditions for the growth for the fungus (e.g., caused by the presence of an antifungal chemical) would seriously affect the plant as well. The debris from dead organisms can either be consumed by living organisms, for example, microorganisms, which in principle will convert it to carbon dioxide, water, and other end products, or be transformed by humification to humic substances (e.g., humic acids) and humin. The humic acids, which may constitute 1–20% of a soil, play an important role, as they form complexes with metal ions because of the presence of free –COOH, –OH and –NH groups. This makes important ions (e.g., Ca^{2+}, Mg^{2+} and K^+) available for the organisms. In addition, because of the polarity of the humic acids, they also retain the water present in soil. They are not defined molecules but instead polymers made up from biochemicals available (e.g., proteins and carbohydrates), and they are constantly formed and degraded by the microorganisms. A final point that should be stressed is that the metabolic capacity of the organisms in the soil is enormous, and will in many cases have a significant impact on pollutants that end up in the soil. This will be further discussed in Chapters 6 and 10.

2.5.4
The Turnover of Chemicals in and between the Spheres

Most important for the possibilities of chemicals to give rise to ecological effects and to be subjected to biological conversions and chemical transformations is their transportation within and between the different spheres. The turnover of chemicals in the lithosphere is of course very slow, while in the pedosphere it is comparable to that of the hydrosphere. As has already been indicated, it is much faster

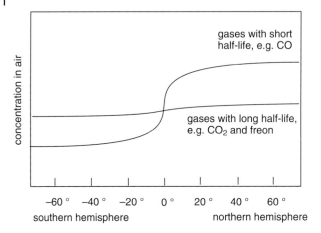

Figure 2.18 The distribution in the atmosphere of gases continuously emitted for years in the northern hemisphere.

and more efficient in the atmosphere, because relatively small amounts of air are transported efficiently by the weather systems. The atmosphere can be divided into two hemispheres, the southern and the northern. The mixing time (the time it takes until the air is homogenous after a discharge) is less than two months within a hemisphere but almost two years between the hemispheres. As most of the emissions into the atmosphere, at the moment, take place in the northern hemisphere, there is a substantial difference between the two hemispheres in the concentration of pollutants with relatively short half-lives (the time it takes for the concentration to fall by 50%) (see Figure 2.18). For example, the concentration of carbon monoxide, a compound that only will stay a couple of months in the atmosphere, is approximately 50 ppb in the southern and 150 ppb in the northern hemisphere. However, the concentration of compounds that are more stable (having a half-life of years, e.g., carbon dioxide and the freons) will be slowly equalized over the whole atmosphere.

Hydrophilic compounds in the atmosphere will be washed away by the rain (water in the atmosphere is rapidly turned over with a half-life of a few weeks) and be transported to the oceans and the soil. From the soil, a major flow for chemicals is with the excess water to rivers and eventually to the oceans, although substantial amounts of chemicals (if they are naturally volatile or become volatile after conversion in the soil) are also transported back to the atmosphere. Also there is a flux of chemicals from the oceans to the atmosphere; approximately 0.03% of the water of the oceans evaporates yearly and with it any volatile chemical. Volatile, lipophilic chemicals (e.g., the freons and methane) will largely stay in the atmosphere, while nonvolatile hydrophilic chemicals will end up in the oceans. Nonvolatile chemicals that are insoluble in water may be adsorbed on various clay particles and deposited in the mud at the bottom of oceans and seas or in the deeper layers of the soil.

3
Chemical Properties

Chemicals that are toxic to organisms may be described, for example, as cytotoxic, carcinogenic, allergenic, etc. The potential of a compound to produce toxic effects depends on its chemical properties, for example, its volatility, solubility, reactivity, and so forth, which in turn depend on the molecular properties of the compound. Important molecular properties are, for example, the size and form of a molecule, the distribution of charge in its different parts (i.e., its polarity), its polarizability (i.e., how easily the electrons can move in the molecule), and the strength of the bonds within the molecule. It is evident that one can discuss and describe hazardous chemicals by their effects on biological systems, by the chemical properties that make them hazardous, or by the molecular properties that determine the chemical properties. To some extent, this text attempts to mix all three approaches and to explain how a toxic effect by a chemical (e.g., carcinogenicity) is caused by a chemical property (such as chemical reactivity) due to the presence of certain functional groups in the molecule (that, for example, enable the chemical to form new bonds with the genetic material). In this chapter, a survey of the most important chemical properties involved in toxicity will be made. The discussion is based on the understanding of how different intermolecular (i.e., between molecules) chemical forces and bonds function, as described in the chemical literature.

3.1
Critical Chemical Properties

3.1.1
Volatility

As will be discussed in Chapter 4, the principal route for exposure to chemicals in the workplace is via the lungs, the chemicals inhaled being present in the air either as a vapor or as an aerosol. A compound's volatility will therefore to a large extent determine the risk that a chemical will be absorbed by persons exposed to it at workplaces, and will also influence its distribution in the environment. Many commercially important chemicals, for example, most organic solvents, are liquid

Chemistry, Health, and Environment. Olov Sterner
© 2010 WILEY-VCH Verlag GmbH & Co. KGaA, Weinheim
ISBN: 978-3-527-32582-5

at room temperature but evaporate readily. A liquid stays liquid because the inter-molecular forces hold the molecules together. However, by increasing the kinetic energy of the molecules, for example, by increasing the temperature, molecules may become energetic enough to break free from the intermolecular forces and pass into the gaseous phase. The energy of the molecules in a liquid is not uniform, but has a Gaussian distribution. Even at temperatures much lower than the boiling point, a small proportion of the molecules will have sufficient kinetic energy to leave the liquid. Therefore, all liquids (and in principle also all solids) have a vapor pressure at all temperatures, and wet clothes will dry at room temperature (and eventually even at temperatures below 0 °C when the water is frozen). When the temperature increases the vapor pressure does the same, and at temperatures when the vapor pressure is the same as the air pressure the liquid will boil.

The volatility of a chemical is influenced by several other factors as well as the temperature and the strength of the intermolecular forces. The air pressure, the vapor pressure of the evaporating compound in the gas phase above the liquid, the composition of the liquid, as well as the molecular weight of the compound in question may also be important. For example, at high altitudes, where the air pressure is lower than it is at sea level, water will boil below 100 °C, and in techni-cal systems it is common to decrease the pressure in order to make a solvent evaporate faster. If the pressure is increased the boiling point increases, as, for instance, in a pressure cooker where water is liquid even at 120 °C. If the gas phase above the liquid contains a lot of the volatile compound this will slow down the evaporation, because condensation (from the gas to the liquid phase) will also take place. Wet laundry will dry faster on a dry day then on a humid day, even if the temperature is the same. When the compound evaporating is present in a mixed liquid consisting of several chemicals, the intermolecular forces experienced by one molecule depend on its neighbors. However, the volatility of the components of a mixture will normally not differ much from the volatility of the pure com-pounds. The molecular weight will influence the volatility because the kinetic energy of a molecule depends on its mass. If two molecules with different molecu-lar weight experience the same intermolecular forces, the lighter will be more volatile (have lower boiling point) because the heavier requires more heat to obtain the same kinetic energy.

The volatility of a pure compound may be characterized in several ways, for example, by the partial pressure of the compound over the liquid at a certain temperature or the boiling point of the pure liquid at a certain air pres-sure. For practical reasons, it is easier to use the boiling point at normal sea level atmospheric pressure, as such values can easily be extracted from the chemical literature. In Table 3.1, the boiling points of a number of compounds with com-parable molecular weights but containing different chemical functionalities are listed.

While it is obvious that butane is more volatile than pentane, which in turn is more volatile than hexane, the difference between the pentane isomers 2,2-dimethylpropane (neopentane), 2-methylpentane (isopentane) and pentane is

Table 3.1 Comparison of the volatility of chemically different compounds.

Name	Structures	Molecular weight	Boiling point (°C)
Butane		58	0
2,2-Dimethylpropane		72	9
2-Methylbutane		72	30
1-Fluorobutane	F	76	32
Diethyl ether	O	74	35
Pentane		72	36
Methyl propyl ether	O	74	38
Propyl chloride	Cl	78	47
Methyl acetate		74	57
1-Propanethiol	SH	76	68
Hexane		86	69
Butanal	H	72	76
1-Aminobutane	NH$_2$	73	78
Butanone		72	80
Nitroethane	NO$_2$	75	115
1-Butanol	OH	74	117
Butyronitrile	CN	69	118
Propanoic acid	OH	74	141
Propanomide	NH$_2$	73	213

not self-evident. However, the van der Waals forces are the only attractive forces acting on these hydrocarbons, and their strength is proportional to the surface area of the molecule. A straight-chained molecule like pentane is elongated and has a larger surface compared to that of 2,2-dimethylpropane, which is more ball-shaped. This effect is general for all molecules in which van der Waals forces make a significant contribution to the intermolecular forces, and 2-chloro-2-methylpropane (*tert*-butyl chloride), for example, boils at 51 °C compared to 77 °C for 1-chlorobutane. It is perhaps surprising to find that compounds containing strongly electronegative atoms such as oxygen and fluorine (1-fluorobutane, diethyl ether (ether), and methyl propyl ether in Table 3.1) do not have higher boiling points than the nonpolar hydrocarbon pentane. However, the dipole moment of a molecule depends not only on the difference in electronegativity between two atoms but also on the length of the bond. Fluorine and oxygen are relatively small atoms that are bound more closely to carbon (bond lengths 1.39 and 1.43 Å, respectively) than, for example, chlorine (bond length 1.78), and the dipole moment of chlorinated hydrocarbons is actually larger than that of the corresponding fluorinated hydrocarbons. The effect of the dipole created by the fluorine–carbon and oxygen–carbon bond is therefore limited, and is comparable with the van der Waals forces. Propyl chloride has a significantly higher boiling point (see Table 3.1), which is due to its larger dipole moment as well as its greater ability to be polarized (giving rise to stronger van der Waals forces), because the three unshared electron pairs of chlorine are less tightly held by the chlorine nucleus than they are in the case of the smaller fluorine. The same effect is responsible for the increased boiling point of 1-propanethiol. Although sulfur has approximately the same electronegativity as carbon it is more easily polarized. The thiol group may in some cases take part in weak hydrogen bonds, although this is not the case in 1-propanethiol. However, the effects are still small, and are comparable to the addition of a methylene group to pentane (to give hexane). The carbonyl groups of aldehydes and ketones are considerably more polar, being electron-withdrawing both because oxygen is more electronegative than carbon and because of the possibility of resonance (discussed later in this chapter). Butanal and butanone therefore have a similar boiling point to that of 1-aminobutane, these being our first examples of compounds that forms hydrogen bonds to themselves. The hydrogen bonds in 1-aminobutane are relatively weak compared to those in 1-butanol because of the difference in electronegativity between nitrogen and oxygen, and the boiling points of the two compounds differ by almost 40 °C. 1-Butanol has a similar boiling point to those of nitroethane and butyronitrile, indicating the polarity of the nitro and the cyano groups, both of which are unable to donate hydrogen bonds. Carboxylic acids and amides are even less volatile because of their polarity and the ability of these functionalities to form hydrogen bonds. Two carboxylic acids (or amides) may form a dimer (by forming two hydrogen bonds to each other, see Figure 3.1), and this may formally be regarded as a polar entity with twice the molecular weight of the monomer (therefore being considerably less volatile). The difference between carboxylic acids and amides (e.g., propanoic acid and propionamide in Table 3.1) is due to the greater tendency of nitrogen to donate its unshared electron pair to the carbonyl group, pushing electrons to the carbonyl oxygen, which will have an even larger negative charge (see Figure 3.1).

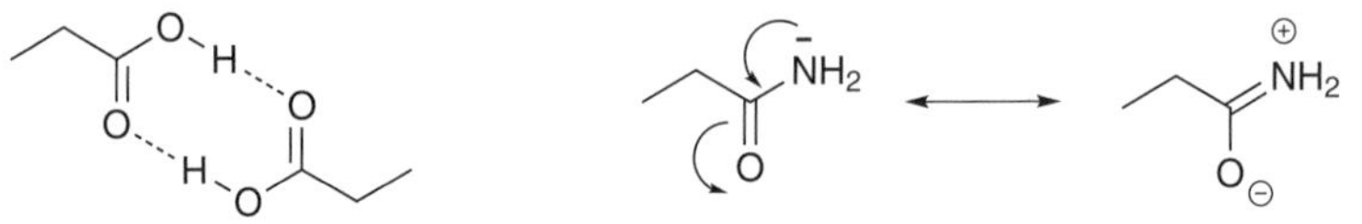

Figure 3.1 A hydrogen bond dimer of propanoic acid (left) and the electron flow from nitrogen to oxygen in propionamide resulting in increased polarity (right).

3.1.2
Solubility

The solubility of a compound is an important property that strongly influences the mobility of a chemical in the environment and between different organisms, as well as its absorption into and distribution in organisms. As will be discussed in the next chapter, cells are surrounded by membranes that are more easily penetrated by nonpolar compounds than by polar or ionic compounds, and most toxic compounds need to get into cells in order to have their toxic effect. In general, a compound (a solute) will be dissolved by another (a solvent) if the intermolecular forces present in the final solution are stronger than those present in the solute and solvent separately. Nonpolar solvents will, in general, dissolve nonpolar compounds easily, while polar solvents will dissolve polar compounds. Because fat is nonpolar while water is polar, the term lipophilic ('fat-loving') or hydrophobic ('water-fearing') is used to indicate that a compound is nonpolar, while hydrophilic and lipophobic indicate the reverse. Amphipathic compounds have both a lipophilic and a hydrophilic part, and prefer surfaces and phase borders (such as cell membranes).

To characterize nonpolar compounds as 'hydrophobic' indicates that there is some kind of repulsion between water and nonpolar molecules that forces them to stay apart, and this is a general misunderstanding. There is no such repulsion: water and fat molecules will be attracted to each other by van der Waals forces. The reason that water and fat do not mix is instead that if fat molecules were to be inserted among the water molecules, strong, stabilizing hydrogen bonds between the water molecules would have to be broken and that would cost the system energy, as indicated in Figure 3.2. This effect is important and useful in biological systems, where it is called the hydrophobic effect or hydrophobic repulsion.

Although many organic compounds can be classified either as hydrophilic (e.g., carbohydrates) or lipophilic (e.g., fats and waxes), most fall between these two extremes. It has therefore been found useful to describe the solubility of a given compound in terms of its distribution between two solvents that are not soluble in each other. If, for instance, 1 mg of a chemical is mixed with 1 mL each of the two inmiscible solvents water and cyclohexane and the mixture is allowed to reach equilibrium, the relative amounts of the compound that dissolve in the two phases (water/cyclohexane) give an indication of its hydrophilicity or hydrophobicity. As octanol to some extent resembles the amphipathic compounds present in biological membranes (see Chapter 2), the two-phase system octanol/water is most frequently used today to characterize a compound's solubility. The proportion of the compound in the two phases is called P, and for practical reasons (for highly lipophilic compounds such as DDT, P is over 1 000 000) the logarithm ($\log_{10}P$, is often used. In Table 3.2, the solubility in water and the P values of some common organic compounds are listed for comparison.

The smallest alcohols, methanol, ethanol, and the two propanols, are all miscible with water in all concentrations at 20 °C because of the strong hydrogen bonding

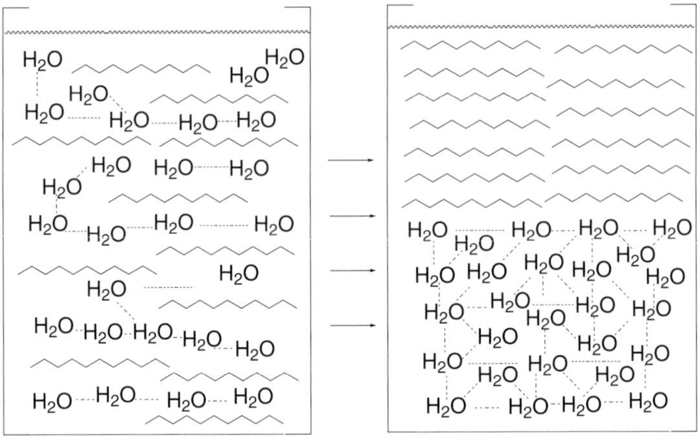

The system is more stable if the maximum number of strong hydrogen bonds between the water molecules can be formed, and will spontaneously go from left to right.

Figure 3.2 Nonpolar compounds will not dissolve in water.

to the hydroxyl groups and the relatively small effect that the alkyl groups have on the water–water interactions. However, for three of the butanols, 1-butanol, 2-methyl-1-propanol (isobutanol) and 2-butanol, the water solubility is limited and the P values are over 4, as the larger alkyl group will interfere with more hydrogen bonds between water molecules and will increase the van der Waals forces between the butanol molecules. There is an upward trend in the water solubility in the butanol series, the fourth member, 2-methyl-2-propanol (*tert*-butanol), being completely soluble in water. This is because of the differences in the total size and surface area of the alkyl groups, which to different degrees will disturb the hydrogen bonds between the water molecules and give rise to attractive van der Waals forces between the organic molecules. For pentanol and higher alcohols the water solubility is low and the P values high. The two cyclic ethers dioxane and tetrahydrofuran (THF) are miscible with water, while diethyl ether, which can be seen as a open form of tetrahydrofuran, and diisopropyl ether have limited water solubility and higher P values. The comparison between tetrahydrofuran and diethyl ether is interesting, as tetrahydrofuran also has a considerably higher boiling point (66 °C) than diethyl ether. In diethyl ether there is free rotation around all bonds, resulting in a partial canceling of the dipole resulting from the carbon–oxygen bonds by the smaller carbon– (carbon-with-oxygen) dipoles, while the dipoles in tetrahydrofuran are forced to work together, making the latter a more polar molecule. The carbonyl function is polar as it is, and may in addition take part in the formation of an hydrate or an enol (see Figure 2.2). Small aldehydes and ketones are miscible with water, but starting with propanal and butanone the solubility is limited (note the similarity in water solubility of 2-butanol and butanone, Table

Table 3.2 The relative solubility in octanol/water (P) compared with water solubility.

Name	Structures	Solubility in water (% w/w, 20°C)	P (octanol/water)
Methanol		∞	0.2
Ethanol		∞	0.5
1-Propanol		∞	1.8
2-Propanol		∞	1.1
1-Butanol		7.8	7.6
2-Methyl-1-propanol		8.5	5.8
2-Butanol		12	4.1
2-Methyl-2-propanol		∞	2.2
1-Pentanol		2.5	36
Dioxane		∞	0.5
Tetrahydrofuran		∞	2.9
Diethyl ether		6.9	7.8
Diisopropyl ether		0.9	33
Acetone		∞	0.6
Butanone		26	2.0
Cyclohexanone		2.3	6.5
Acetic acid		∞	0.7
Methyl acetate		24	1.5
Ethyl acetate		7.9	5.4

3.2). Carboxylic acids are miscible with water up to pentanoic acid (and its isomers), while all esters (even methyl formate) have limited water solubility.

The lipophilicity of organic compounds is a central factor in bioaccumulation (the 'extraction' of chemicals from the environment by organisms and the transportation and concentration from species to species in food chains), and this will be further discussed in Chapter 4.

3.1.3
Adsorption and Absorption

Sorption is the process in which a chemical (a sorbate) becomes associated with a solid material (a sorbent); adsorption refers to sorption on a surface (two dimensions), while absorption is sorption into a material (three dimensions). As most organic compounds emitted to the environment eventually contact solid materials, sorption to organic and inorganic particles in the soil, for example, will influence the fate of an organic pollutant. Adsorption depends on intermolecular forces between the sorbent and the sorbate and results in an equilibrium between the adsorbed and the free state. The free state is normally a solution in the surrounding water, but it can of course also be the gas phase. Adsorption means that a compound is taken out of circulation, at least temporarily, and can be seen as something that protects organisms from pollutants. However, adsorbed compounds are less available to sunlight and microorganisms, and are thereby not degraded at the same rate as they are in a water solution. This means that their lifetime is prolonged, and if the soil is disturbed in some way (for example during construction work) the pollutant may reappear.

Soil is a complex mixture of various inorganic minerals as well as organic matter, and its composition differs from location to location (see also Chapter 2). While the amount of organic matter is substantial at the surface, very little is found deeper down, where, for instance, the groundwater is present. In addition, the size of the particles in soil is an important factor, as small particles have a much larger surface per gram and thereby much better qualities as a sorbent. This makes it difficult to generalize about how efficiently different types of compounds are sorbed. As most minerals contain charged groups at their surface, charged compounds and easily protonated/deprotonated compounds are normally efficiently adsorbed. An example is the herbicide paraquat (see Figure 3.3), which is toxic to the lungs (Section 7.3.5) but has been considered relatively safe to use from an environmental point of view. Paraquat is an efficient herbicide that kills only the plants whose leaves it contacts. The paraquat that ends up in the soil is strongly sorbed and will not wash away.

Most organic pollutants are devoid of electric charge but can still be adsorbed to inorganic materials if they contain polar groups that are attracted by ion–dipole and dipole–dipole forces. However, more important is their sorption to organic matter in soil, and this is strongly correlated with the solubility of the sorbate. Neutral compounds with high water solubility (having small log P values) are poorly sorbed by organic matter, while lipophilic compounds in general are sorbed well. The soil particles are not always retained in the soil, and very small organic

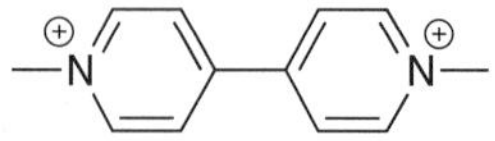

Figure 3.3 The paraquat ion has two positive charges.

particles (approximately 1 µm in diameter), so-called organic colloids (e.g., humic substances), behave as if they are part of the water solution. Colloids do not separate from water by gravitation, and may be seen as microparticles or macromolecules that can still function as sorbents. Organic colloids may in this way substantially increase the 'water solubility' of pollutants.

3.1.4
Persistence

A chemical is called persistent if it stays unchanged for a very long time in the environment. Most inorganic chemicals can be regarded as persistent, and are at most transformed into other inorganic compounds. The lifetime of an organic chemical may vary considerably: many are rapidly degraded by, for example, microorganisms and sunlight, while others are difficult to get rid of. Many compounds developed for specific purposes have been designed to be persistent, as this was considered to be an advantage, and examples are most of the chlorinated pesticides used in the 1950s and 1960s. Persistence is not really a chemical property, it is a lack of other properties that results in the preservation of a compound. The half-life of a chemical in the environment increases if it is poorly soluble in water and if it is strongly sorbed in the soil or at the bottom of a sea, because the microorganisms and sunlight will not get at it. Another important property that persistent chemicals should not have is reactivity. They should in fact be as inert as possible to both chemical and biological degradation. Saturated chemicals are therefore generally more persistent than the corresponding unsaturated analogs, although the aromatic hydrocarbons are the most stable. The persistence of aromatic hydrocarbons increases with increasing number of substituents, especially halogens.

3.1.5
Reactivity

A compound's reactivity is perhaps the most important factor when it comes to the ability of a chemical to cause a toxic effect in an organism. Although volatility and solubility (as well as other properties of a compound) are important for bringing it to the target, it normally has to trigger or precipitate a chemical reaction in some way to induce the toxic effect. However, it should be remembered that chemical reactivity in general can be anything from an explosion to subtle interactions between a ligand and a receptor in a nerve cell. Also, it is not automatically the most reactive chemicals that are the most dangerous, as such compounds are often well known and their hazardous properties may be evident from their behavior. Apparently nonreactive compounds are often considered to be relatively harmless, but may, for example, be slowly converted inside the body to a form that is carcinogenic. It is therefore important to have a basic understanding of the different forms of reactivity and how they can affect an organism as well as be related to the chemical structure of the reactive compound.

3.1.5.1 **Explosives**

Compounds that have an intrinsic instability may be prone to rapid decomposition which converts them into other more stable compounds and releases large amounts of energy. Explosives are compounds or mixtures of compounds that may rapidly react, with the generation of large amounts of heat and pressure, if a certain amount of initial energy (activation energy) is provided. For very sensitive explosives it is enough if they are scraped or struck, while others require an ignition spark or a shock wave from a detonator. If the reaction spreads through the explosive material at less than $1000\,\mathrm{m\,s^{-1}}$ it is called a deflagration, and if the speed is higher it is a detonation. In most cases, explosions can be regarded as an internal combustion without external oxygen. Compounds causing explosions or detonations are not the main subject of this book, but it should be recognized that hazards due to explosive chemicals are of major concern in many workplaces where chemicals are used. In the text, comments will be made about functional groups and combinations of functional groups that are frequently associated with explosives, but not in a comprehensive way.

3.1.5.2 **Acids and Bases**

Acids and bases are reactive and will affect biological systems in various ways depending on their strength and concentration. Strong acids and bases in high concentrations will simply disintegrate and destroy biological tissue, while weaker acids and bases in lower concentrations produce milder effects. Formic acid, hydrofluoric acid, sodium or potassium hydroxide, and nitric acid are examples of acids and bases that are particularly harmful because they break up the skin and make the chemical penetrate deeper. A substantial proportion of all accidents that happen in workplaces with chemicals and that lead to bodily harm involve acids and bases. The damage that exposure to an acid or base will cause to a person depends not only on its strength and concentration but also on the part of the body affected. Some organs are extremely sensitive (e.g., the eyes) while other parts of the body are more resistant. The external effects of acids and bases are fairly well known to anybody working with chemicals and will not be further discussed in this text. However, it should be stressed that accidents with acids and bases are very common and that such compounds should never be handled unless protective measures have been taken. *Always wear protective glasses*. The internal effect of acids will be discussed in Chapter 7.

3.1.5.3 **Reducing and Oxidizing Agents**

Compounds or reagents used for oxidations and reductions will normally react with and damage biological tissue. Strongly oxidizing and reducing agents, for example, sodium dichromate and potassium hydride, are highly reactive and will damage whatever tissue with which they come into contact. During oxidations and reductions an acid or base will be formed as a secondary product, and this will aggravate the effect. However, oxidations and reductions may also be more selective and, for example, change the oxidation state of a metal ion needed for a certain

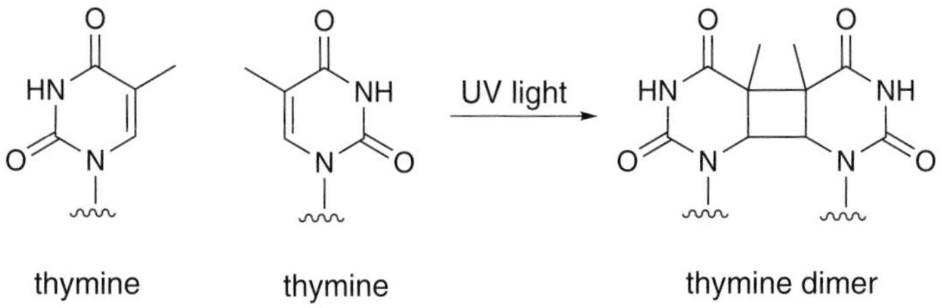

Figure 3.4   The formation of thymine dimers after UV radiation of our cells.

enzymatic reaction (for an example, see Section 7.2.4.2) or generate highly reactive radicals (see below), and we get back to some examples of selective toxicity caused by oxidations and reductions.

3.1.5.4 Excited Molecules

Molecules can be excited to reactive states by the input of energy in some form, most commonly electromagnetic radiation in the form of sunlight. Sunlight consists of a spectrum of wavelengths of which the visible region is only a part. The most powerful is ultraviolet (UV) light, with wavelengths between 10 and 400 nm. At a wavelength of 242 nm, this is able to dissociate molecules (e.g., O_2 + hv → 2 O) or, at a wavelength of 98 nm, ionize them to produce free electrons (H_2O + hv → H_2O^+ + e⁻). Such reactions take place in the upper parts of the atmosphere, and the reactive species formed actually serve to clean the atmosphere of air pollutants. However, the photochemical reactions can also be severely disturbed by pollution, and this will be discussed in Chapter 10. A little UV light penetrates the atmosphere all the way down to the earth, and both humans and other organisms have developed protections against this potentially very dangerous radiation. Nevertheless, exposure to sunlight is associated with an increased risk of skin cancer, and this is caused by photoexcited reactions that take place in the cells of the skin. For example, UV light can excite thymine, a component of our genetic material, to a form that readily reacts with another thymine and forms a thymine dimer (as shown in Figure 3.4).

Another example, which will be discussed in Chapter 7, is the phototoxicity of some plant metabolites (e.g., the psoralenes) and antibiotics (tetracyclines), the toxic effect resulting from the sunlight exciting the chemicals in the skin (even though they may not have been initially applied directly to the skin) to reactive forms.

3.1.5.5 Electrophilic Compounds

The fact that biochemistry is based on molecules containing nucleophilic functionalities will for obvious reasons make us sensitive to electrophilic compounds that react with and bind covalently to, for example, proteins and genetic material. In general, such nucleophilic substitutions or additions will irreversibly transform functional biomolecules to forms that will not work or even be harmful. Important examples are allergens that provoke the immune system to react to the body's own proteins, and carcinogens that transform genes to forms that can no longer control the rate of cell division. Electrophilic compounds are involved in many of the toxic effects that we are most concerned about, for example, allergies, tumors, and teratogenic effects, discussed in detail below.

3.1.5.6 Radicals

Radicals are organic compounds with unpaired electrons, and as a rule these are very reactive. According to fundamental chemical principles (Hund's rule), electrons prefer to be paired, and the search for another electron to pair with is the driving force for the reactivity of radicals. While some radicals, for example, the hydroxyl radical, are so reactive that they will react with the first molecule they encounter, others are less reactive and may even survive the time it takes to diffuse around a cell. (The reactions of radicals are discussed in Section 3.3 below.) Radicals may be formed by the homolytic cleavage of a covalent bond, although this normally requires so much energy (e.g., an open flame) that it is not relevant in biochemical reactions. However, radicals may also be formed by the addition or abstraction of an electron to or from a compound by a reducing or oxidizing agent, for instance by the enzymatic systems that metabolize most of the unnatural chemicals we are exposed to, and this is further discussed in Chapter 5.

3.1.5.7 Noncovalent Binding

A compound may react with other compounds in ways that produce toxic effects by forming noncovalent bonds (electrostatic forces). Such reactions are normally reversible, especially when weak dipole–dipole forces or hydrogen bonds are involved, although their effects on an organism may not be so. Ionic bonds may in rare cases be involved in toxic effects, an example being the reaction between calcium and oxalate (the anion of oxalic acid) yielding an insoluble product that may form stones that block the kidney. Many of the classical toxic compounds acting, for example, on the nervous system bind to receptors in the membranes of the nerve cells and trigger a reaction that may result in, for example, cramps or paralysis. The noncovalent bonds are the forces discussed earlier in this chapter, and are formed by most pharmaceutical agents when they bind to receptors and enzymes, either stimulating or blocking them. A compound that fits perfectly in a receptor, for example, can be seen as a key, the receptor being the lock, and it is evident that the compound must have its functional groups in the right places in space in order to fit. The stereochemistry of the compound is thus extremely important, and for pharmaceutical agents it is well known that one isomer and one enantiomer is the most active. As we shall see in Chapter 7, noncovalent binding is also the mechanism by which some toxic chemicals used in workplaces act, and it should be recognized that it is more difficult to find general relationships between structure and activity for such compounds.

3.2
Reactions between Nucleophiles and Electrophiles

Probably most important chemical reactions causing damage that may, for example, lead to cancer, are nucleophilic substitution and addition. A nucleophile is an electron donor, with at least one unused electron pair that can participate in a chemical reaction. It may be neutral or have a negative charge. An electrophile is an electron acceptor, positively charged or neutral and with the capability in some way to accommodate the electrons donated by a nucleophile. The biochemical components of living organisms, for example, proteins and nucleic acids (the topic of the preceding chapter) are good or reasonably good nucleophiles, whereas living organisms contain virtually no electrophiles, and it is obvious that the exposure of an organism to electrophiles, or chemicals that will be metabolized to electrophiles, will create problems.

3.2.1
Nucleophilic Substitutions

In a nucleophilic substitution or addition, a nucleophile will donate (or 'push') an electron pair and is consequently a Lewis base, while the electrophile acts as an electron sink and will accept (or 'pull') the electron pair, and therefore is a Lewis acid. The actual reaction is the flow of electrons from the source to the acceptor, from nucleophile to electrophile, under the influence of electrical charge, and can be compared with the flow of water in a stream due to the influence of gravity. Although the nucleophilicity and electrophilicity of the reactants are not the only rate-determining factors, as discussed below, a good nucleophile will normally react fast with a good electrophile while a poor nucleophile will react slowly with a poor electrophile. However, fast and slow are relative concepts and, while a very good electrophile may react upon contact with biological tissue and, for example, give rise to chemical burns, poor electrophiles may react selectively with components of genetic material and cause cancer.

The flow of electrons in a nucleophilic substitution or addition is indicated by curved arrows, a full-headed curved arrow indicating the movement of an electron pair while a half-headed curved arrow indicates the movement of a single electron in a radical reaction (see below). An example from inorganic chemistry is the reaction between water and hydrogen chloride (HCl), which takes place if hydrogen chloride gas is bubbled through liquid water and produces hydrochloric acid (Figure 3.5):

Figure 3.5 The reaction between water and HCl.

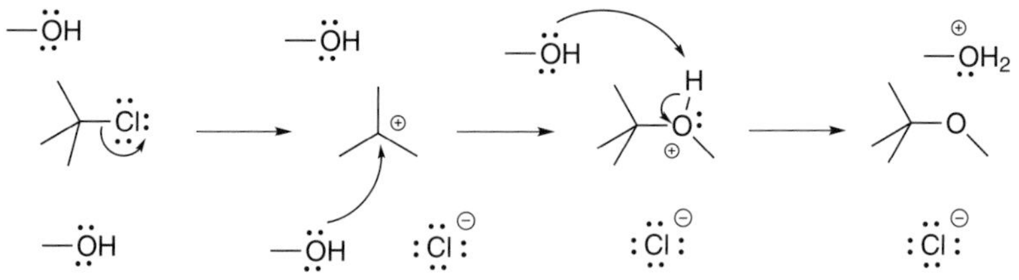

Figure 3.6 The S_N1 reaction between *tert*-butylchloride and methanol.

The base of the arrow always emanates from the site of electron density, either an unshared electron pair or a bond, and the head always points to the site that will accept the electrons. They will either form a new unshared electron pair of an atom or a new bond between two atoms. Nucleophilic substitutions take place with electrophiles that contain a suitable leaving group that will act as the electron sink, parting with the additional electron pair initially introduced by the nucleophile. In nucleophilic additions the electrophile is either positively charged, for example, a carbocation, or has a polarized multiple bond activated by, for example, a carbonyl group. The latter addition includes additions to a carbon–carbon double or triple bond conjugated with electron-withdrawing groups, also called Michael additions. Reactions between nucleophiles and electrophiles normally also include proton transfers – protonations of anions or unshared electron pairs or deprotonations producing anions or unshared electron pairs, which are important parts of the reaction but are often taken for granted and not shown explicitly.

Nucleophilic substitutions may be characterized as S_N1 or S_N2, depending on the number of molecules that are involved in the rate-determining step of the reaction. In S_N1 (Substitution Nucleophilic monomolecular) reactions, the electrophile reacts alone in an initial slow step to form a highly reactive cation that is extremely electrophilic and immediately reacts with the nucleophile. The reaction rate is therefore determined by the concentration of the electrophile alone; the concentration of the nucleophile does not matter (as long as it is there). S_N1 reactions are favored by factors that stabilize the reactive intermediate formed, for example, the presence of electron-donating groups. An example is the substitution of a chlorine atom by a methoxy group during the transformation of *tert*-butylchloride to methyl-*tert*-butylether in methanol (Figure 3.6):

The rate-determining step involves only one molecule, *tert*-butylchloride, and as soon as the trimethylmethyl cation is formed it directly adds the nucleophile methanol. After the addition, the oxygen of the added methanol has a positive charge, because it has donated one electron pair to the cation. This is taken care of by the abstraction of the proton from the same oxygen by another methanol molecule, acting as a base, and the two electrons of the oxygen–hydrogen bond make a second unshared electron pair on the ether oxygen. Methanol is not known to be a good nucleophile, but that does not matter in an S_N1 reaction. Any nucleophile will do, simply because the true electrophile (the trimethylmethyl cation) formed from the substrate is so very reactive. In that sense one can say that electrophiles that react via an S_N1 mechanism are less dangerous, because they are not choosy when it

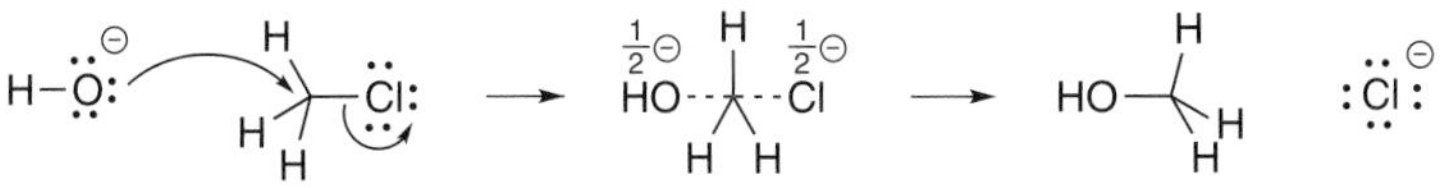

Figure 3.7 The S_N2 reaction between methyl chloride and a hydroxide ion.

comes to selecting a nucleophile. Organisms always consist of large amounts of water that will be sufficiently nucleophilic to take care of such electrophiles.

In an S_N2 reaction the reaction rate is instead determined by the concentration of two molecules, both the electrophile and the nucleophile, and such a reaction is said to be bimolecular. The two have to collide and undergo a reaction in which both participate. An example is the reaction between a hydroxide ion and methyl-chloride (Figure 3.7):

The hydroxide ion (the nucleophile) is attracted by the positively charged carbon in the electrophile, which can accommodate the electrons donated by the nucle-ophile by pushing the electrons between the carbon and the chlorine (in the car-bon–chlorine bond) toward the chlorine. In the middle of the reaction, one reaches a high-energy transition state that appears to have five bonds to the central carbon. This collapses to the products, in which a new bond has been formed between the oxygen and the carbon while the carbon–chlorine bond has been broken. The chlorine atom leaves the electrophile as a negative ion, together with the two electrons that bound it to the carbon. It is called the leaving group in the reaction, and as such it ultimately takes care of the electrons donated by the nucleophile. As the nucleophile takes part in the rate-determining step of S_N2 substitutions, the nucleophilicity of the nucleophile will be an important factor for the reaction. An S_N2 electrophile will react fast with a good (but perhaps not at all with a poor) nucleophile, and may therefore be highly selective for nucleophilic groups in proteins and genes. It should be mentioned that nucleophilic substitutions seldom proceed via pure S_N1 or S_N2 mechanisms, and it is common for mixed mecha-nisms to be observed in which one or the other predominates.

3.2.2
Nucleophilic Additions

Nucleophilic additions, for example to carbon–carbon double bonds activated by electron-withdrawing groups (EWGs), are very common among the toxic chemi-cals. A carbon–carbon double bond is normally not electrophilic, but if it is polar-ized by its substituents and one of them has the ability to accept an extra pair of electrons, it becomes a good electrophile. The most common EWG that activates carbon–carbon double bonds for nucleophilic additions is the carbonyl group, conjugated with the double bond, as for example in α,β-unsaturated aldehydes and ketones (α and β because the unsaturation is located between the carbons α and β to the carbonyl group). Nucleophilic additions to α,β-unsaturated carbonyl com-pounds as well as α,β-unsaturated nitriles and nitro compounds are called Michael additions.

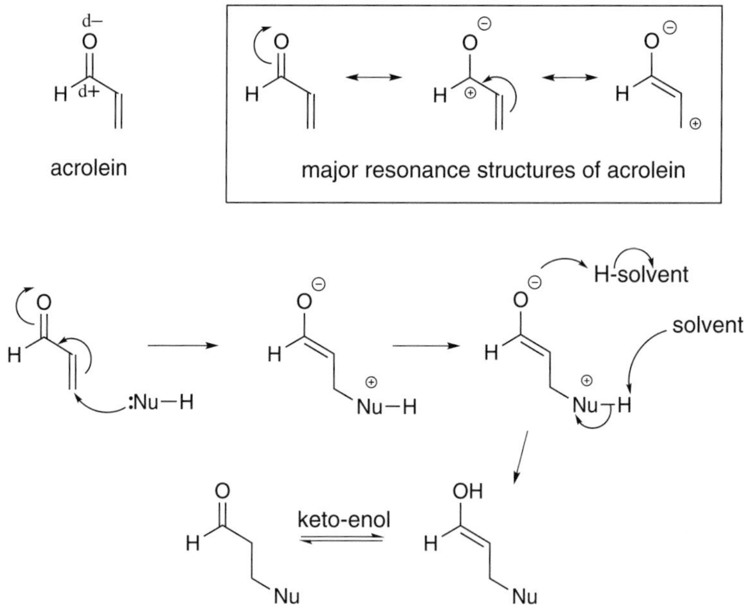

Figure 3.8 The Michael addition to acrolein.

The addition of, for example, esters of acetoacetic acid to unsaturated aldehydes, amides, esters, ketones, and nitriles was described as early as 1887 by the American chemist Arthur Michaels, who has given the reaction its name.

The simplest α,β-unsaturated aldehyde is acrolein, a powerful electrophile and a highly toxic chemical. As can be seen in Figure 3.8, the carbonyl group is polarized because it contains an electronegative oxygen atom, which actually causes the electrons in the carbon–carbon double bond to be slightly drawn toward the carbonyl group. Three reasonable resonance structures can be drawn for acrolein; the one without electric charge is, of course, the most stable, but the other two are fairly stable because the negative charge is positioned on the electronegative oxygen. This indicates that both the carbonyl carbon and the β-carbon have a substantial positive charge, and it is to these carbons that nucleophiles will add (see Figure 3.8).

As well as S_N1 and S_N2 substitutions and nucleophilic additions there is an array of other reactions that electrophiles may undergo, although this lies outside the scope of this text.

3.2.3
The Electrophile

Electrophiles involved in nucleophilic substitutions and additions relevant to toxicity contain functionalities of the general types discussed in this section. In addition, other features (e.g., steric and electronic factors) will influence the reactivity of an electrophile, and this subject is discussed in Section 3.4.

3.2.3.1 Strained Cyclic Hydrocarbons and Heterocycles

In epoxides, aziridines, and thiiranes (three-membered ring with oxygen, nitrogen, and sulfur), there is a combination of polarized bonds that will attract a nucleophile, and a ring strain that would be relieved if the ring opened up. The ring strain is caused by the fact that saturated hybridized carbon atoms have an angle of 109.5° between their sp³ orbitals, while the angles in an equilateral triangle are 60°, so the bonds are somewhat 'banana shaped', and this increases the energy of three-membered rings. The energy that the system will gain by opening the ring is the driving force for the reaction (see Figure 3.9). Cyclopropanes are in most cases not electrophilic, because they lack the heteroatom to polarize a bond in the ring. In some cases, electron-withdrawing substituents on the cyclopropane ring may activate it and facilitate a nucleophilic attack. An example is the microbial antibiotic duocarmycin A, which possesses potent biological activities (e.g., antitumour activity) because the cyclopropane ring can react with certain positions in the genetic material. In four-membered rings, for example, oxetanes, the ring

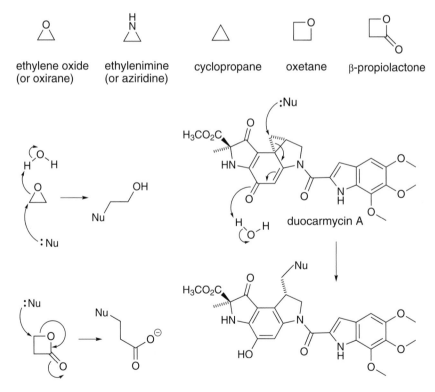

Figure 3.9 Electrophiles containing strained cyclic hydrocarbons and heterocycles.

strain is smaller and normally not enough to make these compounds electrophilic. However, if the polarization of the carbon–oxygen bond increases because of the presence of an EWG, as is the situation in β-propiolactone, oxetanes will also be good electrophiles (β-propiolactone is a well-known chemical carcinogen that reacts with genetic material).

3.2.3.2 Weak Single Bonds between an sp³ Carbon and a Leaving Group

Atoms or groups (neutral or positively charged) bound to an sp^3 carbon with a weak, polarized bond have a tendency to leave together with the electron pair of that bond, either as an anion or as a neutral species. Such atoms or groups are called leaving groups, and we have already observed that a chlorine atom may act as a leaving group in S_N1 and S_N2 reactions. In general, the quality of a leaving group is correlated with the acidity of its corresponding acid, and good leaving groups are the basis of relatively strong acids having negative pKa values. (The pKa value of an acid is a measure of its acidity, at the pH corresponding to the pKa at which the acid is ionized to 50%. For example, the pKa of acetic acid is 4.8, and in a water solution having pH 4.8 acetic acid is 50% present as CH_3COOH and 50% as CH_3COO^-. If the pH is 3.8 it is 90% CH_3COOH and 10% CH_3COO^-, while if the pH is 5.8 it is 10% CH_3COOH and 90% CH_3COO^-.) More basic leaving groups (having weaker corresponding acids with positive pKas) can be regarded as fair (up to pKa 10) to poor (pKa over 10). In Figure 3.10, some examples of leaving groups of different qualities are shown.

Dimethyl sulfate is obviously a better electrophile than methyl sulfate, because methyl sulfate is a better leaving group than sulfate. (Dimethyl sulfate is an efficient methylating agent used in chemistry to put methyl groups on various nucleophiles.) However, methyl sulfate is not a poor electrophile, and electrophiles having sulfate as a leaving group in activated positions may be reactive enough to cause toxic effects. Examples of this are given in Chapter 5. The hydroxyl group in alcohols is a poor leaving group, the pKa of the corresponding acid (H_2O) being 16. However, if the oxygen is protonated (the methyl oxonium ion in Figure 3.10 is protonated methanol) it becomes an excellent leaving group. This trick can be performed in the laboratory, and also by enzymes carrying out reactions that one would not expect to proceed with reasonable speed. Acetate is normally not a leaving group in electrophiles relevant to chemical toxicology, but if it is attached to a saturated nitrogen instead of carbon it will become a good leaving group because the nitrogen–oxygen bond is so much weaker from the beginning.

For electrophiles with a weak single bond between an sp^3 carbon and a leaving group, substitution competes with elimination, and the ratio depends on the position of the leaving group (primary, secondary, or tertiary sp^3 carbon). There are few electrophiles in biochemistry, and those that are there are used in a safe way. One example of a fair leaving group used in nature is pyrophosphate, present in, for example, dimethylallyl pyrophosphate and isopentenyl pyrophosphate, which take part in the enzyme-catalyzed biosynthesis of terpenoids as shown in Figure 3.11. The fully protonated acid pyrophosphoric acid is a strong acid, but the anions formed after deprotonation are also acidic. The geranyl pyrophosphate formed in

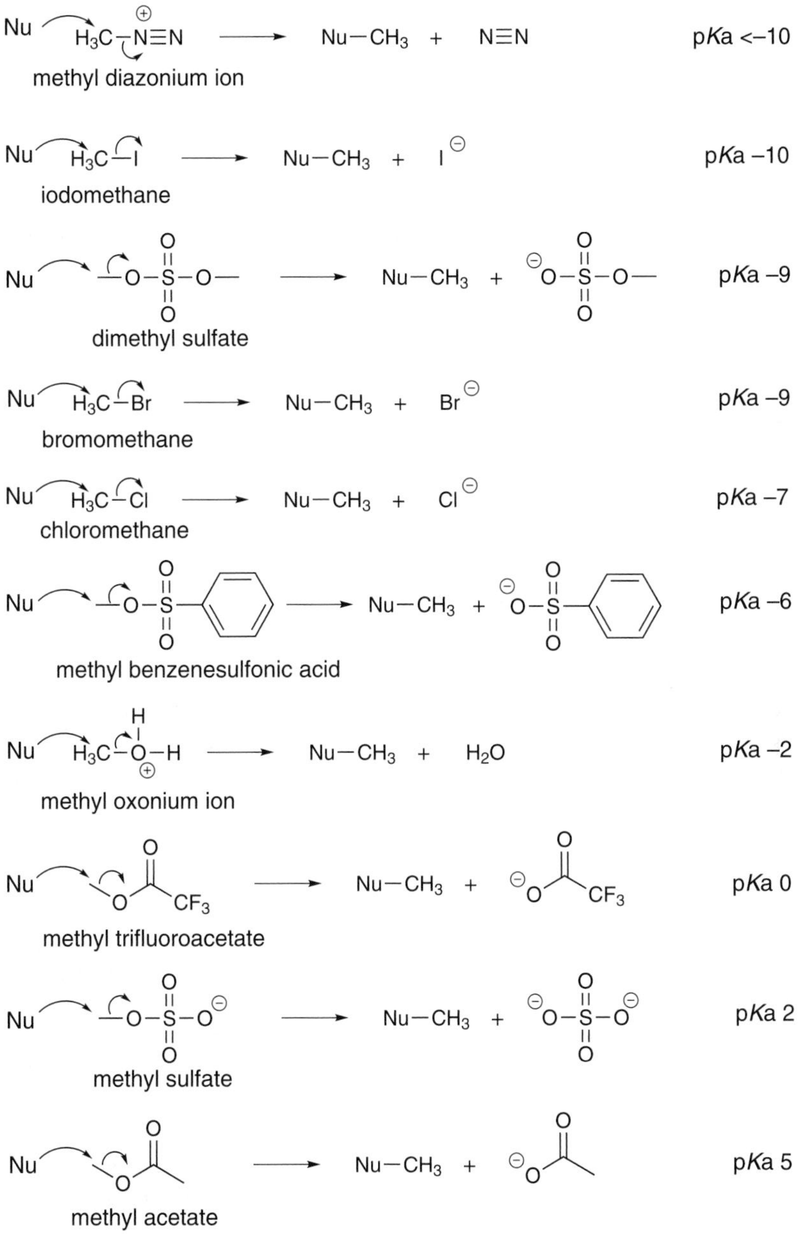

Nu H₃C⁺N≡N ⟶ Nu—CH₃ + N≡N   pKa <−10

methyl diazonium ion

Nu H₃C—I ⟶ Nu—CH₃ + I⁻   pKa −10

iodomethane

Nu —O–S(=O)₂–O— ⟶ Nu—CH₃ + ⁻O–S(=O)₂–O—   pKa −9

dimethyl sulfate

Nu H₃C—Br ⟶ Nu—CH₃ + Br⁻   pKa −9

bromomethane

Nu H₃C—Cl ⟶ Nu—CH₃ + Cl⁻   pKa −7

chloromethane

Nu —O–S(=O)₂–C₆H₅ ⟶ Nu—CH₃ + ⁻O–S(=O)₂–C₆H₅   pKa −6

methyl benzenesulfonic acid

Nu H₃C—O(H)⁺–H ⟶ Nu—CH₃ + H₂O   pKa −2

methyl oxonium ion

Nu —O–C(=O)–CF₃ ⟶ Nu—CH₃ + ⁻O–C(=O)–CF₃   pKa 0

methyl trifluoroacetate

Nu —O–S(=O)₂–O⁻ ⟶ Nu—CH₃ + ⁻O–S(=O)₂–O⁻   pKa 2

methyl sulfate

Nu —O–C(=O)–CH₃ ⟶ Nu—CH₃ + ⁻O–C(=O)–CH₃   pKa 5

methyl acetate

Figure 3.10 Some important leaving groups and the pKa values of their acids.

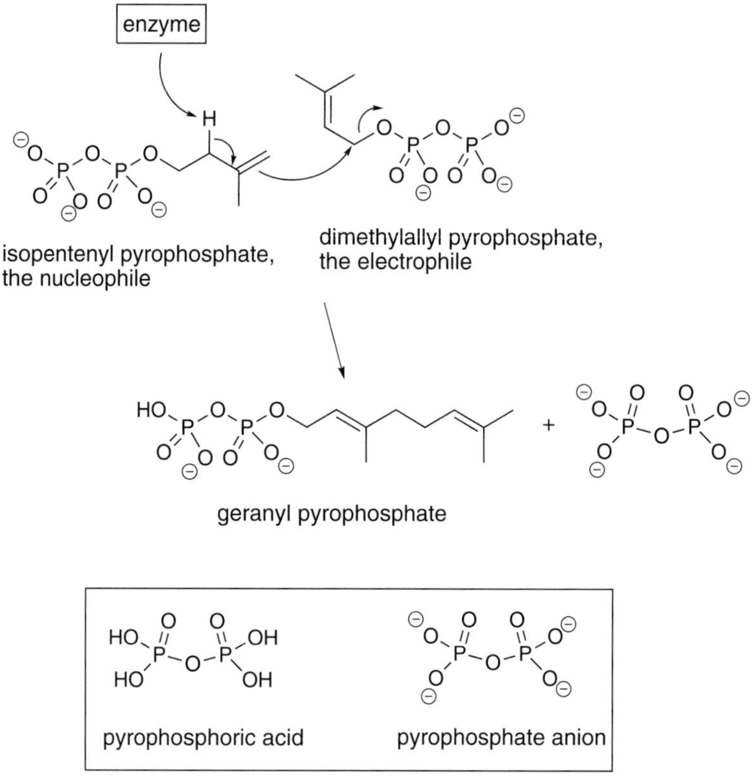

Figure 3.11 Pyrophosphate is a leaving group in nature.

the reaction shown in Figure 3.11 is a monoterpenoid, which can undergo the same reaction again to form a sesquiterpenoid. Another example of an electrophile used by organisms is the cofactor SAM, discussed in Chapter 5 (Section 5.4.5).

3.2.3.3 Polarized Double or Triple Bonds

We have already noted that carbonyl groups may react with water to form hydrates (Section 2.2), and this is in fact a nucleophilic addition. The same reaction could in principle also take place with a polarized triple bond, for example in nitriles, although these are much less reactive than ketones. If the positive end of a polarized double or triple bond has a suitable leaving group, the addition may be followed by an elimination, making the overall reaction look like a substitution. Carboxylic acid chlorides (see Figure 3.12), bromides, and anhydrides are examples of such electrophiles, which are often very reactive. The product is an acyl derivative, and the reaction is often called an acylation (in analogy to alkylations).

Also, polarized double bonds in heteroallenes (two adjacent double bonds with at least one heteroatom) can react with nucleophiles, as shown by the toxic methyl

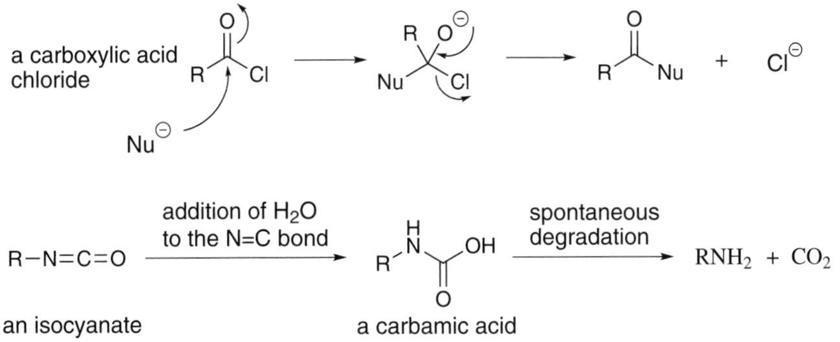

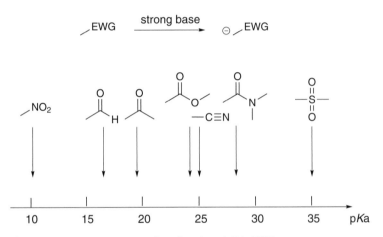

Figure 3.12 Additions to polarized multiple bonds.

Figure 3.13 Approximate pKa values for selected CH$_3$–EWGs.

isocyanate. Other isocyanates are used for the preparation of polyurethanes, used, for example, in upholstery. The bubbles formed in this material are caused by water reacting with the isocyanate to form an unstable carbamic acid that decomposes into an amine and gaseous carbon dioxide (see Figure 3.12).

However, most important are the conjugated additions discussed above (Figure 3.8), with an EWG activating a carbon–carbon double or triple bond. In general, the more efficiently the electrons are withdrawn from the carbon–carbon multiple bond by an EWG, the more reactive it is toward nucleophiles. Because EWGs attract electrons, they will stabilize a negative charge on an adjacent carbon, and compounds with the general formula CH$_3$–EWG are therefore slightly acidic and can be ionized by a base (see Figure 3.13). By comparing the acidity (pKa value) of the methyl protons of various CH$_3$–EWGs, one can get an impression of how efficient they are in activating carbon–carbon multiple bonds (see Figure 3.13).

The combination of two EWGs on the same unsaturated carbon will naturally increase the electrophilicity of the carbon–carbon multiple bond.

3.2.4
The Nucleophile

The nucleophilicity of a nucleophile depends on its ability to donate electrons and to form a new covalent bond with the electron sink. In general, the best nucleophiles have a negative charge, which not only provides it with the free electron pair necessary for the reaction but also enhances the attractive force to the partially positively charged center of the electrophile. However, nucleophilicity does not depend on the presence of a full negative charge, the partial negative charge caused by polarized bonds and conjugations is sufficient for turning many organic compounds into good nucleophiles. Most nucleophiles contain a heteroatom, in most cases oxygen, nitrogen, or sulfur, which provides the unshared electron pair that forms the new covalent bond to the electrophile. Table 3.3 lists a number of nucleophiles and their relative nucleophilicity toward the electrophile iodomethane (in methanol as solvent).

Obviously, a heteroatom is a better nucleophile when it is negatively charged. Compounds having the same heteroatom with the same charge are more nucleophilic the more basic the unshared electron pair is. The methoxide ion is a stronger base than the phenoxide ion or the acetate ion (methanol is a weaker acid than phenol or acetic acid) and is consequently a better nucleophile. Triethylamine is a stronger base than ammonia and therefore a better nucleophile. When comparing nucleophilic atoms in the same column of the periodic table, nucleophilicity increases with increasing atom number. This is due to the fact that the electrons in a larger atom (e.g., sulfur) are less tightly associated with the nucleus than are

Table 3.3 The relative nucleophilicity of various nucleophiles toward iodomethane in methanol.

CH_3OH	(methanol)	1
F^-	(fluoride ion)	500
CH_3COO^-	(acetate ion)	20 000
Cl^-	(chloride ion)	23 000
$(CH_3CH_2)_2S$	(diethyl sulfide)	220 000
NH_3	(ammonia)	320 000
PhO^-	(phenoxide ion)	560 000
Br^-	(bromide ion)	620 000
CH_3O^-	(methoxide ion)	1 900 000
$(CH_3CH_2)_3N$	(triethylamine)	4 600 000
CN^-	(cyanide ion)	5 000 000
I^-	(iodide ion)	26 000 000
$(CH_3CH_2)_3P$	(triethyl phosphine)	520 000 000
PhS^-	(thiophenoxide ion)	8 300 000 000

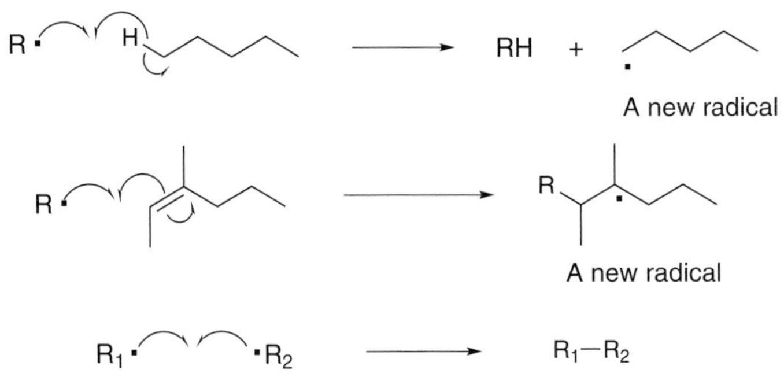

Figure 3.14 Principle reactions of radicals.

those of a smaller atom (e.g., oxygen): the electrons are more polarizable and may interact more efficiently with the types of electrophile that we are mainly concerned with.

3.3
Radical Reactions

As has been mentioned, most radicals are highly reactive because of the presence of an unpaired electron. They will react with other compounds by abstraction of an atom, by addition to a multiple bond, or by combination with another radical (see Figure 3.14). The first and the second reactions generate one radical for each one consumed, so that radical reactions are often called chain reactions. As the concentration of radicals is low the third reaction is unlikely to happen, but it is important because it will effectively take radicals out of circulation. The fishhook arrows used in Figure 3.14 show the change in position of single electrons, not electron pairs.

Radicals and radical reactions have attracted considerable attention from toxicological chemists because molecular oxygen readily participates in radical reactions, forming reactive and toxic oxygen species. This will be further discussed in Chapters 5 and 7.

3.4
Factors Modulating Reactivity

The reactivity of a compound naturally depends on a number of factors, and large differences in reactivity may be observed in a group of compounds that contain a particular reactive functional group. Most important are the electronic and steric effects, which are associated with the reactive chemical, but the conditions under which the reaction is carried out may also influence the outcome. In the human body, the temperature and pH normally do not change, but a compound dissolved in the water solution of a cell would react differently it were present inside the lipophilic membranes (i.e., different solvents). In addition, a large number of enzymatic conversions in organisms, as well as spontaneous chemical transformations, will take place with most compounds. As the products formed may be more reactive and more toxic, it is important to have an idea of which conversions and transformations are likely to take place.

3.4.1
Electronic Effects

The electronic effects are divided into inductive (or field) effects and resonance effects. Consider, for example, compounds that are reactive because they are acidic. Weak organic acids include carboxylic acids and phenols. As a class of compounds, carboxylic acids are depicted as R–COOH, and phenols as Ar–OH, but the acidity of individual carboxylic acids and phenols depends greatly on the nature of R and Ar. Compare the acidity of acetic acid and trichloroacetic acid (see Figure 3.15). The latter is almost 10000 times stronger because the three electronegative chlorines makes the α-carbon partially positively charged, and this stabilizes the negative charge of the carboxylate group. Trichloroacetic acid is therefore more willing than acetic acid to give up its acidic proton and become an anion because of the inductive effect of its α-substituents.

A similar difference is observed between phenol and 4-nitrophenol (see Figure 3.16). To some extent this is caused by the inductive effect of the nitro group, but the main contribution comes from its ability to accommodate an extra pair of electrons by resonance (i.e., the resonance effect). In Figure 3.16 the additional resonance form of the anion of 4-nitrophenol (compared to phenol itself) is shown, and this ability to distribute the negative charge over several (electronegative) atoms will stabilize the anion and facilitate its formation. (In Figure 3.16, the

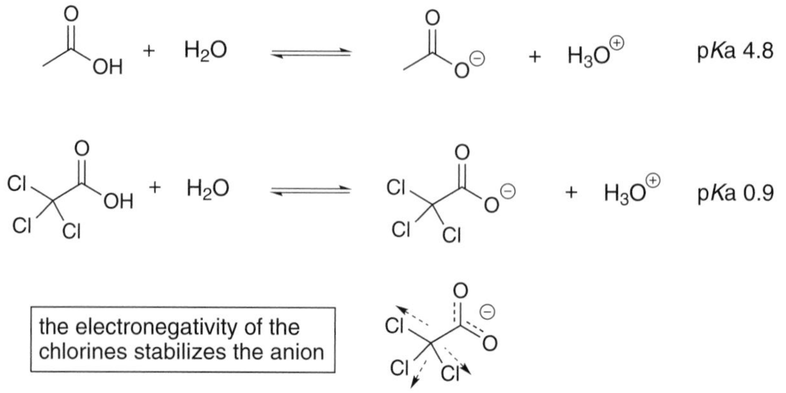

Figure 3.15 The acidity of carboxylic acids depends on the α-substituents.

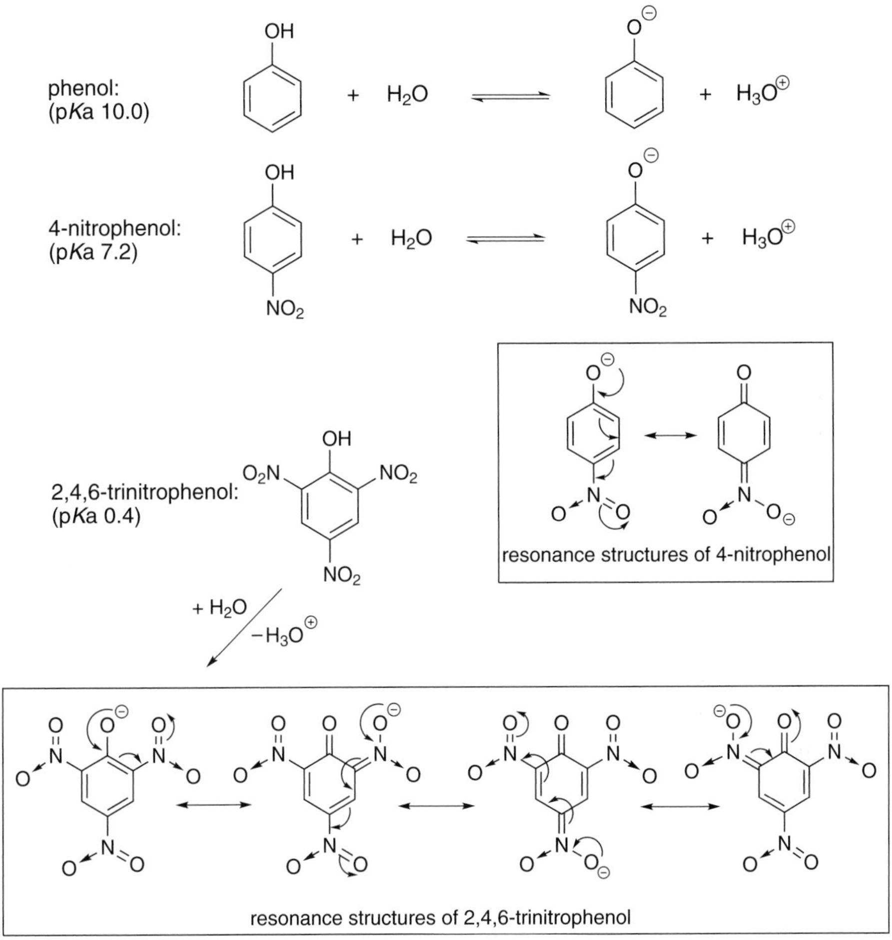

Figure 3.16 Nitro groups will increase the acidity of phenols.

nitrogen atom in the nitro group is shown with a double bond to one of the oxygens and a single bond, made from the unshared electron pair of the nitrogen, to the other. In reality, the two nitrogen–oxygen bonds are identical and both oxygens have a half negative charge, similarly to the situation in carboxylates.) If the phenol is substituted with several nitro groups in suitable positions, the resonance effect is additive. 2,4-Dinitrophenol is a considerably stronger acid with a pKa value of 4.0 (more acidic than acetic acid), while the pKa of 2,4,6-trinitrophenol is 0.4. The latter is a strong organic acid, and in Figure 3.16 it is indicated how the charge of the anion is distributed on four oxygen atoms thanks to resonance. Note that the nitro groups have to be in the correct positions (2, 4 and 6). In the 3 or 5 position they cannot participate in any resonance structures and will only give a weaker inductive effect (the pKa of 3-nitrophenol is 9.3).

The nitro group has a strong electron-withdrawing effect, but also, for example, aldehyde, keto, and cyanide substituents will have an effect in the same direction. Other substituents connected via a heteroatom (e.g., ethers, amines, or halogens) may instead donate electrons and have an opposite effect.

Electronic effects also influence the reactivity of electrophiles and radicals, although things get a bit more complicated. Electrophiles must have an electron sink, and are polarized by definition. Electronic effects by substituents can influence the polarization, and also stabilize or destabilize the high-energy intermediates or transition states formed during the reaction. An example is the difference in electrophilicity between bromoacetone, 2-methyl-3-bromopropene, and 2-methyl propylbromide (see Figure 3.17). Bromoacetone is a highly reactive chemical and a violent lacrimator that has been used as a war gas, 2-methyl-3-bromopropene is a good electrophile that is carcinogenic, while 2-methyl propylbromide is a poor electrophile. The first two have a double bond adjacent to the carbon that holds the bromine, the leaving group, and the p orbitals of the double bond will interact with and stabilize the p-like orbital holding both nucleophile and leaving group in the transition state, thereby lowering the energy for the transition state of the reaction (see Figure 3.17). In contrast, the 2-methyl propyl radical is more reactive than the corresponding radicals of acetone and 2-methylpropene (see Figure 3.18),

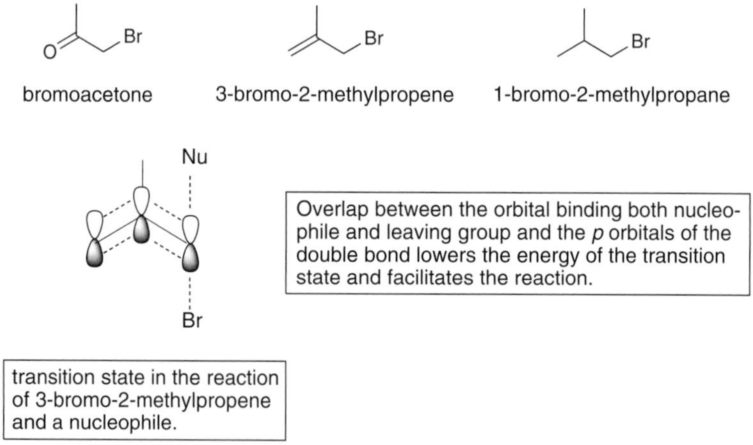

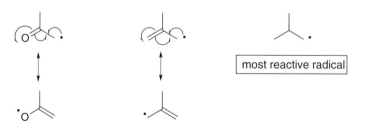

Figure 3.17 — content:

bromoacetone   3-bromo-2-methylpropene   1-bromo-2-methylpropane

Nu

Overlap between the orbital binding both nucleophile and leaving group and the *p* orbitals of the double bond lowers the energy of the transition state and facilitates the reaction.

Br

transition state in the reaction of 3-bromo-2-methylpropene and a nucleophile.

Figure 3.17 Electronic effects may increase electrophilicity.

most reactive radical

Figure 3.18 Resonance-stabilized radicals are less reactive.

because the unpaired electron in a radical is stabilized by the electrons of an adjacent double bond, as indicated.

It should again be stressed that it is not necessarily the most reactive species that is the most dangerous, especially in limited amounts, as such compounds will react with, for example, the water of our bodies. Instead, it is the less reactive species that are stable enough to get around and react selectively with targets such as genetic material that we should be most concerned about. However, it depends on the amount of material that one is exposed to as well as how one defines 'dangerous'.

3.4.2
Steric Factors

In addition to electronic factors, nucleophilic substitutions are very sensitive to steric effects that block the nucleophile from coming in contact with the electrophile. The major effect is observed with sterically hindered electrophiles, but nucleophiles may also be too big. For S_N2 reactions, the general and important rule is that an electrophile reacts faster the less substituted is the carbon with the leaving group; for example, bromomethane reacts faster than bromoethane, which in turn is a better electrophile than 2-methyl bromopropane and 2,2-dimethyl bromopropane. However, steric effects not only affect the proximate positions, but may also have an effect further away. An example is shown in Figure 3.19, showing the relative rates for the reaction between ethanol (the nucleophile) and various bromides (the electrophiles) in the solvent ethanol. For comparison, the rate of the reaction with ethyl bromide has been set at 1.

Although the steric effect in electrophiles may appear to be easily predicted, this is not the case. The problem how to compile all the different chemical properties of an electrophile in a way that the risks associated with it can be estimated is most efficiently addressed by the computational techniques available today.

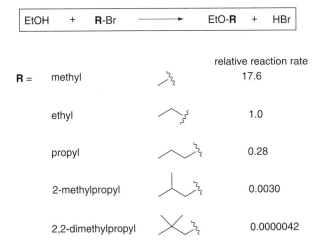

Figure 3.19 The reactivity of primary bromides is also modulated by steric effects.

3.4.3
Transformations and Conversions that Increase Reactivity

Although transformations and conversions are not inherent in a molecule, there are in principle no compounds that are inert in an organism or in the environment. The multitude of chemical and enzymatic reactions that may take place will sooner or later change the structure of the compound and convert/transform it to another compound. If the new compound formed is more toxic, we can say, from the point of view that is interesting in this text, that a chemical or enzymatic activation has taken place. If one aims at predicting the toxic and ecotoxic effects of a chemical, it is crucial to understand and be able to foresee the reactions that a compound is likely to undergo inside organisms subjected to various enzymatic systems, or in the abiotic environment subjected to, for example, irradiation by the sun. By definition, a conversion of one chemical to another in the biosphere is enzymatic (catalyzed by enzymes, see below) and takes place inside an organism, while a transformation is chemical reaction and can take place both inside and outside organisms.

4
Uptake, Distribution, and Elimination of Chemicals

In order for a toxicant to give rise to a systemic toxic effect, it has to be taken up by the organism and distributed before it reaches its target organ. There are a number of different routes for chemicals to enter our bodies, but two organ systems, the digestive (intestines) and the respiratory (lungs), have been specifically designed for the uptake of chemicals. In addition, chemicals can in principle enter the body wherever they contact it, through the skin or the mucous membranes in the nasal cavities and the mouth, in the stomach or the rectum, and even via the ear (as the father of Hamlet so tragically experienced). In places of work, people are mainly exposed to chemicals as air contaminants, and the lungs normally account for 80–90% of the total uptake.

4.1
Transport through Biological Membranes

Any uptake of a chemical from the outside into the body requires it to pass through biological membranes, which one way or another consist of layers of cells. Cells are surrounded by a cell membrane, briefly described in Chapter 2, and the cell membranes constitute a major barrier for the chemicals that are transported.

4.1.1
Diffusion

Diffusion is the transport of molecules in the gas or liquid phase (solution) due to the continuous and rapid movement of the molecules. The efficiency of diffusion as a general transport mechanism depends on the temperature and the weight of the molecule. At 37 °C the average velocity of water molecules is 2500 km h^{-1}, while a glucose molecule, being approximately 10 times heavier, only moves at 850 km h^{-1}. In spite of its velocity, a molecule will not get far (especially in solution) before it has a collision with another molecule, which changes its direction, so that it will mainly move back and forth within the same small volume. However, with time this movement will result in a spreading in all directions, driven by concentration differences in the system and with a homogenous liquid or gas as

Chemistry, Health, and Environment. Olov Sterner
© 2010 WILEY-VCH Verlag GmbH & Co. KGaA, Weinheim
ISBN: 978-3-527-32582-5

the end point. The amount of transport caused by diffusion is therefore extremely sensitive to distance, and this will in practice limit the size of cells. An average cell is 10–20 μm in diameter, and for most compounds the transport from the cell membrane to the center of the cell by diffusion takes place in seconds. However, if the cell was the size of a football (20 cm) it would instead take years!

The transport by diffusion through the membrane of a cell in our bodies also depends on two additional things: the difference in concentration between the two sides of the membrane and the resistance of that particular membrane to the passage through it of that particular molecule (or the permeability of the membrane to those molecules). A substantial concentration gradient will always promote diffusion, whereas no net diffusion will take place if the concentration is equal on both sides. The diffusion resistance of the membrane is linked to both the membrane itself, which is similar in all cells as long as the double layer (see Section 2.4.1) of amphipathic molecules is considered, and to factors associated with the diffusing species. If one compares how easily chemicals in a water solution diffuse through a synthetic membrane consisting only of amphipathic compounds and no proteins, it is obvious that very small neutral molecules (e.g., molecular oxygen and water) are not much affected by the membrane but diffuse straight through even if they are polar. They are believed to simply fit into cavities in the membrane created by its continuous movement. The bigger and the more polar a compound, the more difficult it is to diffuse through a membrane. Ions will not be able to pass through a synthetic membrane (see Figure 4.1) simply because they are covered by several layers of stabilizing water molecules (formed by strong ion-dipole and dipole-dipole intermolecular forces). In order to leave the water solution and move into the lipophilic core of the membrane, an ion would have to get rid of this water, and this costs a lot of energy. Compounds containing, for example, carboxylic acid or amine functions may be ionized at physiological pH, although most organic acids and bases are not completely ionized, which makes it easier for them to pass through membranes. Most lipophilic organic compounds, also reasonably large (up to approximately 500–1000 g mol^{-1}), will

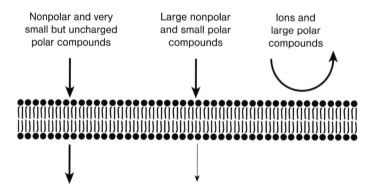

Figure 4.1 Diffusion of different species through a synthetic membrane.

rapidly pass through membranes, while hydrophilic compounds, if they are not as small as water or methanol, will not.

Two additional things are important to remember concerning the diffusion of exogenous compounds through biological membranes, namely that (a) it is by far the most important mechanism for transportation, and (b) with time almost anything will diffuse through a membrane.

4.1.2
Passive Transport by Proteins

For hydrophilic compounds that can only diffuse through biological membranes with difficulty, there are in all cells special proteins, namely membrane transport proteins, that may facilitate the transport. There are two principal mechanisms for protein-assisted transport: passive transport, that only takes place with a concentration gradient (or an electrochemical gradient for ions) and active transport (see Section 4.1.3), that can also operate against a concentration gradient. Certain membrane transport proteins, namely carrier proteins, may transfer hydrophilic compounds that the cell either needs (e.g., sugars, amino acids, and ions) or wants to dispose of, in and out of the cell. Carrier proteins only transport one type of compound, which must fit into the protein as a key into its lock, and the specificity that this type of transport shows is characteristic and important, but not absolute. As the transporting proteins have a limited capacity, it is also characterized by its ability to become saturated. The principal function of these transmembrane proteins is that they can adopt two conformations, which are available for the transported molecule on different sides of the membrane. In both conformations, the solute can bind to the protein as well as be released, and if there is a conformational change in the protein between the binding and the release, the solute will have been transported across the membrane (see Figure 4.2). As there is no preference for binding or releasing on either side, it is the concentration of the solute on the different sides of the membrane that will determine the flow in each direction. Most mammalian cells use this mechanism to transport glucose from the

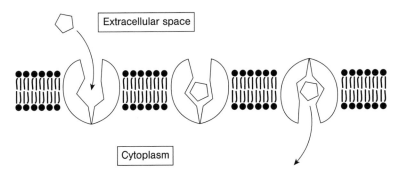

Figure 4.2 Passive protein-assisted transport.

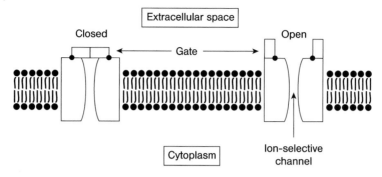

Figure 4.3 Passive transport with channel proteins.

extracellular fluid in which the concentration of glucose is relatively high. Note that the same system also transports the sugars mannose, ribose, and xylose, but not, for example, fructose.

Other membrane transport proteins, channel proteins, form transmembrane pores in the cell membranes through which hydrophilic molecules may pass. The pores are gated by other proteins attached to the cell membrane so that they can open and close as a response to some stimulus, for example, an external signal. In their open form, they provide channels of water through the membrane through which hydrophilic compounds may diffuse from one side or the other without coming into contact with the inner part of the membrane, or may be passively transported by the flow created by hydrostatic or osmotic pressure (see Figure 4.3). Another name for this transport is filtration. The size of the pores may differ in different cell types; in most cells they only let small molecules pass. In mammalian cells it is mainly ions that are transported in this way. The pores are then called ion channels, and they are to a large degree selective for a certain ion by the size of the pore and its internal properties. As the regulation of the flow of certain ions through the membrane is very important for cells, especially nerve cells, the interference of exogenous chemicals with the pores may give rise to toxic effects (see Chapter 7).

Exogenous compounds may also facilitate diffusion, as demonstrated by the ionophores. These increase the membrane's permeability for ions, either by disguising the ion in a lipophilic envelope and helping it to pass through the membrane as a mobile ion carrier (as the insecticide and nematicide valinomycin, Figure 4.4) or by forming unnatural channels through which an ion can diffuse (as the gramicidins, used as antibiotics).

As mentioned above, passive transport only takes place in the direction of a concentration gradient (or the electrochemical gradient for ions), and can only result in dilution of a chemical. On the other hand, it shows a high degree of specificity and it is relatively simple, not requiring the input of energy for its function (although the gating of the ion channels is an active process). Passive transport was not developed for exogenous chemicals, so when they are transported through membranes by this mechanism it is simply because they happen to fit into the carrier proteins or the channel proteins.

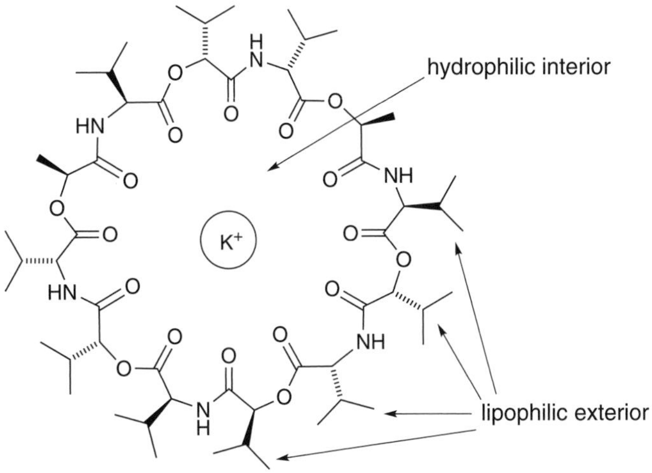

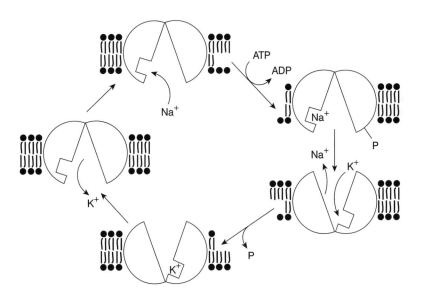

Figure 4.4 Valinomycin.

Figure 4.5 The plasma membrane Na^+–K^+ pump.

4.1.3
Active Transport by Proteins

In contrast to passive transport, active transport requires energy in the form of ATP (primary active transport) or an ion gradient (secondary active transport) and it is able to concentrate a chemical or an ion against a concentration gradient (see Figure 4.5). The most important example of a primary active transport is the plasma membrane Na$^+$–K$^+$ pump, which generates and maintains the sodium (Na$^+$) and potassium ion (K$^+$) gradients across plasma membranes.

The Na$^+$ and K$^+$ gradients serve several functions, for example, to maintain the osmotic balance and regulate the cell volume, and we shall come back to this in Chapter 10. An example of a secondary active transport is the resorption of glucose in the kidney tubules (see Section 4.5.3), where the glucose concentration is low. This is coupled with and driven by the simultaneous uptake of Na$^+$, of which the concentration is low inside the cell and high outside, and the carrier protein will only transport glucose and Na$^+$ together. The Na$^+$ that gets into the cell in this way is pumped out again by the Na$^+$–K$^+$ pump, so a low concentration of Na$^+$ is maintained inside the cell.

Besides the absorption of chemicals through cell membranes by active transport, compounds are also excreted from the body by the same mechanism. The brain has active transport systems for getting rid of organic acids and bases, and so have the liver and kidneys (see below).

4.1.4
Exocytosis and Endocytosis

The need for transportation also of larger molecules, such as proteins, polysaccharides, and polynucleotides, or even particles, is satisfied by the transport mechanisms exocytosis and endocytosis. As depicted in Figure 4.6, they involve the formation of vesicles surrounded by a membrane, like a miniature cell. These are formed either inside a cell and can by fusing with the plasma membrane be emptied into the extracellular space (exocytosis), or by budding a piece of the plasma membrane with a volume of extracellular fluid (containing whatever the cell desires) into the cell where it is dissolved (endocytosis). Endocytosis can be

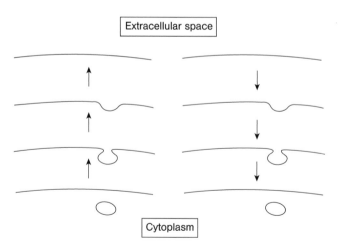

Figure 4.6 Exocytosis (left) and endocytosis (right).

divided into pinocytosis, when a volume of fluid is ingested (pino = drink), and phagocytosis, when particles are ingested (phago = eat). Almost all mammalian cells make frequent use of pinocytosis for their transport needs, and the terms endocytosis and pinocytosis sometimes replace each other, while only specialized cells use phagocytosis.

Insulin-producing cells export the insulin by exocytosis. Cells that need cholesterol for membrane synthesis produce and place receptors for LDL (low-density lipoproteins) in the plasma membrane. LDL, which essentially is a complex between one protein and 2000 cholesterol molecules, is attracted by the receptors, concentrated on the outside of the cell, and engulfed by pinocytosis. The newborn obtain immunity to various diseases by absorbing immunoglobulins (proteins that recognize, for example, pathogenic microorganisms). Phagocytosis is used by the macrophages and the neutrophiles of the immune system. They 'eat' bacteria and other cells recognized as alien as well as damaged and dead cells of their own. Although exocytosis and endocytosis do not play a central role for the transport of exogenous compounds, they are involved in, for example, the absorption of the botulin toxins (toxic proteins, discussed in Section 9.3.9), and we shall also came across a few other examples.

4.2
Absorption of Chemicals in Man

4.2.1
Absorption via the Skin

The skin is involved in many important tasks. It constitutes a barrier and helps to protect the organism, and it regulates the body temperature, but it is not involved in the absorption of nutrients. Therefore, it does not contain any specialized transport systems, and the only way for exogenous compounds to be absorbed by penetrating the skin is to diffuse through it. In addition to diffusion directly through the cells that make up the biological membrane, it is in some cases also possible to diffuse between the cells. The space between the cells contains fat, which can be dissolved by organic solvents leaving a skin that is considerably more permeable to chemicals. The skin consists of several layers, but for our discussions a division into dermis and epidermis is adequate. The principal difference between the two is that the epidermis, which is the outermost, is an epithelial tissue containing no blood vessels, while the dermis is a connective tissue rich in blood vessels (see Figure 4.7). For an exogenous chemical, therefore, it is sufficient to penetrate the epidermis; arriving at the dermis it will be efficiently swept away by the blood.

The epidermis is composed of 10–20 layers of cells, of which the innermost layer is constantly dividing, generating new cells that push those above closer to the surface. However, as soon as the cells lose contact with the basal layer they stop dividing and begin to differentiate. They fill themselves with keratin, a protein,

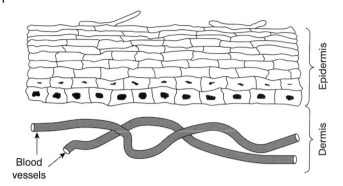

Figure 4.7. The skin.

dissolve the cell nucleus, and dry as they are slowly pushed upwards. When they approach the surface they are dead, flattened, highly keratinized, and relatively dry cells, containing only 5–10% water. Eventually they are sloughed off by wear (we all lose approximately 600 000 skin cells per day). The process of renewing the epidermis takes a few weeks, and for products that penetrate the epidermis very slowly this continuous shedding serves as protection. The thickness of the skin may vary substantially in different parts of the body and will adjust to the actual need, caused by different amounts of keratin in the cells of skin experiencing different workloads. In addition, the skin contains hair follicles and different glands which make up approximately 1% of the surface of the skin.

The absorption of an exogenous chemical through the skin depends on both the skin, its general condition and thickness, and on the chemical, and how efficiently it can diffuse through the epidermis. The epidermis of the skin of the arms and legs is more than 10 times thinner than that of the palms, and exogenous chemicals will consequently be absorbed faster in such areas. However, even more important is the water content of the epidermis as the diffusion rate increases rapidly in moist skin, and the water content may increase dramatically if, for example, rubber gloves or boots are worn. In addition, if the skin is diseased or damaged by reactive chemicals, it may be easier for exogenous compounds to penetrate. As discussed above, molecules that are small and nonpolar will diffuse rapidly through biological membranes, and the epidermis is no exception. Many organic solvents, for example, methanol, will diffuse through the skin so efficiently that toxic amounts may be absorbed as a result of accidents when solvents are spilled. Although water molecules are normally able to diffuse through the membrane of a living cell rapidly, water will not diffuse through the epidermis as efficiently. However, it will make it swell, something that everyone can observe after having taken a bath, and as mentioned above this makes the skin more permeable to other chemicals.

Many other chemicals also increase the permeability of the skin, and facilitate more rapid absorption of others. An example is the polar organic solvent dimethyl sulfoxide (DMSO), which is used in veterinary medicine to increase the absorption of certain remedies administered dermally. The mechanism is again that DMSO makes the outermost cells in the epidermis swell and hence easier to diffuse through, and in addition it will loosen the cells from each other.

It is important to remember that there is no sharp line dividing compounds into one group that efficiently diffuse through the epidermis and are absorbed and another group that is not able to penetrate the skin. Most compounds will eventually get through, although it takes time. Synthetic gloves and boots are often used as a protection against chemical spills when larger amounts of chemicals are handled, and although some synthetic materials are efficient barriers for certain types of compounds, they should not be regarded as fail-safe. Just as chemicals can penetrate the skin, they will eventually also penetrate synthetic materials, and whatever diffuses through a rubber glove will encounter skin that is humid and very easily penetrated. Good advice is therefore not to wear such equipment for too long, and to wash it thoroughly if it has been in contact with hazardous chemicals. A special risk with synthetic gloves and boots is the possibility that a chemical is spilled inside a glove or a boot, and in such cases also quite polar compounds may be absorbed efficiently. This happens frequently, for example during the preparation and use of pesticides, and is responsible for many unnecessary accidents that could easily be avoided.

Most absorption of exogenous chemicals via the skin depends on accidental spills combined with carelessness. If one is observant and immediately takes care of spills on the skin or on the clothes by removal of the contaminated garments and careful washing of the skin, and one uses protective equipment in a sensible way, such incidents could essentially be eliminated.

4.2.2
Uptake by the Gastrointestinal Tract

Even though the gastrointestinal tract only accounts for a small part of the uptake of chemicals in workplaces, this route is of course of interest in other contexts like accidents, suicides, and so forth. The gastrointestinal tract, all the way from the mouth to the anus, can be regarded as an approximately 8 m long tube that passes through the human body. Its contents are still exterior to the human organism, but its task is to convert whatever is put in at one end to something that can be taken up and used. The whole construction is controlled by the nervous system, although hormones regulate some functions. After mechanical treatment in the mouth, the food and drink passes the esophagus and reaches the stomach where it is degraded both chemically (by hydrochloric acid) and enzymatically (by pepsin, which will hydrolyze peptide bonds in proteins). The resulting mush (called chyme) is then passed into the small intestine (divided into the duodenum, jejunum, and ileum), and the low pH in the stomach (approximately 2) is neutralized with bicarbonate secreted by both the pancreas and the liver. The pancreas also secretes enzymes that degrade proteins, polysaccharides, fats, and nucleic acids, while the liver in addition secretes bile that emulsifies fats, and the cells of the small intestine also secrete a variety of enzymes (see Figure 4.8). A continuous conversion of the nutrients to absorbable units takes place, and when they reach the end station, the rectum, via the large intestine, only non-digestible material is left. Also important for this process is the fact that the intestines harbor hundreds of different species of bacteria, which in number of cells exceed the number of cells of a human being by a factor of 10. In some ways, the bacterial flora of the intestines can be regarded as a separate organ in the body, with a huge and different metabolizing capacity as well as the ability to produce toxins of its own, and we shall see how this can create problems.

Chemicals are absorbed all the way from the mouth to the rectum, although most of the absorption takes place in the first part of the small intestine. The total internal surface of the small intestine available for absorption is approximately $300 \, m^2$ (as big as a tennis court!), created by numerous villi (see Figure 4.9) that stick out like arms into the contents of the intestine, the villi in turn being covered

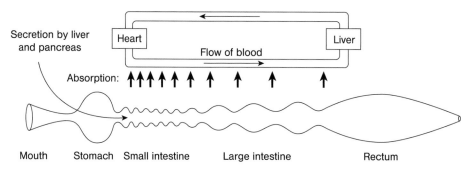

Figure 4.8 A schematic view of the gastrointestinal tract.

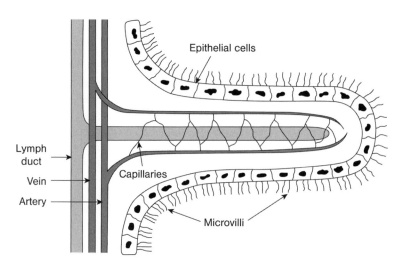

Figure 4.9 The structure of a villus.

by microvilli (small projections that increase the surface even more, see Figure 4.9). The small intestine absorbs approximately 8.5 L fluid material per day in an adult. The inside of a villus is vascular, and the blood vessels are only separated from the outside by a thin monocellular layer of endothelial cells with microvilli. It is obvious that any molecule that passes through the intestine and is able to diffuse a little will be adsorbed. This absorption is so efficient because the surface is large, the diffusion barrier is low (compared with the skin), and the time spent in this environment is relatively long. In addition, the endothelial cells lining the intestines are packed with membrane transport proteins that pick up hydrophilic compounds (e.g., monosaccharides and free DNA bases) and ions (e.g., amino acids and minerals) that cannot diffuse by themselves. Even endocytosis is used to engulf some proteins, for example, the special protein (called intrinsic factor) that binds vitamin B_{12} and helps it to pass through the membrane, and the extremely toxic botulin toxins.

Exogenous compounds that for any reason end up in the intestines can be expected to be absorbed, most likely by diffusion in the small intestine but also by being structurally similar to the endogenous compounds and fitting into one of the many specific membrane transport systems. For example, thallium, cobalt, and manganese all use the same carrier as that for iron, while lead is sufficiently similar to calcium to make use of its transport system. Poorly absorbed are chemicals that are more or less insoluble in the contents of the intestines, because absorption in the intestines requires that the chemical should be dissolved. Examples are paraffin and metallic mercury, which both pass through the intestines almost unaffected. Mercury is well known to be a very toxic heavy metal, and we shall return to it several times. If mercury vapor is inhaled or its salts are ingested it gives severe toxic effects, but ingestion of comparable amounts of metallic mercury is relatively harmless. In this context we can also note a case of dispositional antagonism, which is taken advantage of when intoxications are treated with active carbon. Toxicants are adsorbed to the carbon, which is insoluble and stays in the intestines.

As indicated above, the environment in the gastrointestinal tract shows several distinct features. One is the low pH in the stomach, which is not found anywhere else in the body. Although absorption directly from the stomach into the blood is small, it is evident that acidity will affect this uptake. Many carboxylic acids are ionized at pH 7 but are protonated and neutral in the stomach, while amines are protonated and ionized. It has actually been shown that some amines may be transported from the blood, where they are neutral, to the stomach, as the reverse cannot take place. Besides being corrosive, acid is also an efficient catalyst for many chemical reactions, and the pH in the stomach is optimal for the transformations of nitrite (encountered for example as a preservative in meat) and amines or amides to *N*-nitrosamines or *N*-nitrosamides. Some of these, for example, the dimethylnitrosamine formed from dimethylamine (see Figure 4.10), are stable and will be absorbed in the intestines, but convert enzymatically to potent carcinogens in, for instance, the liver, and we shall return to *N*-nitrosamines in Chapters 5 and 9.

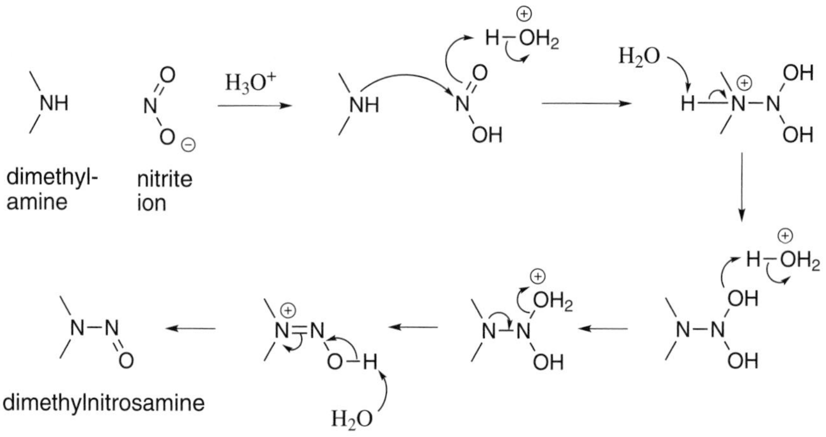

Figure 4.10 The formation of dimethylnitrosamine.

The most dangerous nitrosamines are those formed from secondary amines (with two alkyl groups and one hydrogen attached to the nitrogen). Secondary amines are found in many foodstuffs (amino acids are secondary amines), and several have found important uses in industry, so it is not unlikely that people will be exposed to nitrite and secondary amines at the same time. However, it is not known whether the formation of nitrosamines in the stomach of humans has any significant impact on tumor frequency.

The metabolizing capacity of the bacterial flora of the intestines is enormous, and has a different profile compared to our own metabolic system. While our biochemistry is oxidative (aerobic), the bacterial flora in the intestines, especially in the colon, can be regarded as a reductive (anaerobic) enclave separated from us by a protective intestinal mucous layer. Across the colonic mucous layer, which is approximately 1 mm thick, there is a redox potential of several hundred meV, positive in our cells and negative on the outside, and this adds another element to the absorption of exogenous compounds in the intestines, because reactive radicals are easily generated in such an environment. Some years back scientists were able show that extracts of the feces of some people had quite high mutagenic activity, and the compounds responsible were isolated and characterized. The mutagenic principles turned out to be fecapentaenes, for example, fecapentaene 12, which is the major component (see Figure 4.11), and it is suspected that they are involved in the genesis of colon tumors.

It has been shown that the fecapentaenes are formed from the plasmalopentaenes (a mixture of different fatty acid esters, the stearic acid ester shown in Figure 4.11 being a major component) by various bacterial strains of *Bacteroides* in the colon. The plasmalopentaenes are believed to be protective compounds, produced by the epithelial cells of the intestines to act as a scavenger of radicals formed at the redox border. They are therefore important endogenous metabolites, and their conversion to fecapentaenes depends on the presence of the *Bacteroides*

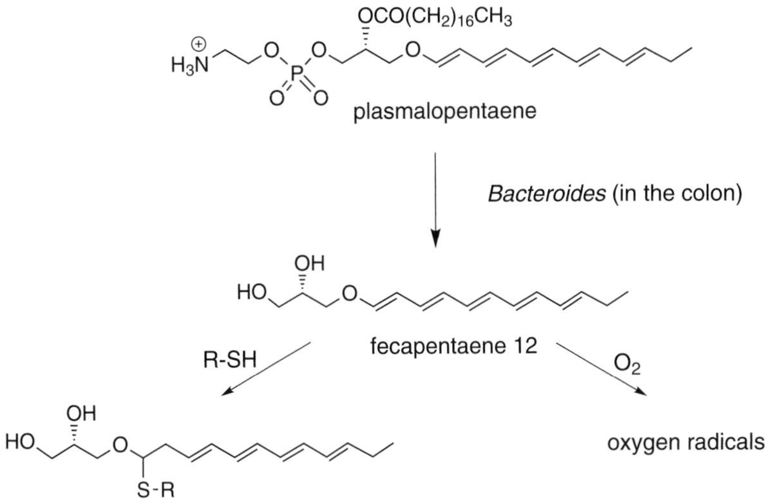

Figure 4.11 Fecapentaene 12 and its precursor plasmalopentaene.

strains in the colon. As nutritional factors are known to influence the composition of the bacterial flora, it is possible that the formation of the potentially carcinogenic fecapentaenes is a piece in the cost–cancer puzzle (see Section 9.3). It is not understood in detail how the fecapentaenes exert their mutagenic activity. Fecapentaene 12, a very potent mutagen, can react with nucleophiles (e.g., thiols, see Figure 4.11) that add to the first carbon of the enol residue, but it has also been shown that the fecapentaenes generate reactive oxygen radicals. The hydrolysis of the enol moiety would, after keto–enol tautomerism, yield an unsaturated aldehyde that could be suspected to be the electrophilic form of the fecapentaenes, but this is apparently not the case.

From the intestines, the blood flows directly to the liver via the portal vein, and the liver is the first organ that compounds absorbed in the intestines contact. This is 'intentional', and the liver converts nutrients into suitable forms before they are distributed to the rest of the body (the so-called first-pass effect). Also, exogenous compounds absorbed in the intestines (but not necessarily exogenous compounds absorbed in the skin or the lungs) will primarily be metabolized in the liver, which is perfect if they thereby are rendered less toxic. However, as we shall see in the next chapter, this is not always the case.

4.2.3
Uptake via the Lungs

As already noted, the lungs are the principal absorption organs when it comes to exogenous chemicals that people are exposed to in workplaces. The lung is constructed for the exchange of oxygen and carbon dioxide between the air and the blood, and the contact surface between the two elements is large (approximately

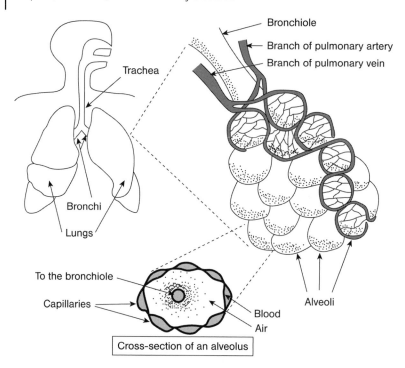

Figure 4.12 The respiratory system with the alveoli.

$70 \, m^2$, a badminton court!). The exchange takes place in the functional unit called the alveolus, but before reaching this region of the lungs the inspired air passes through the pharynx, the larynx, the trachea, and the branching bronchi (one per lung) which end in the alveoli (see Figure 4.12).

In the alveoli, the distance between the air and the blood is very small, approximately $1 \, \mu m$, and this ensures that the diffusion of oxygen and carbon dioxide (as well as most compounds that end up in the alveoli) is very rapid and efficient.

An important variable for absorption in the lungs is the working load of an exposed individual. At rest, an adult inhales approximately $5 \, L$ air per minute, which increases to $20 \, L \, min^{-1}$ for a person working in a normal manufacturing industry. At extreme workloads, this will of course increase even more. As more air passes through the lungs per minute, more of the air contaminants will be absorbed, and this will be further discussed in Section 4.4.3.

4.2.3.1 Air Contaminants in the Form of Aerosols

Dust (solid particles), fume (particles formed by combustion in general), smoke (particles formed by combustion of organic material), fog and mist (liquid droplets) and smog (particles and gas formed from car exhausts) contain particles and/or droplets that are sufficiently small to hover in the air, and are called aerosols. The absorption of aerosols in the lungs depends on the weight of the particles or

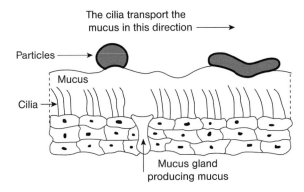

Figure 4.13 Mucociliary escalator.

droplets. If they are light enough and, as a consequence, small enough, they can follow the inspired air all the way down to the alveoli, while the bigger (and heavier) particles will not get that deep. The reason for this is that inhaled air passes through a number of turns and windings at rather high speed, and the bigger particles are thrown, simply by inertial force, into the mucous membranes of the nose, the larynx, and the bronchi. This protects the lungs from particles, and especially from bacteria that are adsorbed on dust. The mucus is transported by small hairs on the epithelial lining to the pharynx, where it is swallowed [this is called the mucociliary escalator (see Figure 4.13)], and the mucus covering the inside of the respiratory system is continuously renewed. Anything that goes down with the mucus to the stomach may of course be absorbed in the intestines, so the mucociliary escalator does not excrete the aerosol from the body. However, as a protective device for the lungs it is very important, and chronic exposure to hazardous air contamination (i.e., smoking) may decrease and eventually inhibit its function.

Very small particles and droplets stay in the air, and although there is no sharp limit it is normally considered that only particles/droplets with a diameter of less than 1 μm are light enough to escape the mucociliary escalator to a significant extent. Down in the alveoli, such a small particle will through Brownian diffusion (very small particles 'diffuse' because they are constantly bombarded by gas molecules) come into contact with and be adsorbed on the moist inner walls. The dissolvable constituents will be absorbed quickly, but anything macroscopic that does not dissolve can be left here for some time, because in the alveoli there is no ciliary escalator. Eventually, special cleaner cells of the immune system called macrophages enter the alveolus from the blood, engulf the particle by endocytosis, and degrade it both chemically and enzymatically with the contents of their lysosomes. However, particles of some materials, for instance quartz and asbestos, cannot be degraded, and the macrophage will succumb from its efforts. This leads to the stimulation of cell (fibroblast) division in the area, and new layers of connective cells are formed on the inside of the alveolus. As a result, the distance between air and blood increases and the diffusion rate for oxygen decreases. Individuals exposed chronically to quartz dust (working in a quarry for example) or asbestos (used as insulation in older buildings) may develop an irreversible and disabling fibrosis called silicosis and asbestosis, respectively. Asbestos is also carcinogenic, in synergy with smoking, but the mechanism for that effect is not known.

4.2.3.2 Air Contaminants in the Form of Gases and Vapors

Chemicals that are present in air as single molecules will not be affected by any inertial forces, but stay in the air until they are condensed on a surface or absorbed by an adjacent solid or liquid. In principle, airborne molecules therefore have no problem to reach the alveoli, but on their way down they also come into contact with and can be absorbed by the mucous membranes. In that sense, the upper respiratory tract can be compared with a scrubber that purifies the air from water-soluble and reactive contamination. Hydrophilic molecules therefore do not reach the alveoli as efficiently as lipophilic molecules, although they do get there as well. Highly reactive gases, such as hydrogen chloride (HCl), ammonia (NH_3) and sulfur dioxide (SO_2), are strongly irritating on the mucous membranes in the nose. Respiration is blocked, and air containing such gases in significant amounts cannot be inhaled in any larger quantities.

Chemicals with intermediate reactivity and lipophilicity, for example, phosgene, ozone, isocyanates, the halogens, nitric oxide, and nitrogen dioxide will be irritating in high concentrations, but it will be possible to inhale them in lower concentrations. Such compounds can give injuries all the way down, depending on their reactivity and concentration. When they reach the alveoli, their reactivity may harm the cell walls, so that the epithelial cells start to leak. The additional liquid in the alveoli will increase the distance between air and blood and hinder the diffusion of oxygen. This condition, which can be compared to internal drowning, is called pulmonary edema and is believed to be caused by lipid peroxidation of the membranes of the epithelial lung cells (lipid peroxidation is described in Section 7.2.2). Any person accidentally exposed to such gases should be kept under observation for 48 h, as the pulmonary edema takes time to develop.

Lipophilic compounds with little or no reactivity reach the alveoli without problems, and are efficiently absorbed by the blood. Examples are the common organic solvents, which are volatile and whose vapors we can inhale and absorb without difficulty.

4.3
Bioaccumulation

There are no principal differences between man and other organisms when it comes to the uptake of exogenous chemicals, because there are always biological membranes and cell membranes to traverse. (The uptake of chemicals in the roots of plants, on the other hand, is rather complicated.) The rule that lipophilic compounds are taken up easily therefore applies to any creature, and organisms will continuously 'extract' lipophilic contaminants from the environment. The process by which chemicals are absorbed and concentrated by organisms is often called bioconcentration. The bioconcentration factor (BCF), that is the ratio between the concentration of the compound in an organism and its concentration in the surrounding environment, depends largely on the chemical properties of the compound; for example, for nonvolatile compounds in the hydrosphere the log P value is strongly correlated with the BCF value. It also depends on the time that the organisms are in contact with the polluted environment. For many compounds it will take days for equilibria to be established, and the ability of the organism to convert/excrete the compound will of course also play a role.

Chemicals that are relatively stable in an organism and will not be rapidly converted or excreted, for example, polyhalogenated aromatic hydrocarbons, may be absorbed by another organism that uses the first as food. In this way chemicals may enter and be concentrated in food chains, a process called biomagnification or ecological magnification. Especially during the period when halogenated hydrocarbons were used in large amounts there were several examples of how the higher organism at the end of a food chain became affected by pollutants as a result of biomagnification. Birds of prey that fed on marine organisms could contain such large amounts of DDT and PCB that their reproduction was inhibited, while birds of prey that fed mainly on plant-eating animals contained less. However, such birds would instead get more of the seed desinfectants used in agriculture (e.g., hexachlorobenzene). Note that bioconcentration and biomagnification do not only apply to lipophilic organic compounds; inorganic chemicals such as metal ions (see Chapter 10) may also be strongly bioconcentrated as they are bound by proteins.

The term bioaccumulation is used for the combination of bioconcentration and biomagnification, especially in natural environments when it is not possible to separate the two processes. In summary, the accumulation of a chemical in an organism can take place if it is extremely lipophilic and resistant to chemical and/ or biological degradation, or if it is complexed and stored out of circulation by the organism. In the next section we shall see some examples of this.

4.4
Distribution of Chemicals in Organisms

4.4.1
Chemical Properties

Systemic effects of chemicals require that the compound is absorbed by the blood and, if it is not the blood itself that is the target, distributed so that it can reach its target organ. The initial distribution of a compound absorbed into the blood via any of the routes discussed above depends mainly on two things: the ability of the chemical to pass through biological membranes and the blood flow in various tissues and organs. As already discussed, molecules that are small and/or lipophilic will be rapidly distributed to all vascular tissues. However, with time, this initial distribution may well change because a number of equilibria will be established. Some organs and tissues, notably the body fat, the nervous system, and the bone marrow, can be considered to constitute a more 'lipid' environment, while others, for example, the blood, the muscles and the eyeballs, are more 'watery'. In the blood, both endogenous and exogenous chemicals with different polarity may be bound to plasma proteins (e.g., albumin), which may serve as a temporary storage for such compounds. The competition between two compounds for the same binding site on albumin may even lead to a dramatic and dangerous release of one compound as another takes its place.

In general, lipophilic compounds will eventually be distributed to lipophilic tissues like the adipose tissue, but as there are few blood vessels in our fat this will take time. The opposite is, of course, also true: it takes time for lipophilic compounds to leave the adipose tissues, which in practice function like a long-term storage system that in extreme cases can be compared to accumulation. In spite of what is generally believed, even lean and athletic individuals have body fat. Approximately 20% of the weight of such persons is fat. If body fat is consumed, during extreme slimming or starvation, significant amounts of exogenous lipophilic compounds may suddenly be mobilized (depending on the amount present from the beginning and the rate of fat burning), and this may lead to intoxication.

In addition, many exogenous compounds are distributed in somewhat unexpected ways. The liver and the kidneys, as we will see later, are important excretion organs, and they frequently come into contact with compounds that should be expelled. Among other things they take care of many metals, notably toxic heavy metals such as cadmium and lead. If a dose of lead (as Pb^{2+}) is given intravenously, the concentration in the liver is 50 times higher than in the blood 30 min after administration. In the liver and the kidneys the heavy metals are complexed by special proteins called metallothioneins, which are small (approximately $6500 \, g \, mol^{-1}$) and contain large amounts of the amino acid cysteine (25–30%). In this way the metallothioneins are also thought to function as antioxidants, as metal ions (e.g., Cu^+) that may participate in the generation of reactive oxygen species (Section 7.2.2) are inactivated. Lead, together with strontium (Sr), is also distributed to the skeleton, as both may change places with calcium, and the fluoride ion may take the place of a hydroxyl group in bone. Lead has no known toxic effect in bone, although it will slowly be released as our bones are continuously degraded/rebuilt. A radioactive isotope of strontium (^{90}Sr) is formed, for example, during nuclear fission in atom bombs, and the distribution of a radioactive element to the skeleton where it will stay for a long time is of course serious. Fluoride depositions in bone will also give chronic effects.

4.4.2
Biological Barriers

Although the term 'barrier' may be misleading, the passage of chemicals between the blood and the brain is especially difficult, and this resistance is called blood–brain barrier. In the brain, the endothelial cells of the blood capillaries are held together particularly tightly, not leaving any intercellular pores, and they are surrounded by glial cells that obstruct this passage. In addition, the interstitial fluid (the fluid between cells and tissues) of the brain has a low protein content, which hampers the diffusion of both lipophilic and hydrophilic chemicals (e.g., metal ions) that prefer to diffuse attached to a protein. Only freely soluble molecules that readily diffuse through the membranes, or are transported by carrier proteins, will pass through the blood–brain barrier efficiently. TCDD is an example of a compound that is distributed to all organs but the brain, due to its strong binding to plasma proteins. The blood–brain barrier protects the brain from many chemicals, but it may also act like a cage for compounds that are able to get in and then are transformed into something that cannot get out. We have briefly discussed the absorption of mercury; liquid mercury is poorly absorbed in the intestines, but elemental mercury vapor (mercury is a liquid at room temperature and slightly volatile) is readily absorbed in the lungs and distributed throughout the body. When it arrives at the brain, dissolved elemental mercury [Hg(0)] is lipophilic enough to diffuse through the blood–brain barrier, but if it is oxidized (Hg(0) will be oxidized everywhere in the body) to divalent mercury ions (Hg^{2+}) inside the brain it cannot get out because of the blood–brain barrier. Chronic exposure to mercury vapor will therefore, besides other effect on the kidneys for example, also give CNS depression. Most organometallic compounds, for example, tetraethyllead and methylmercury (see Chapter 10), readily enter CNS. The blood–brain barrier develops during the first years of childhood, and thus there is a significant difference between infants and adults, which explains some of the differences in sensitivity to toxicants (e.g., the neurotoxicity of lead in children).

The placenta (or placental barrier as it also is called) is only present during a special period, when it performs the extremely important task of regulating the flow of chemicals between the blood of the mother and the blood of the fetus (which are not mixed, see Figure 4.14). The nutrients are passed through the placenta by active transport, while exogenous compounds may pass through it just as they pass through other biological membranes, and to call placenta a barrier is therefore wrong. The functions of the placenta are also discussed in Chapter 7.

4.4.3
Uptake as a Function of Distribution

The way a chemical is distributed in a body may affect its absorption, especially via the lungs. Volatile compounds can be inhaled and passed into the blood via the alveoli, but they can of course also diffuse in the other direction from the blood to the air. As the balance between uptake and excretion in the lungs depends on

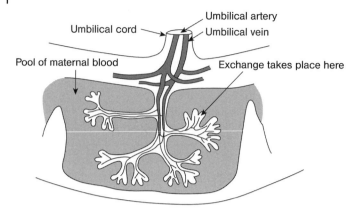

Figure 4.14 Placenta.

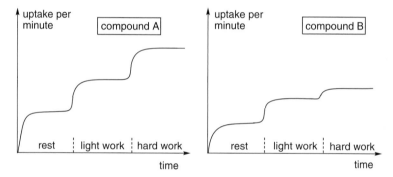

Figure 4.15 The uptake of compounds with large (a) and small (b) blood/air coefficients. See text for explanation.

the amount of the chemical in the air and in the blood, the most efficient absorption is obtained when the blood, each time it passes through the body, is depleted in the chemical by metabolism, distribution to other organs, and so forth. The distribution can be described by partition coefficients between the different media and organs (or similar substitutes) involved, for example between blood and air, blood and brain, blood and body fat, blood and oil, and so forth. If the blood/air coefficient for a chemical is 10, the brain/blood coefficient is 30 and the fat/blood coefficient is 100, and the person is breathing air containing $1\,\mathrm{mg\,L^{-1}}$ ($1000\,\mathrm{mg\,m^{-3}}$), the concentrations in the different tissues at equilibrium will be $10\,\mathrm{mg/kg}$ blood, $300\,\mathrm{mg/kg}$ brain tissue and $1000\,\mathrm{mg/kg}$ body fat. If two compounds with different blood/air coefficients, >30 for compound A (e.g., styrene) and <5 for compound B (e.g., 1,1,1-trichloroethane) are compared, it is clear that the compound with higher solubility in blood will be absorbed more efficiently. Figure 4.15 shows how much of the two compounds will be absorbed by persons breathing air containing similar (and relatively high) concentrations of the compounds, at different work-loads. At rest, the uptake of compound A is approximately 80% of the inhaled amount, and the proportion taken up is more or less independent of the person's workload (i.e., how much air that is inhaled per minute). For compound B, with its smaller blood/air coefficient, the uptake at rest is lower (less than 50% of the inhaled amount), but it decreases even more when the person works harder. This is caused by a fairly rapid saturation of the blood. Compound B is not only absorbed from the air but is also excreted from the blood back to the air in the lungs.

Saturation can also take place further away in the body. If the coefficients for distribution from the blood to other organs are high, the blood coming back to the lungs will have low concentrations of the compound and new amounts can be absorbed from the respiration air. Toluene, having the intermediate blood/air coefficient of 12 and a high oil/blood coefficient of 1200 (a high oil/blood coefficient indicates that a compound is readily absorbed by tissues containing a lot of lipophilic material) will be absorbed in the lungs to approximately 60% at rest. The amount present in blood leaving the lungs will be efficiently distributed to other organs, and when the blood returns it is almost free from toluene. Even if the workload is increased gradually, 50–60% of the inhaled amounts are always absorbed (see Figure 4.16, left part). This can be compared with the solvent trichloroethylene, having a similar blood/air coefficient (16) but a much lower coefficient for oil/blood (15). (Remember that blood contains large amounts of proteins that may 'dissolve' lipophilic compounds, so the partition between oil and blood cannot be compared with that between oil and water.) It will at rest also be absorbed by the blood in the lungs to 60%, but its poor solubility in the rest of the body makes it difficult for the blood to get rid of the compound. Therefore, the blood coming to the lungs will contain increasing amounts of trichloroethylene as the exposure continues, and the exposure for high concentrations of trichloroethylene may in the end saturate the body completely, resulting in the absorption of 0% of the inhaled amounts (see Figure 4.16, right part).

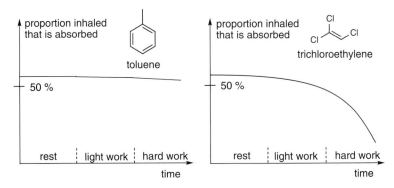

Figure 4.16 The uptake of compounds with similar blood/air but different oil/blood coefficients. See text for explanation.

It should be noted that the experiments discussed in this section were carried out with healthy volunteers, and that while acute exposure to the high concentrations of the solvents discussed above is not considered harmful, chronic exposure for similar concentrations certainly would be.

4.5
Excretion of Chemicals from Man

As accumulation of any chemical in one organism in principle is a bad thing, it is important for organisms to be able to convert chemicals. This can be done in several different ways, for example, by making use of its nutritional value and metabolizing it to water and carbon dioxide, or by excreting it as it is or as a metabolite.

Just as there are several routes for chemicals to get into the body, there are also a number of ways out. While the three principal ways out, via the lungs, via the liver, and via the kidneys, will be discussed in some detail, a few minor routes also deserve to be mentioned. Compounds can in limited amounts be excreted by the sweat, in hair, nails and teeth, by salivary and pancreatic secretion, and in other ways. In addition, and more importantly because this route is coupled with the exposure of an extremely sensitive individual, some chemicals may be excreted via the mother's milk. The milk is characterized by its high content of fat and calcium and a relative low pH (6.5), and it can therefore be suitable for the elimination of lipophilic compounds, metals that can change place with calcium, and basic amines. It has been shown that the milk is the major excretion route for DDT in breast-feeding women, and if she has recently been exposed to lead this will also be present in the milk. Excretion via the saliva also takes place, but, as the saliva is normally swallowed, reabsorption can take place in the intestines and nothing has really changed.

4.5.1
Excretion via the Lungs

Volatile compounds that have been absorbed by any route and that are present in the blood will readily diffuse from the blood to the air in the alveoli of the lungs and be excreted by leaving the body with the exhaled air. This is contrary to absorption via the lungs (see Section 4.2.3), and the direction of the net flow of a volatile compound in the lungs depends on the concentration in the air, the concentration in the blood, and the blood/air coefficient. So, if the air contains nothing of a volatile compound present in the body it will purify the blood coming to the lungs, and the blood may in its turn extract the compound from the tissues it passes. For very volatile and lipophilic compounds (e.g., diethyl ether) this is the major excretion route, and, as it is relatively rapid, only small amounts of such compounds will ever be metabolized in the body. The concentration in the exhaled air is directly proportional to the concentration in the blood, notably for less volatile

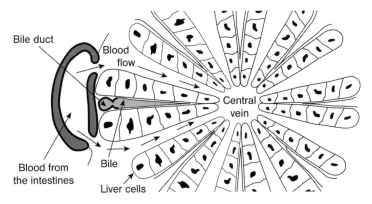

Figure 4.17 The liver.

solvents such as ethanol, only a small proportion of which are excreted by the lungs, and this is taken advantage of in breath analysers.

4.5.2
Excretion via the Liver

The liver is fed by blood coming from the intestines, and anything absorbed there will immediately come into contact with the liver cells. These will absorb all kinds of compounds, convert them in various ways, and pass them back to the blood or to the bile (see Figure 4.17). After passing through the liver, blood containing the metabolites produced in the liver is collected in veins and pumped to the heart. The bile produced by liver cells is secreted in bile ducts, which are eventually emptied in the small intestine. Bile contains mainly amphipathic compounds (bile salts, cholesterol, and lecithin), which are responsible for the emulsification of fats in the intestines, but it also serves as a vehicle for the compounds that the liver excretes.

The liver will excrete many metals as ions, for example, lead, manganese, and also to some extent mercury. Lead is concentrated (approximately 100 times compared to the blood) in the bile by active transport, while mercury is not. Nevertheless, excretion via the bile is still the major route of excretion for mercury, indicating that the half-life of this highly toxic metal in the body is long. In addition, other organic compounds are deposited in the bile, especially acidic and basic compounds but also some neutral ones having a high molecular weight. There is no sharp limit for how big they should be, but excretion via the liver is normally the major route for compounds with a molecular weight exceeding $300\,\mathrm{g\,mol^{-1}}$ in rats and $500\,\mathrm{g\,mol^{-1}}$ in humans. Although this may appear to be a high figure, we shall see in Chapter 5 that many smaller organic molecules are metabolized in the liver by conjugation with fairly large endogenous compounds, and the resulting conjugates are considerably heavier than the original compound. An example of an organic compound that is excreted via the liver is the disodium salt of

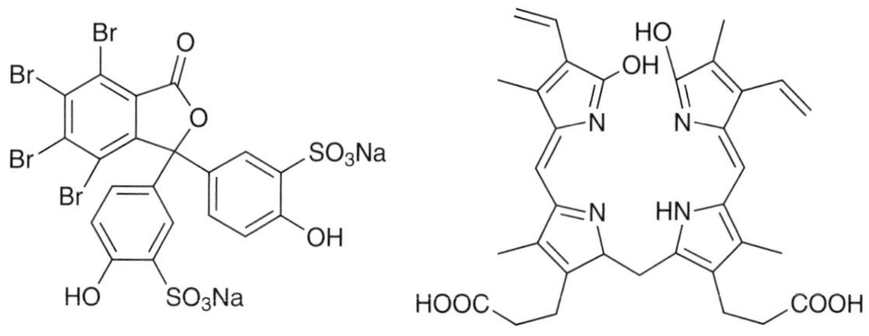

Figure 4.18 Sulfobromophthalein (left) and bilirubin.

sulfobromophthalein (see Figure 4.18), with a molecular weight of $838 \, g \, mol^{-1}$, which is used to test the function of the liver. It is injected into the blood, and should disappear rapidly if excretion through the liver is functioning normally. Another compound excreted in this way is the endogenous compound bilirubin, a colored breakdown product of hemoglobin, and a person whose liver is malfunctioning is jaundiced because bilirubin is retained in the body.

The exact mechanism for transporting chemicals that are to be excreted from the liver cells to the bile is poorly understood, but it appears to be by active transport. It can also be noted that this route for excretion has not matured in infants (another example of a physiological difference between adults and infants). Excreting organic compounds into the intestines may not appear to be a very good idea, because we have already noted that few organic molecules will escape absorption in the intestines. An example is the highly toxic organometallic compound methyl mercury, well-known as the Minamata poison (Section 10.7), which is absorbed in the intestines and transported with the blood to the liver where it is excreted. In addition, the bacterial flora have the ability to reverse some of the metabolic conversions carried out in the liver and re-convert the excreted metabolite to an absorbable compound. The process by which an excreted compound is reabsorbed in some form is named the enterohepatic circulation, and although it is not a perpetual machine with no losses it still prolongs the time that an exogenous chemical spends in the body. We shall return to the enterohepatic circulation in Chapter 5. Besides being an organ involved in the metabolism and excretion of exogenous compounds, the liver is also the target organ in many intoxications, and this will be discussed in Chapter 7.

4.5.3
Excretion via the Kidneys

The kidneys are the most important excretion organs. Although the two kidneys only comprise 1% of the body weight they receive approximately 25% of the blood coming directly from the heart, and 20% of this is filtered in the functional unit of the kidneys, the nephron (see Figure 4.19). The kidneys contain approximately a million of these units, each with a glomerulus (a set of fine capillaries with rela-

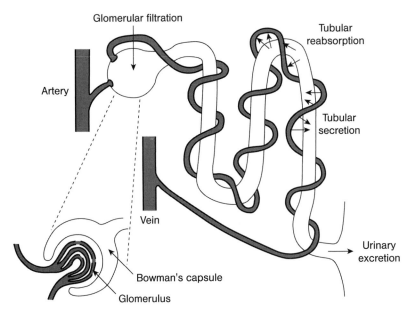

Figure 4.19 The nephron.

tive large pores that allow fairly big molecules, up to small proteins, to leave the blood) surrounded by a sac (Bowman's capsule) that collects the filtered fluid. As the blood passes through the glomerulus, it loses part of its water together with any dissolved relatively low-molecular-weight compounds. Albumin and other plasma proteins (with adsorbed lipophilic compounds) are retained. The liquid produced by the glomerular filtration, approximately 200 L/24 h, is called the primary urine, and this passes through the tubili, which are surrounded by blood vessels. (The glomerulus and the tubule together make up a nephron.) Evidently, a person cannot afford to lose 200 L water per day, and most of the water, as well as nutrients and minerals, are reabsorbed back into the blood again (tubular reabsorption). Besides simple diffusion, several specific transport systems are involved to make the loss of valuable and useful chemicals as limited as possible.

What is left in the concentrated urine (approximately 1.5 L is produced per day) is a solution of anything originally present in the blood but filtered in the glomerulus, unable to diffuse back to the blood and not fitting any of the transport systems. If an exogenous compound that cannot be used and should be excreted is too lipophilic to stay in the urine, it must be further converted in the body to a more hydrophilic form. In addition, there are also transport systems in the tubili that actively excrete (tubular secretion) certain compounds (organic acids and bases) from the blood to the urine. In the beginning of the era of penicillin as an antibiotic drug it was expensive, and its rapid elimination via the kidneys was a problem. Penicillin was therefore administered together with an acid (probenecid, see Figure 4.20), which occupied the transport system normally pumping penicillin

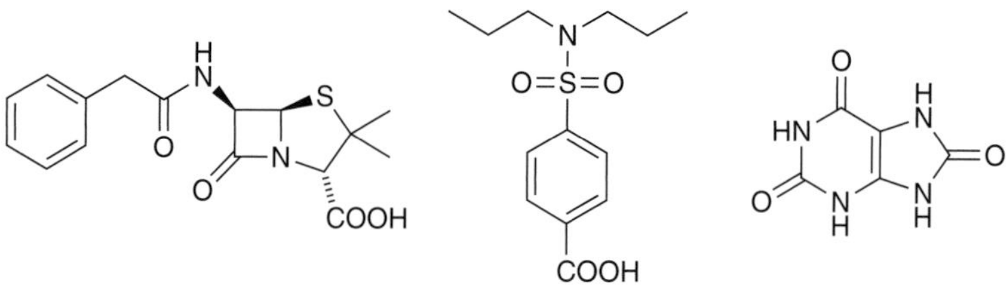

Figure 4.20 Penicillin G (left), probenecid and uric acid (right).

from the blood to the urine, and this prolonged the time the penicillin stayed in the body (a synergistic effect!). Today penicillin is relatively cheap, and higher doses are given instead. Another compound that is excreted in this way is uric acid (see Figure 4.20), present in our urine and associated with gout. If the transport of uric acid is inhibited, the concentration of it goes up and the risk of gout increases.

5
Metabolism of Exogenous Compounds in Mammals

Although the presence of exogenous chemicals in our environment has increased dramatically during the last century, the chemical soup that life on earth has evolved in has always contained numerous compounds that are of little use and are toxic to any organism. Consequently, life on earth has always needed ways to handle such chemicals, and it is not surprising that enzymatic systems that exclude/degrade/detoxify/expel exogenous chemicals have evolved. One way of achieving this is by having specific membrane transport proteins that are selective in their choice of compounds to be taken up or excreted (discussed in the preceding chapter). Such systems only protect us to a certain degree, and we have seen how easily lipophilic exogenous compounds pass through biological membranes and enter our bodies. In some rare cases compounds can be converted from being exogenous to become endogenous, and thereby in principle be useful. However, as we shall learn in this chapter, such conversions, even if in the end they produce a harmless compound, often generate more toxic intermediates and by-products. The excretion of compounds is facilitated if they have chemical properties that make them suitable for excretion, for example, are volatile (excretion with the exhaled air), have a molecular weight over 500 (excretion with the bile) or, most importantly, have high solubility in water (excretion with the urine). If the compounds do not possess the required properties to begin with, they can obtain them by being converted enzymatically by the metabolism. The volatility of a compound may be increased if the metabolism decreases its molecular weight (by breaking bonds) and/or converts a functional group. A high molecular weight can be achieved by combining the exogenous compound with an endogenous compound, while increased water solubility can be accomplished by converting chemical functionalities to more polar forms. There are many examples of compounds that are transformed into more volatile forms or into compounds with higher molecular weight and at least partially excreted by the lungs and the liver, but the kidneys are the main excretion organs, and the conversion of exogenous compounds into highly water-soluble derivatives is the most important way for the body to get rid of such compounds.

In this chapter, the principal enzymes involved and the most important conversions are discussed, and we also describe how a number of well-known organic compounds are metabolized in mammals. The survey of the principal conversions

Chemistry, Health, and Environment. Olov Sterner
© 2010 WILEY-VCH Verlag GmbH & Co. KGaA, Weinheim
ISBN: 978-3-527-32582-5

concentrate on how one functionality at a time may be changed chemically, while the practical examples at the end of the chapter demonstrate that it may be difficult to predict the outcome of the metabolism of a compound containing several groups that participate in conversions.

5.1
Overview

The conversion of exogenous compounds is commonly divided into two principal reaction types, the phase I and the phase II reactions. The result of a phase I reaction (or, in some cases, a combination of several phase I reactions) is the introduction of a suitable functional group that increases the water solubility and can be used in a phase II reaction. This is achieved either by the enzymatic introduction of such a functionality into the framework of the molecule or by exposing a group already present but masked by other groups. The phase II reactions, which are often called conjugations, take advantage of this chemical handle and attach polar hydrophilic groups to it, producing products with considerably higher water solubility. In addition, substitutions are also included among the phase II reactions. Obviously, it is not always necessary to go through both phases if the phase I reaction produces a compound that is sufficiently water soluble to be excreted with the urine or if a compound already contains suitable functionality for the phase II reactions. An example of a compound that is metabolized by both phase I and phase II reactions is benzene (see Figure 5.1). Benzene is oxidized to phenol (a phase I reaction introducing the hydroxyl group as a chemical handle) which subsequently is conjugated with glucuronic acid. The latter is a carbohydrate derivative that has a carboxylic acid functionality that is ionized at physiological pH. It also has several hydroxyl groups, and its size and polarity will facilitate the excretion of any glucuronic acid conjugate via the kidney or the liver.

Benzene is a cheap chemical obtainable in unlimited amounts from oil. It has several useful properties and has been an important industrial solvent as well as an additive in petrol. However, it is also toxic, and chronic exposure may lead to the development of leukemia, so that its use is today strictly regulated. The mechanism by which benzene is carcinogenic is still not completely understood, although neither benzene itself nor its two major metabolites phenol and phenol-*O*-glucuronide are directly involved. Instead, it is one or several other metabolites that are responsible for this specific effect, and the example of benzene illustrates an important aspect of the metabolism of exogenous compounds. It is always to be expected that a compound will be metabolized by several parallel routes, and a minor route, perhaps only responsible for the conversion of one or a few percent of the original compound, may generate the metabolites that finally turn out to be responsible for the toxic effect. We therefore return to benzene at the end of this chapter and identify some minor metabolites that are involved in its toxicity. This is also a common reason for the difference in sensitivity observed between differ-

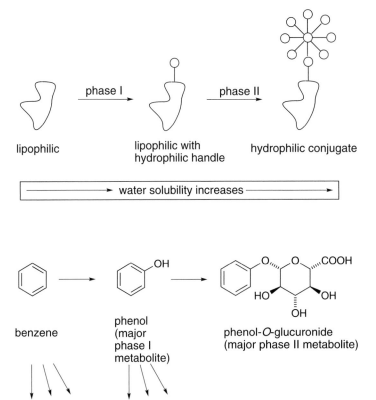

Figure 5.1 The two phases of the metabolism of exogenous compounds, exemplified by the major conversion of benzene.

ent species (discussed in Chapter 7), as the balance between different metabolic pathways may also be slightly different in closely related species.

It is not unusual for the term 'detoxification' to be used as a synonym for the metabolism of exogenous compounds, but this is not correct. For example, most carcinogenic compounds require enzymatic conversion (metabolic activation), and it is actually the metabolites that transform the normal cells into tumor cells. Metabolic conversions will always change the toxicity of a compound. The metabolites may be more or less toxic than the original compound but they are never equally toxic, although in some cases a metabolite can be more toxic in one way and less in another.

Another important concept that we shall come back to is the induction of the enzymes that are required for a certain conversion, for example by the presence of the substrate for that conversion in certain amounts, and induction is very important for the enzymatic systems involved in the conversions of exogenous compounds. Many exogenous compounds are powerful inducers, and, as is further discussed in Chapter 7, the induction of metabolizing enzymes is often the reason for potentiating and synergistic toxic effects as well for differences in sensitivity among different species.

5.2
Enzymatic Systems Involved in Oxidations

Enzymes that will metabolize exogenous compounds are present in all tissues in the body, although the liver has the highest metabolic capacity of all organs. However, for a special enzymatic conversion other organs and tissues may have a similar or even higher metabolic capacity, and differences in the metabolic profiles between various organs and tissues may be responsible for organ-specific toxic effects of compounds that require metabolic activation (which for natural reasons often affect the liver). In the cells, most of the phase I reactions take place in the endoplasmic reticulum, a relatively lipophilic organelle to which lipophilic compounds would be attracted, while most of the phase II reactions are carried out in the cytoplasm. As discussed later in this section, the phase I reactions are dominated by one group of enzymes, the cytochrome P450s, which in principle are oxidative but also can perform reductions under special circumstances. Several important phase II systems are normally active and often compete for the same types of substrates. Again, there is reason to remember the existence of the '11th organ system' composed of bacteria in the intestines (Section 4.2.2), with its huge and different metabolic capacity.

5.2.1
Primary and Secondary Metabolism

The metabolism of all organisms can be divided into a primary and a secondary part, although they are not completely separate. The primary part deals with all nutrients, their degradation into suitable units (amino acids, saccharides, nucleotides, etc.) that can be used for the construction of macromolecules and for the production of energy, and so forth. It takes care of all metabolic conversions necessary to keep a cell (organism) alive, and if the primary metabolism fails or is obstructed, a life-threatening situation results. It is also noteworthy that the primary metabolisms of different species are more or less identical; even organisms as different as man and bacteria share many of the principal primary metabolic pathways. The secondary metabolism is less well understood. It emanates from primary metabolites and converts them to metabolites with other functions, and the secondary metabolites (e.g., alkaloids and terpenoids) produced in various organisms are often completely different from each other. Examples of possible roles that secondary metabolites may play are as substances used for signaling, within an organism as hormones and between organisms as pheromones, or as substances that taste bad or are toxic and are used for defensive reasons. The toxic compounds from natural sources that we discuss later are secondary metabolites, and so also are many antibiotics (e.g., the penicillins), the cardiac glycosides (e.g., digitoxin), and many antitumor agents (e.g., taxol).

Exogenous compounds will be metabolized by enzymes that belong to both the primary and the secondary metabolism. The principal conversions of the primary metabolism are covered by textbooks in biochemistry and will not be discussed here. In this chapter we confine ourselves to an overview of other important conversions, mainly belonging to the secondary metabolism, that take place with exogenous compounds in mammals.

5.2.2
Cytochrome P450

The cytochrome P450 system (discussed as a system, but in reality the enzyme cytochrome P450 exists as a large number of variants, see below) catalyzes oxidations mainly but also some reductions, and is the most important metabolic system for the conversion of exogenous compounds. It is not a uniquely mammalian enzyme system, having also been found (in identical or similar forms) in many other organisms. In mammals it is found in most tissues, although its function does not appear to be the same everywhere, reflecting the role of cytochrome P450 in the conversion of steroid hormones. The highest concentrations are naturally found in liver cells, but its presence has also been demonstrated in, for example, the lungs, kidneys, placenta, and testes. The cytochrome P450 system is also called the polysubstrate monooxygenase system or the mixed-function monooxygenase system, reflecting the fact that it is not specific for one type of substrate but instead accepts a surprisingly broad range of structures. It is composed of two enzymes, cytochrome P450 and a reductase (NADPH-cytochrome P450 reductase) providing the electrons necessary for the reactions in the form of the reduced coenzyme NADPH. Interestingly, there are approximately 20 cytochrome P450 enzymes for each reductase, placed in a ring around the reductase, and the whole complex is embedded in the phospholipid matrix of the endoplasmic reticulum. The cytochrome P450 system will not work outside this membrane, being dependent on the membrane for its function, and the phospholipids improve the contact between the two enzymes. However, when liver cells are disrupted by homogenization the endoplasmic reticulum is fragmented into vesicles, diameter 0.1 μm, and these are fairly easily purified from the rest of the cell debris. The vesicles are called microsomes and can be used when the effects of cytochrome P450 conversions are to be studied *in vitro*. The cytochromes are proteins that readily take part in electron transfers because they contain a heme group with an iron atom that easily goes between the ferric [Fe(III)] and ferrous [Fe(II)] states, and we shall return to cytochromes when we discuss toxic effects on the respiration chain and energy production in Chapter 7. Similar heme groups are found in the hemoglobin of the blood and in the chlorophyll of plants, although in the latter the iron is exchanged for magnesium. The fact that the cytochrome P450

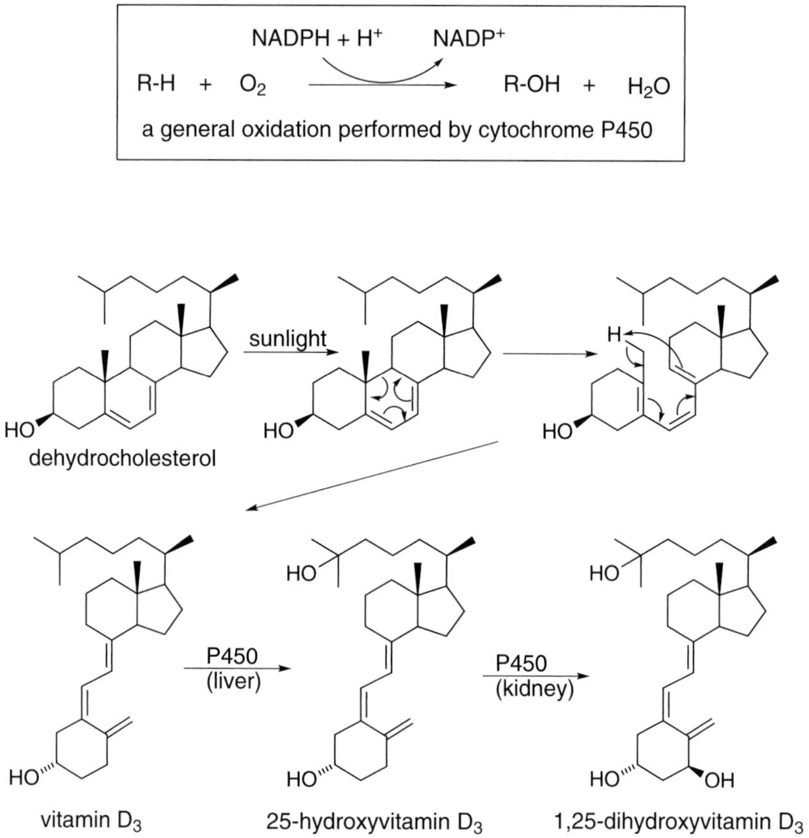

Figure 5.2   The oxidation of vitamin D₃ by cytochrome P450.

system is a monooxygenase reveals that it uses molecular oxygen as an oxidation agent and that one of the two oxygen atoms forms water. The overall reaction for the oxidation (a hydroxylation in this case) of a substrate RH can be written as shown in Figure 5.2. An example of the physiological role of this system, to prepare various variants of vitamin D, is also showed in Figure 5.2.

The molecular mechanisms by which the cytochrome P450 system oxidizes various substrates have been intensely studied over the years, but they are still not understood in all details. In this section the general mechanism is discussed, while the details linked to the oxidation of certain functional groups will be given in Section 5.3. Cytochrome P450 oxidations are initiated by the association of a lipophilic substrate molecule with the cytochrome P450 enzyme in its oxidized form [Fe(III)] in the endoplasmic reticulum (a lipophilic environment). The complex then accepts an electron from the reductase via the coenzyme NADPH, reducing the iron to Fe(II). Just as hemoglobin, with a similar Fe(II)-containing heme group, binds oxygen, the reduced form of cytochrome P450 will immediately

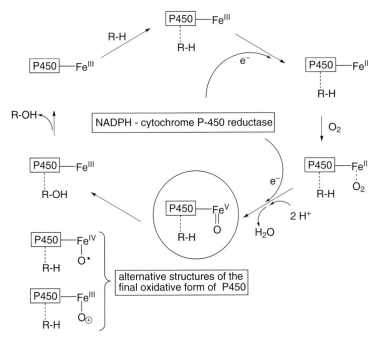

Figure 5.3 The oxidation cycle of cytochrome P450.

attract a molecule of oxygen (which under normal circumstances is present in all tissues) to the iron atom. Now the stage is set for the acceptance of a second electron from the reductase: two protons are picked up from the surrounding water and a molecule of water is split off from the very unstable complex. The second oxygen of O_2 has now combined with the iron to the final oxidation reagent, a complex (circled in Figure 5.3) containing what could be a perferryl group with the iron in an extremely oxidized state [Fe(V)]. How this final form looks is not understood, and the oxidation states of iron and oxygen in the final reactive enzyme could also be Fe(III)–O$^+$ or Fe(IV)–O$^•$. However, it is known that other perferryl compounds prepared synthetically are very efficient oxidizing agents and perform similar oxidations to those performed by cytochrome P450, because Fe(V) is eager to accept electrons and be reduced to something more normal [Fe(II) or Fe(III)]. The substrate is now oxidized, for example, by the insertion of an oxygen into a carbon–hydrogen bond, so that hydroxylation has taken place, as shown in Figure 5.3, whereupon the substrate can leave the enzyme and the cytochrome P450 returns to its original state.

Just as hemoglobin is blocked by carbon monoxide (CO) (discussed in Chapter 7), the reduced form of cytochrome P450 is also sensitive to this toxic gas. This effect can be utilized when cytochrome P450 is studied *in vitro*, but it will not be relevant *in vivo* because of the more serious effect of CO on the blood. In certain situations, for example, during anaerobic conditions, cytochrome P450 can make use of a molecule other than molecular oxygen as the oxidation agent. That compound, whatever it is, will be reduced, and this is one way that metabolic reductions of exogenous compounds take place. Although the oxidation cycle of cytochrome P450 appears to be secure, several complexes are not very stable, especially that formed after oxygen has added to the reduced heme group (in the lower right-hand corner of Figure 5.3), which has a tendency to get into (chemical) trouble.

5.2.3
Generation of Reactive Oxygen Species

Instead of accepting an electron from the reductase, an electron may be transferred internally from the iron (which will be oxidized) to the O_2 (see Figure 5.4). The negatively charged molecular oxygen possessing an additional electron (called superoxide, superoxide anion, or superoxide radical) will leave the complex, while the cytochrome P450 with its substrate will re-enter the oxidation circle, and the effect of this extra loop is that it generates superoxide, which, as an anion radical, may participate in radical reactions.

The generation of superoxide during electron transfer to molecular oxygen highlights a problem that any life-form that uses oxygen for the production of energy has to cope with. Molecular oxygen itself has a certain reactivity: in its ground state it is a di-radical (with two unpaired electrons) which is fairly stable (because the two unpaired electrons have parallel spin in the ground state of O_2) but which readily participates in and promotes radical reactions. Oxygen would therefore be expected to give synergistic effects with anything that generated radicals in the body, for example, exposure to ionizing radiation or to any exogenous compounds that are metabolized to radicals. This is believed to be a major cause of the toxic effects observed in mammals exposed to high concentrations or high

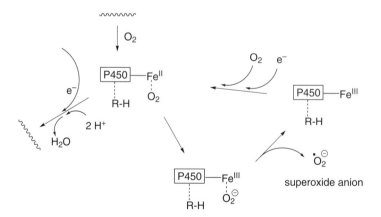

Figure 5.4 Reactive oxygen species are generated during cytochrome P450 oxidations.

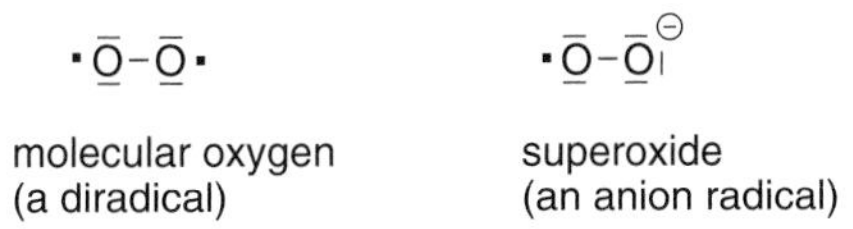

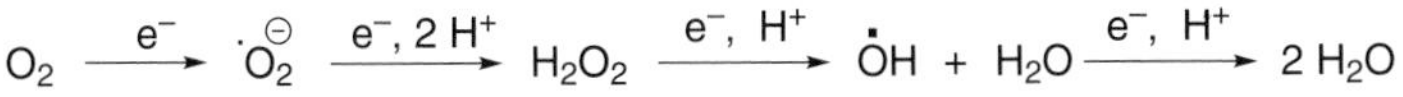

$$\cdot\overline{\underline{O}}-\overline{\underline{O}}\cdot \qquad \cdot\overline{\underline{O}}-\overline{\underline{O}}\overset{\ominus}{\underline{}}$$

molecular oxygen   superoxide
(a diradical)   (an anion radical)

$$O_2 \xrightarrow{e^-} \cdot\overset{\ominus}{O_2} \xrightarrow{e^-,\,2\,H^+} H_2O_2 \xrightarrow{e^-,\,H^+} \overset{\cdot}{O}H + H_2O \xrightarrow{e^-,\,H^+} 2\,H_2O$$

Figure 5.5   The reduction of O_2 to H_2O via reactive oxygen species.

pressures of oxygen, as some radicals are always generated in our bodies by, for example, the normal background radiation. In its ground state, molecular oxygen can only function as an oxidation agent by accepting one electron at the time, the second electron transforming superoxide anion to the dianion of hydrogen peroxide, which at physiological pH is protonated. The hydroperoxyl radical (the corresponding acid of the superoxide anion) has the pKa value 4.8 and is consequently ionized at pH 7.4. Hydrogen peroxide is a well-known oxidation agent, used to bleach hair, as a rocket propellant and as a disinfectant, and although it is not limited to one-electron reductions, like molecular oxygen it is in some situations reduced to the hydroxyl radical on the reduction path to water (see Figure 5.5). The hydroxyl radical is one of the most reactive compounds known; it will immediately initiate radical chain reactions, and its involvement (and that of other radicals) in various toxic effects will be discussed in Chapter 7.

Other oxidizing enzymatic systems will also leak superoxide, as will oxyhemoglobin (the oxygenated form of hemoglobin), and in addition we will encounter several examples of how superoxide is generated by the autoxidation of exo- and endogenous compounds with molecular oxygen. It is believed that several percent of the oxygen consumed in the mitochondria (the cell organelle where energy is produced and most oxygen is consumed) is reduced to superoxide and hydrogen peroxide, and this continuous generation of reactive oxygen species is a threat to any organism. It has been counteracted by the evolution of several protective enzymatic systems, which are present in tissues where oxygen is being reduced and take care of the reactive oxygen species. Examples are superoxide dismutase (SOD), catalase (see below), and glutathione peroxidase, and the reactions they catalyze are shown in Figure 5.6. The dismutation of two molecules of superoxide may also take place spontaneously, but is much more efficient when SOD is present. The reduction of hydrogen peroxide by glutathione peroxidase is coupled with the oxidation of the thiol groups of two molecules of glutathione (GSH, discussed in detail later in this chapter) to glutathione disulfide (GSSG) (two thiol groups are oxidized to a disulfide). To regenerate glutathione, which is an important endogenous compound in several ways, the disulfide is reduced back to the thiols by glutathione disulfide reductase with the coenzyme NADPH.

Glutathione peroxidase depends on both vitamin E and selenium as cofactors for its function. Shortage of these two essential substances, for example, because

superoxide dismutase (SOD): $2\ \overset{\cdot\ominus}{O_2}\ +\ 2\ \overset{\oplus}{H} \longrightarrow H_2O_2\ +\ O_2$

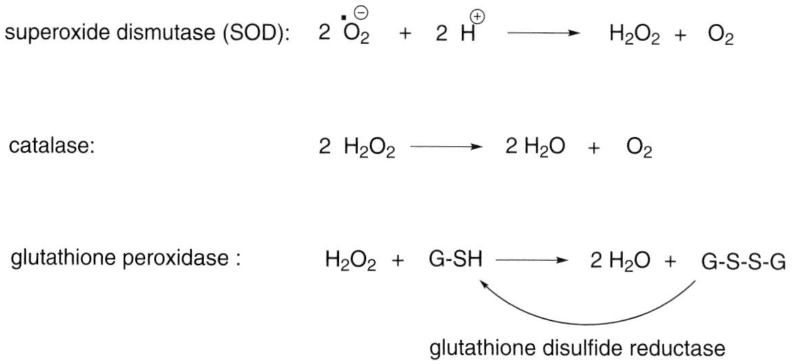

catalase: $2\ H_2O_2 \longrightarrow 2\ H_2O\ +\ O_2$

glutathione peroxidase : $H_2O_2\ +\ G\text{-}SH \longrightarrow 2\ H_2O\ +\ G\text{-}S\text{-}S\text{-}G$

glutathione disulfide reductase

Figure 5.6   The enzymatic disarmament of superoxide and hydrogen peroxide.

$\overset{\cdot\ominus}{O_2}\ +\ H_2O_2\ +\ \overset{\oplus}{H} \longrightarrow \overset{\cdot}{O}H\ +\ H_2O\ +\ O_2$

the Haber Weiss reaction

$H_2O_2\ +\ Fe^{2+}\ +\ \overset{\oplus}{H} \longrightarrow \overset{\cdot}{O}H\ +\ H_2O\ +\ Fe^{3+}$

the Fenton reaction

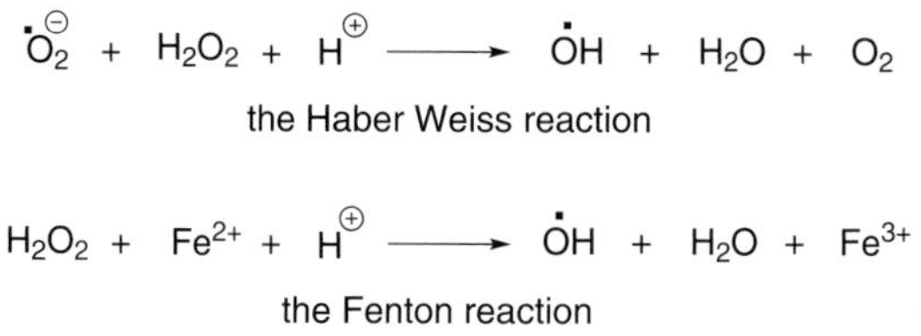

Figure 5.7   Left unattended, superoxide and hydrogen peroxide may generate hydroxyl radicals.

of malnutrition, will weaken the defense against reactive oxygen species and increase the risk of toxic effects. Investigations have show that there is a correlation between insufficient intake of vitamin E and selenium and an increased incidence of tumors, and it is suspected that at least part of this is due to their role in hydrogen peroxide reduction. Another possible disturbance of this defense, which has a limited capacity, is that a sudden generation of large amounts of reactive oxygen species will saturate the enzymes and increase the possibility for these compounds to take part in other reactions. Examples of some deleterious reactions are shown in Figure 5.7, namely the reaction between superoxide and hydrogen peroxide and that between hydrogen peroxide and Fe(II), both of which generate the extremely reactive hydroxyl radical. Although most of the 4 g of iron that an adult contains is not available for the latter reaction, the small part that is makes it biologically relevant. Iron, like all other essential elements and compounds, is toxic in high concentrations, and the effects observed in iron overload suggest that they are caused by increased formation of hydroxyl radicals.

Interestingly, reactive oxygen species are also put to good use by our immune system. They are deliberately produced by phagocytes and used to degrade foreign material. In addition, it has been shown that in certain plants lipid peroxidation generates reactive oxygen species as an efficient defense against parasites.

The number 450 in the name cytochrome P450 comes from its absorption of visible light with an absorption maximum at 450 nm in its reduced form with carbon monoxide bound to the Fe(II) ion, while the letter P stands for pigment. However, it does not consist of one single enzyme but rather of a range of isoenzymes that all have slightly different properties, and the exact composition of a cytochrome P450 subset varies with species, organ, sex, age, health, stress, and so forth. They all absorb light of approximately the same wavelength, and the exact value of the wavelength is sometimes used to identify individual enzymes (e.g., cytochrome P448, etc.). It is not understood why so many isoenzymes are produced, for example, whether they are intended to carry out slightly different reactions, and in this text we will simply call them all cytochrome P450. An important aspect of the cytochrome P450 system is its ability to be induced, so that the number of enzymes increases in situations where the need for cytochrome P450 conversions is high. The presence of exogenous compounds is a common cause for induction, and different compounds will induce different isoenzymes (or sets of isoenzymes). Potent inducers of cytochrome P450 are, for example, polyhalogenated aromatic and alicyclic hydrocarbons, for example, TCDD, PCBs, PBBs, and the pesticides aldrin, dieldrin, lindane, and the chlorinated phenoxy acids. These compounds are all discussed in detail later.

5.2.4
Other Oxidative Enzymes

Besides cytochrome P450, there are a number of other enzymes involved in phase I oxidations. These are less general than the cytochrome P450 system with which they in most cases compete, and they appear to be intended for more specific conversions. Only a brief overview is given in this section; further details about the conversions can be found in Section 5.3.

The microsomal FAD-containing monooxygenase is, like the cytochrome P450 system, localized in the endoplasmic reticulum and depends on NADPH and molecular oxygen for its action, and it is obvious that it is not easy to distinguish between the two. It oxidizes mainly heteroatoms: nitrogen (primary, secondary, and tertiary amines), sulfur (thiols, thioethers, and thiocarbamates) and phosphorus (phosphines and phosphonates). Many of these compound classes are also believed to be substrates for cytochrome P450, especially before the existence of the microsomal FAD-containing monooxygenase had been established, and the balance between the two has not yet been determined. In contrast to cytochrome P450, it does not appear to be inducible; instead, it has been suggested that its activity is regulated by hormones.

The dioxygenases, which, for example, are important in the primary metabolism of mammals for the degradation of amino acids, will use molecular oxygen for their oxidations, just like the monooxygenases. The difference between the two is that the dioxygenases insert both oxygen atoms, as indicated in Figure 5.8. Dioxygenases are also present in bacteria, and, as we shall discuss in Chapter 6, they are very important for the degradation of organic compounds by microorganisms.

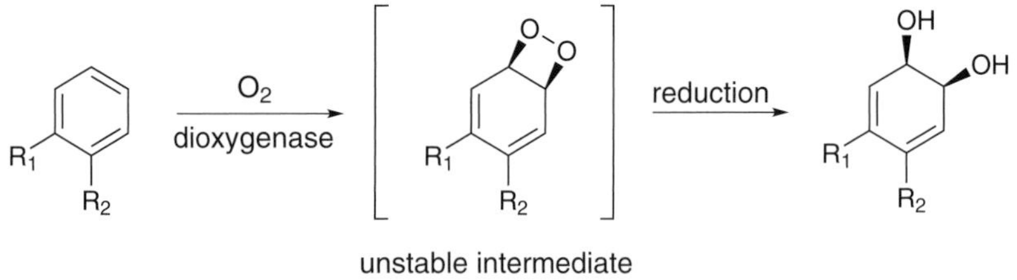

Figure 5.8 Oxidation of an aromatic hydrocarbon with a dioxygenase.

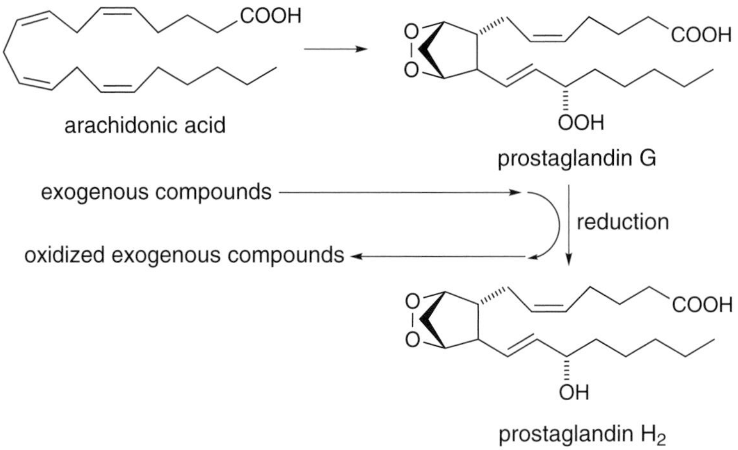

Figure 5.9 The cooxidation of an exogenous compound coupled with the reduction of a hydroperoxide, catalyzed by a peroxidase.

Amine oxidases (monoamine oxidases) typically oxidize primary amines to aldehydes, but also secondary amines to ketones (compare with the transaminases of the primary metabolism that oxidize α-amino acids to α-keto acids). They are not located in the endoplasmic reticulum but instead in the mitochondria or are dissolved in the cytoplasm of the cell. They are present in different tissues, and in CNS for example they are involved in the conversion of neurotransmitters.

Peroxidases are enzymes that couple the reduction of hydrogen peroxide or a hydroperoxide to the oxidation of a substrate. Metabolic activation by peroxidases has received attention recently, and it is believed that they play a significant role in the generation of several types of electrophilic compounds. Several examples of this will be given in the next section. An interesting example is prostaglandin synthase, which actually generates its own hydroperoxide. It will oxidize arachidonic acid (an essential fatty acid) to prostaglandin G, which is subsequently reduced by the same enzyme to prostaglandin H_2 (see Figure 5.9). During the second step, exogenous compounds can be cooxidized, and a number of examples of this with various types of substrates have been reported.

Another example is catalase, which uses hydrogen peroxide produced by, for example, SOD (see above) to oxidize various substrates, for example, phenols, formic acid, formaldehyde, and alcohols according to the general formula $RH_2 + H_2O_2 \rightarrow R + 2H_2O$. Actually, a substantial part of the ethanol consumed by humans is converted to acetaldehyde by catalase. The conversion shown in Figure 5.6 is only carried out when excess hydrogen peroxide accumulates in the cell.

Alcohol dehydrogenases, which exist in several similar forms, will oxidize primary alcohols (to aldehydes) and secondary alcohols (to ketones), although some amounts of alcohols will also be oxidized by cytochrome P450 and catalase/hydrogen peroxide. Tertiary alcohols will normally not be oxidized at the alcohol function. Aldehydes are considerably more toxic than primary alcohols, and also more lipophilic, so this oxidation is clearly a metabolic activation. The alcohol dehydrogenases are soluble enzymes found especially in the liver, the kidneys, and the lungs, and they are responsible for most of the conversion of exogenous alcohols, although alcohols are also oxidized to some extent by monooxygenases in the endoplasmic reticulum. It should be noted that these are reversible reactions: the aldehydes and ketones may be reduced back to alcohols by the same enzyme (but with the reduced form of the coenzyme instead of the oxidized, see below), and the equilibrium depends on whether other conversions or any excretions take place with the different forms.

Aldehyde dehydrogenases oxidize both aliphatic and aromatic aldehydes to carboxylic acids, a reaction that shows all the features (i.e., less toxic and more hydrophilic products are formed) of a metabolic detoxification. The reaction is irreversible, and the carboxylic acid is a good substrate for the phase II reactions, which in practice will be a strong driving force for pushing the equilibrium between primary alcohols and aldehydes to the right. Aldehyde dehydrogenases exist both in a general form (as many isoenzymes) and in one specific for formaldehyde. Other enzymes that oxidize aldehydes have also been identified, including the molybdenum-containing flavoprotein aldehyde oxidase and xanthine oxidase.

Acyl coenzyme A dehydrogenase takes part in the β-oxidation pathway which is responsible for the conversion of fatty acids to acetyl units, which are subsequently oxidized in the citric acid cycle, a part of the primary metabolism. It converts the coenzyme A thioester of a straight-chain saturated carboxylic acid to an α,β-unsaturated thioester, and exogenous carboxylic acids (possibly formed by the oxidation of the corresponding alcohol or aldehyde) may of course also be activated as a thioester and oxidized.

5.3
Phase I Conversions

Although the classification of certain metabolic conversions of an exogenous compound as a phase I or a phase II reaction is not always unambiguous, phase I reactions can be considered to be either oxidations, reductions, or hydrolyses.

5.3.1
Oxidations

5.3.1.1 Epoxidations of Carbon–Carbon Multiple Bonds

Carbon–carbon multiple bonds are readily epoxidized by cytochrome P450, and as the reaction increases the electrophilicity of the compound this is a metabolic activation from a toxicological point of view. However, although most of the exogenous compounds that we are exposed to contain a carbon–carbon multiple bond, this is not as serious as it may appear, and this is for two main reasons. Firstly, alkenes and aromatic hydrocarbons also undergo other conversions which do not generate epoxides. Secondly, the epoxides formed in a phase I oxidation are in most cases either chemically unstable and will spontaneously form other products or are good substrates for phase II enzymes that will hydrolyze the epoxide to a relatively harmless 1,2-diol (see below). Assuming that the mechanism for cytochrome P450 oxidations presented in Figure 5.3 with the perferryl complex as the true oxidizing agent is correct, the epoxidation of an alkene can be imagined as either a one- or a two-electron transfer (see Figure 5.10). The intermediate complex formed may then either cyclize to an epoxide, rearrange via a 1,2-alkyl (-hydride if R_4 is a hydrogen) shift to a carbonyl compound, or react with a nucleophilic atom (e.g., a nitrogen of the heme group) of the enzyme and form a covalent bond to it (the two latter possibilities are only

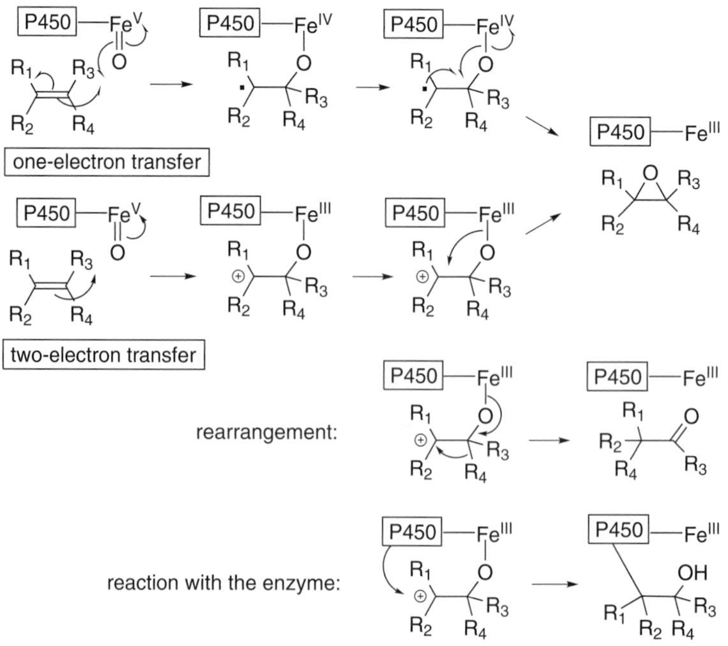

Figure 5.10 Possible mechanisms for the oxidation of alkenes.

shown for the two-electron transfer route in Figure 5.10). The reaction with the enzyme, which has been shown to take place with an intermediate and not with the epoxide end product, will inactivate the cytochrome P450. This has been called a suicide mechanism, as the enzyme activates a substrate to a form that also destroys the enzyme.

Aromatic compounds may also be epoxidized by an electrophilic attack of the perferryl complex (or via the corresponding radical). If the cation is formed as an intermediate, as shown in Figure 5.11, it may yield the corresponding products (via routes a and b in Figure 5.11) to those obtained when alkenes are epoxidized (see Figure 5.10). However, epoxides of aromatic compounds are unstable in general and will spontaneously rearrange to the corresponding ketone, as indicated in Figure 5.11, and the ketone (formed directly by route a or via the epoxide by route b) would immediately be transformed to the corresponding phenol (as a result of the keto-enol equilibrium). A hydroxylated aromatic compound (a phenol) is therefore to be expected to be the initial product after cytochrome P450 oxidation of aromatic compounds, if the aromatic nucleus is attacked. Heterocyclic aromatics may also be epoxidized, as the example with furan in Figure 5.11 indicates. The transient epoxide formed will rearrange, in the case of a furan

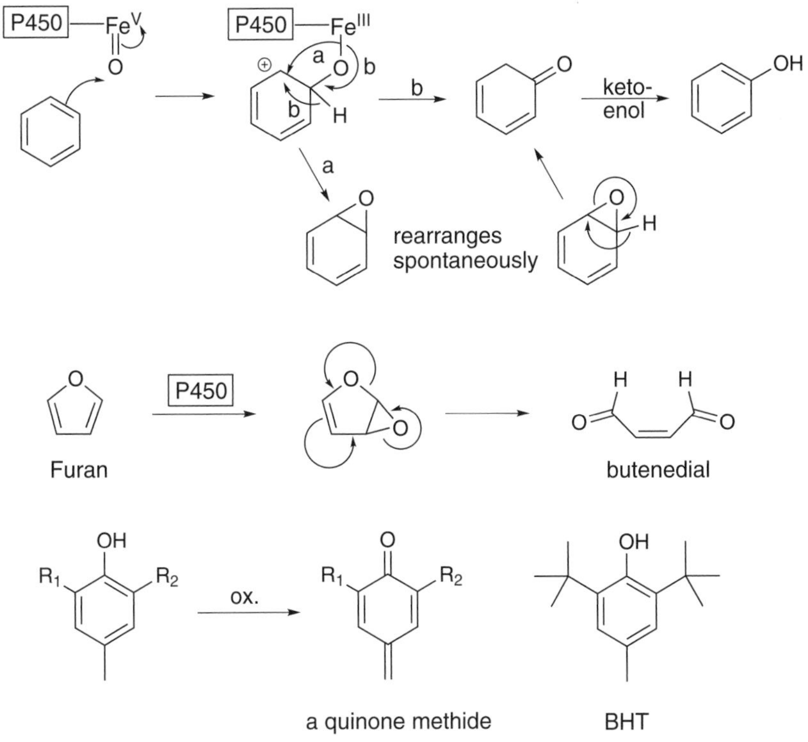

Figure 5.11 Oxidation of aromatics.

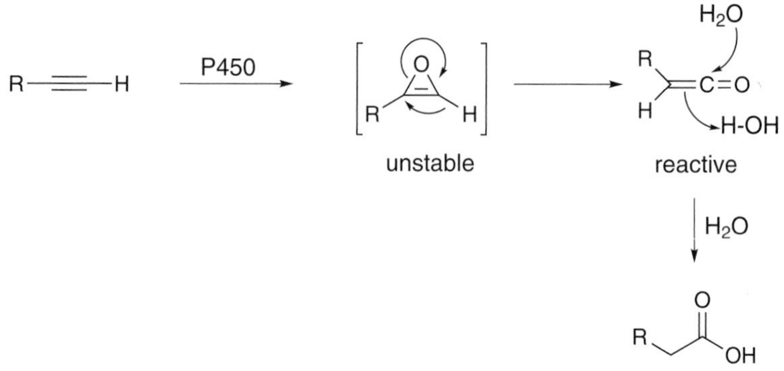

Figure 5.12 Oxidation of alkynes.

with no substituents in positions 2 and 5, to an unsaturated 1,4-dialdehyde, which is reactive. Furan is used as a solvent and as an intermediate in chemical synthesis. It is known to be toxic to the liver and kidneys, and exposure to furan has been associated with liver tumors. The furan ring, which is aromatic, is a common functionality in natural products, and we contact furans in many different ways.

When aromatics containing suitable substituents (e.g., phenols with a *para*-hydroxy or -methyl group) are oxidized, reactive benzoquinone derivatives are frequently formed. In the example shown in Figure 5.11, a *p*-hydroxytoluene is oxidized to a quinone methide, which has electrophilic properties. BHT, butylated hydroxytoluene (3,5-di-*tert*-butyl-4-hydroxytoluene) is a synthetic antioxidant widely used as an additive in foods, cosmetics, and drugs, and has been shown to damage the lung tissue of experimental animals fed large doses. The toxic effect of BHT is due to its conversion to a reactive quinone methide.

It is believed that alkynes also are epoxidized, although they have not been so thoroughly investigated, as the use of alkynes (except for acetylene gas) is not widespread. The product formed initially from the epoxidation of a terminal alkyne is too unstable to be isolated, and rearranges to the ketene. This in turn is reactive and may add water to form a carboxylic acid, and experiments with labeled compounds have supported the mechanism shown in Figure 5.12. Cytochrome P450 is also destroyed to some extent during the oxidation of alkynes, and it is reasonable to believe that this is due to the formation of a reactive ketene (which of course may react with another nucleophile instead of water) at the surface of the enzyme.

5.3.1.2 Hydroxylations of Saturated Carbon

Hydroxylations will insert an oxygen atom into the bond between any saturated atom and a hydrogen atom. It is possible that unsaturated atoms could also be hydroxylated, although the normal pathway for oxidizing unsaturated compounds appears to be via the epoxide, as discussed above. However, as epoxides

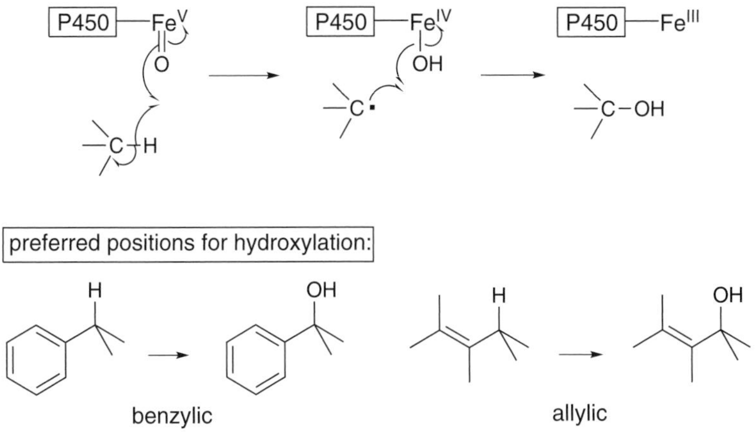

Figure 5.13 Hydroxylation of hydrocarbons.

of aromatic compounds will rearrange to phenols, this oxidation could formally be regarded as a hydroxylation. Hydroxylations of carbon–hydrogen bonds are the most common, resulting in alcohols (for example, the endogenous hydroxylations of steroids), although nitrogen–hydrogen hydroxylations (producing hydroxylamines) are important because of the toxicity of the products. In principle it is possible to hydroxylate any X–H bond, although in practice it is the position that for one or another reason is activated that is hydroxylated at a reasonable rate. Important positions that are hydroxylated relatively easily are the allylic and the benzylic positions, that is, a saturated carbon with at least one hydrogen adjacent to a double bond or an aromatic system (see Figure 5.13). The reason for this is obvious when the most likely mechanism for the hydroxylation is examined, as it involves the formation of a relatively stable allylic or benzylic radical after the abstraction of a hydrogen atom from the substrate as the rate-determining step (or the corresponding allylic or benzylic cation if the oxidation proceeds with a two-electron transfer). This will then collapse to form the hydroxylated substrate, and the regenerated cytochrome P450 is then ready for the next oxidation cycle.

Another position that is activated for hydroxylation is the C–H α to heteroatoms, that is, a carbon atom with at least one hydrogen and a heteroatom bound to it. In the case where the heteroatom is N, O, or S, this hydroxylation is often called α-hydroxylation or oxidative dealkylation. In Figure 5.14 a reasonable mechanism for the α-hydroxylation of an amine is shown. The heteroatom is thought to take part in the initial step by being oxidized, and the α-hydrogen is picked up by the perferryl complex. The unpaired electron on the heteroatom together with one of the electrons of the C–H bond forms a π-bond between the heteroatom and the α-carbon, whereupon the hydroxyl group is transferred from the iron to the carbon and the electrons in the double bond move back to the heteroatom so that it recovers its lone pair. As mentioned in Chapter 2, the

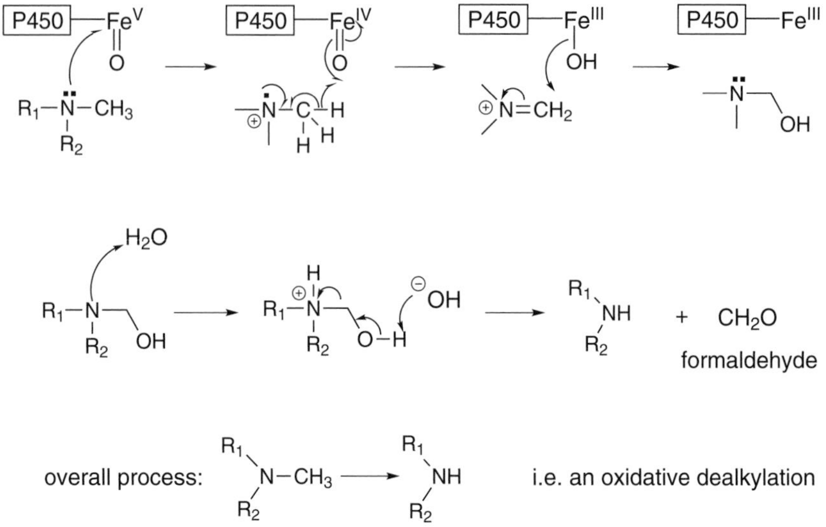

Figure 5.14 α-Hydroxylation, leading to oxidative dealkylation.

product formed after α-hydroxylation, with two heteroatoms bound to the same saturated carbon, is in principle unstable in contact with water. The initially formed N-hydroxymethyl derivative formed in Figure 5.14 will therefore spontaneously generate the corresponding amine and formaldehyde. The overall result of the hydroxylation is therefore that the bond between the nitrogen and carbon that is oxidized is cleaved, explaining why this hydroxylation is also called oxidative dealkylation.

The same hydroxylation may also take place if the heteroatom is chlorine, bromine, or iodine, although this reaction has been given a different name: oxidative dehydrohalogenation. As the name implies, HCl, HBr, or HI is eliminated after oxidation, the hydrogen comes from the hydroxyl group added by the hydroxylation, and the double bond formed by the elimination is a C=O bond. A primary halide would yield an aldehyde, a secondary a ketone, while a tertiary halide is not able to undergo oxidative dehydrohalogenation as it lacks α-hydrogens. A compound with two halogens attached to the same carbon is hydroxylated faster, and the result of the dehydrohalogenation is a carboxylic acid halide, which normally is quite reactive. Chloramphenicol (see Figure 5.15), a natural antibiotic used to treat infections, has been shown to deactivate cytochrome P450, presumably by reacting irreversibly as the corresponding carboxylic acid chloride derivative with the enzyme. A compound with one hydrogen and three halogens on the same carbon may also be hydroxylated, and chloroform will generate phosgene (the dichloride of carbonic acid), which is highly toxic and at least partly responsible for the toxicity of chloroform.

Saturated hydrocarbons that completely lack functional groups are not readily oxidized in mammals. The smaller members are often volatile enough to be

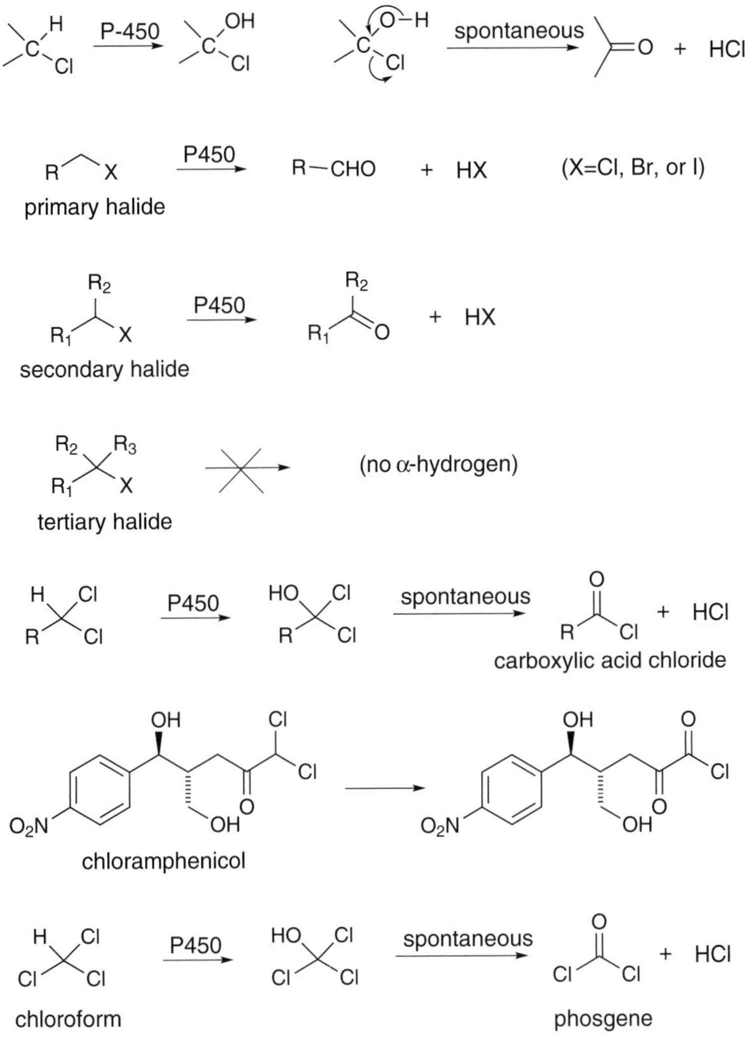

Figure 5.15   Oxidation of halides.

excreted by the lungs, but the larger members cannot be excreted as they are. As we have already indicated, the mechanism for hydroxylation of saturated carbon is via the carbon radical, and the more a molecule is able to stabilize an unpaired electron the more likely it is to be formed. The more branched a hydrocarbon radical is, the more stable it will be, and the more likely it is that this carbon will be hydroxylated. For unbranched (linear) alkanes it has been noted that the most favored (but not the only) position for hydroxylation is the carbon adjacent to the terminal methyl group (called ω-1 hydroxylation, as the terminal carbon is called the ω-carbon). We shall return to this topic later in this chapter.

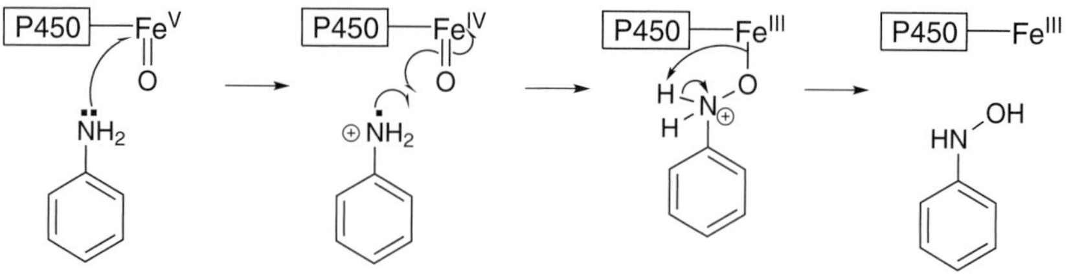

Figure 5.16 N-Hydroxylation of an aromatic amine.

5.3.1.3 Hydroxylation of Amino Groups

Also nitrogen in activated positions may be hydroxylated, especially if α-hydroxylation (see above) is not possible. Aromatic amines are a class of compounds that has received a lot of attention over the years, because they are toxic and because they have been used in several industrial chemical processes. In an aromatic amine the carbon α to the amino group has no hydrogen and consequently cannot be oxidized by α-hydroxylation. Instead, oxidation may take place in the aromatic system or in the amino group. As the nitrogen is benzylic and thereby activated, nitrogen hydroxylation takes place fairly easily. Hydroxylation of an amino group can be carried out by both cytochrome P450 and by the microsomal FAD-containing monooxygenase, and peroxidases have also been shown to be involved in the oxidation of aromatic amines in some tissues (e.g., in the epithelial cells of the urinary bladder). A possible mechanism for the hydroxylation of an aromatic amine to the corresponding hydroxylamine by cytochrome P450 is shown in Figure 5.16.

5.3.1.4 Oxidation of a Single Bond to a Double Bond

Although this is less common, several examples are known. One obvious possibility is that carboxylic acids with two CH_2 groups next to the carboxylic acid functionality are attacked by the β-oxidation of the primary metabolism. In this, an activated (as the coenzyme A thioester) fatty acid is oxidized by the enzyme fatty acyl CoA dehydrogenase with the coenzyme FAD, thereby becoming unsaturated. As mentioned, this is intended to oxidize fatty acids to acetic acid units (acetyl CoA) that can be used for the production of energy in the mitochondria. The oxidation is carried out according to Figure 5.17, via the oxidation of the α,β-single bond to a double bond.

Exogenous carboxylic acids may also undergo β-oxidation. However, the reaction will normally not stop after the first oxidation but proceeds via the β-hydroxyl and the β-keto derivatives to the new carboxylic acid lacking two carbons (see Figure 5.17). In addition, the oxidation of carbon–carbon single to carbon–carbon double bonds (presumably by dehydrogenases) has also been observed in other situations, and one case where the oxidation appears to be an important activation step is with valproic acid (an anti-epileptic drug). The oxidized derivatives of valproic acid shown in Figure 5.18 are responsible for the liver toxicity of the drug.

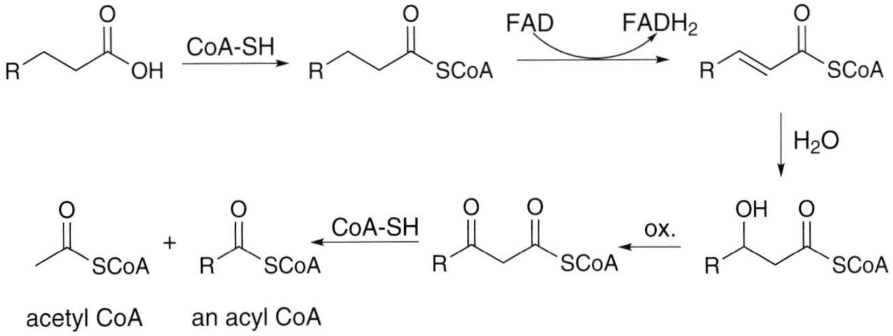

Figure 5.17 The oxidation of a single bond to a double bond during β-oxidation.

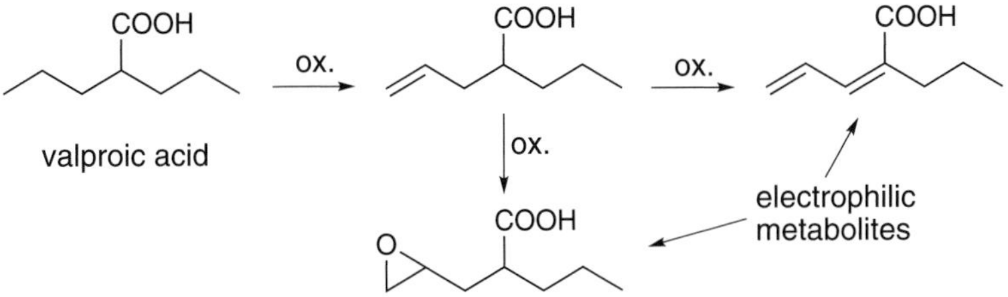

Figure 5.18 The metabolic activation of valproic acid.

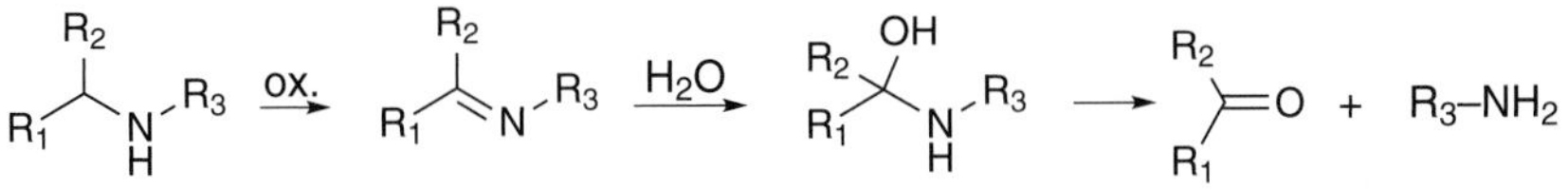

Figure 5.19 The oxidation of an amine to an imine, which can be hydrolyzed to a ketone.

Amines (primary, secondary, or tertiary, as long as they have an α-hydrogen) have been shown to be oxidized to imines by amine oxidases (see Figure 5.19). The imines are not stable in contact with water and will be hydrolyzed to the free amine and an aldehyde or a ketone (depending on the nature of R_1 and R_2 in Figure 5.19). The hydrolysis proceeds via the α-hydroxylamine, formed after the addition of water to the imine double bond. Consequently, α-oxidation of amines by cytochrome P450 and oxidation of amines to imines by amine oxidases both yield the same end products.

5.3.1.5 Oxidation of Heteroatoms

Different heteroatoms, for example, N, S, P, Se, and I, may be oxidized by cytochrome P450, by the microsomal FAD-containing monooxygenase, and by amine oxidases. The hydroxylation of nitrogen has already been discussed, but nitrogen (in tertiary amines) can also be oxidized to the corresponding aminoxide (e.g.,

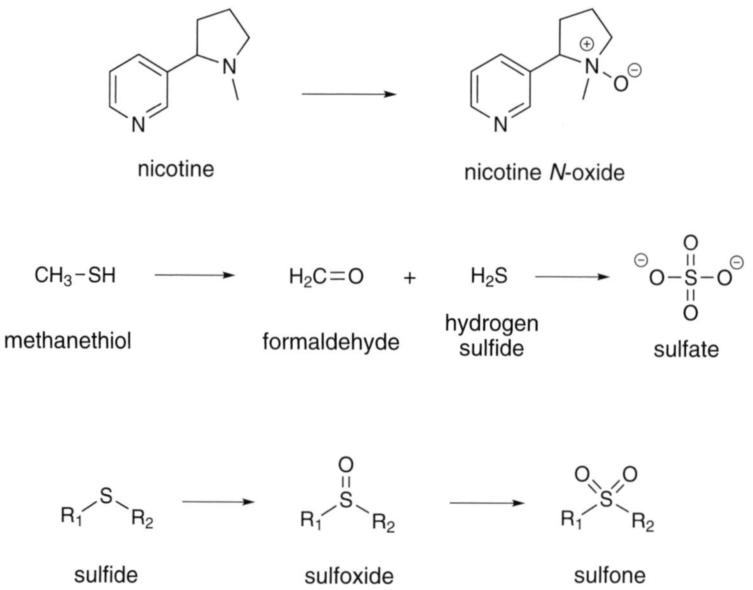

Figure 5.20 Oxidation of nitrogen (in tertiary amines) and sulfur.

nicotine to nicotine *N*-oxide, see Figure 5.20), which has high water solubility and can be excreted without any conjugation. Sulfur can be oxidized in several steps, and some examples are given in Figure 5.20.

5.3.1.6 Oxidation of Alcohols and Aldehydes

The oxidation of alcohols to aldehydes or ketones is a reversible reaction catalyzed by alcohol dehydrogenases that use NAD^+ as a coenzyme, and the equilibrium depends on the further destinies of the oxidized/reduced forms. In contrast, the oxidation of aldehydes to carboxylic acids is irreversible. Aldehydes, formed from primary alcohols, are in most cases oxidized to carboxylic acids so efficiently that only negligible amounts are reduced back to the alcohol. For secondary alcohols a true equilibrium can often be observed. The reduced form has a good handle (the hydroxyl group) for conjugations, and this facilitates its excretion by the kidneys, while ketones are more volatile than alcohols and may, at least those of lower molecular weight, be excreted by the lungs. In addition, ketones are in some instances further oxidized on the carbon α to the keto function. Acetone, for example, is oxidized to pyruvic acid, which enters the citric acid cycle and is converted to carbon dioxide and water (Figure 5.21).

5.3.2
Reductions

Aldehydes and ketones can be reduced to alcohols by the same enzymes that oxidize alcohols but with (the reduced form of) the coenzyme NADPH

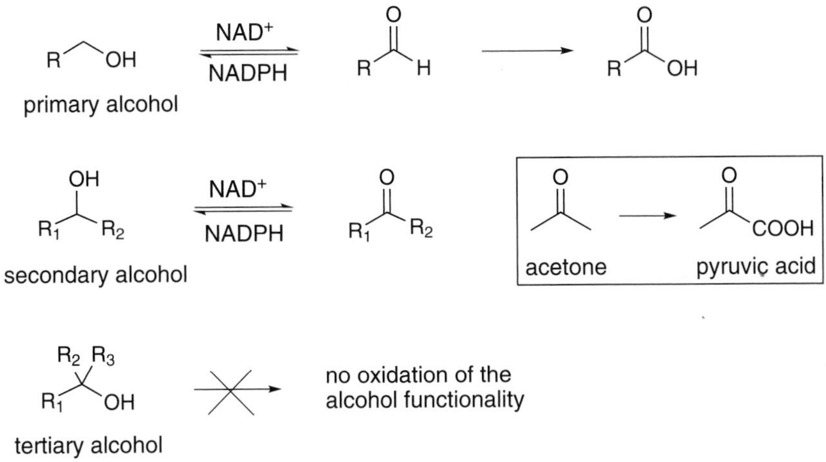

Figure 5.21 Oxidation of alcohols.

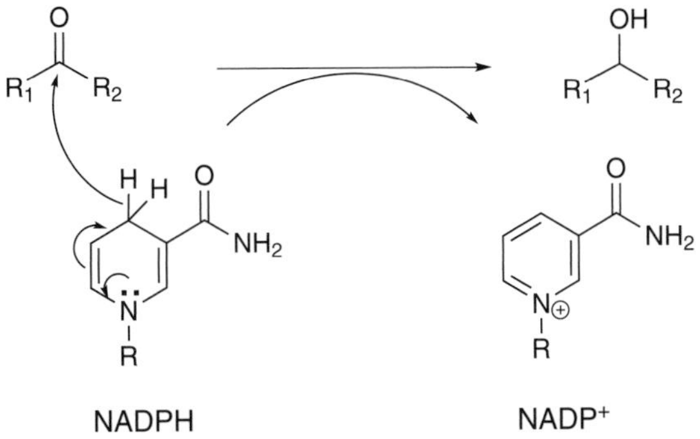

Figure 5.22 Reduction of a ketone with the coenzyme NADPH.

(see Figure 5.22). The oxidation of alcohols is favored by the fact that the NAD$^+$/ NADH ratio in mammalian cells is normally high, while reductions are favored by a low NADP$^+$/NADPH ratio. Aldehydes are, as discussed above, irreversibly oxidized to carboxylic acids and are therefore less likely to be reduced.

Cytochrome P450 will to a small extent also carry out reductions of exogenous compounds, in spite of its character as an oxidase. Here, the substrate will replace oxygen as the electron sink and be reduced, and the substrate and molecular oxygen will compete for the electrons supplied by the NADPH-cytochrome P450 reductase. Low oxygen concentrations will promote the reduction of such a substrate, while high concentrations will inhibit it. Reduction is, of course, most important under anaerobic conditions, which we primarily find in the distant parts of the intestines, and here bacteria will carry out reductions of exogenous compounds with enzymes that show similarities to cytochrome P450. Nitro heteroatom reduction:

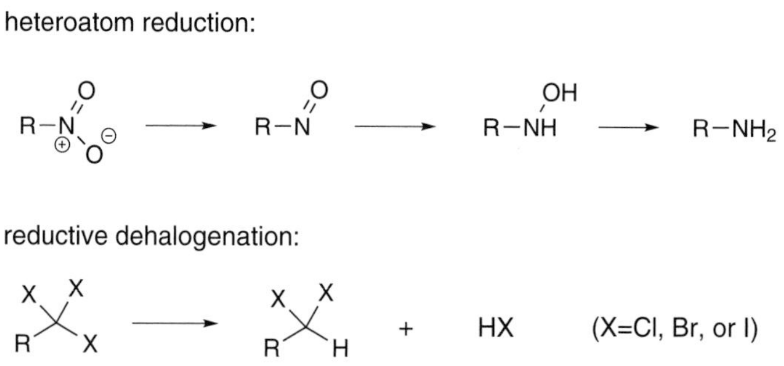

$$R-\overset{\overset{O}{\parallel}}{\underset{\underset{O^{\ominus}}{\oplus}}{N}} \longrightarrow R-\overset{\overset{O}{\parallel}}{N} \longrightarrow R-\overset{\overset{OH}{\vert}}{NH} \longrightarrow R-NH_2$$

reductive dehalogenation:

$$\underset{R}{\overset{X}{\diagup}}\overset{X}{\underset{X}{\diagdown}} \longrightarrow \underset{R}{\overset{X}{\diagup}}\overset{X}{\underset{H}{\diagdown}} \quad + \quad HX \qquad (X=Cl, Br, or I)$$

Figure 5.23 Examples of reductions that may take place in mammals.

compounds, for example, will be reduced to the corresponding nitroso compounds, which in turn may be further reduced to the hydroxylamine and the amine (see Figure 5.23). Azo compounds may also be reduced, initially to hydrazines, but also, after cleavage of the nitrogen–nitrogen bond, to two amines, and other oxidized heteroatoms may also be reduced. Halogenated organic compounds may be reduced via reductive dehalogenation (further discussed later in this chapter), by which, in principle, a halide ion is replaced by a hydride ion. The banned insecticide DDT is metabolized by this route (discussed in Chapter 9), although very slowly.

5.3.3
Hydrolyses

Oxidations and reductions are chemically more complicated reaction, involving the transfer of electrons. They are carried out with the help of coenzymes (e.g., NADPH in cytochrome P450 oxidations) which have been prepared by the primary metabolism at the expense of chemical energy. Hydrolyses do not involve the transfer of electrons, and are reactions between a substrate and a molecule of water that are energetically favorable but hampered by an activation energy that makes them slow at body temperature. The enzymes that catalyze hydrolyses take this energetic threshold down and speed up the reaction considerably.

5.3.3.1 Hydrolysis of Epoxides
Some epoxides are used as industrial chemicals, for example, ethylene oxide, which is used for the production of certain polymers and for the sterilization of medical equipment, and the handling of such chemicals is very hazardous. In addition, we have seen that epoxides are formed inside our bodies as a result of phase I conversions, and this is of course a serious threat to our health. However, as we have already indicated, most epoxides formed from aromatic compounds are unstable and generally rearrange to phenols within seconds (although some are more stable). In addition, the majority of the reasonably stable but still reactive epoxides that are formed as a result of phase I conversions

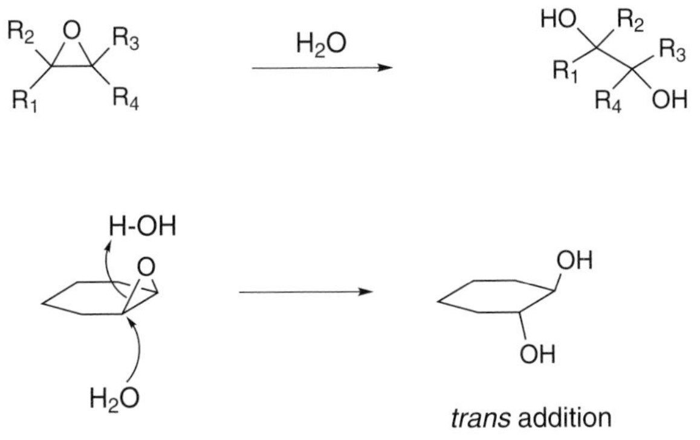

Figure 5.24 The hydrolysis of epoxide by epoxide hydrolase.

are detoxified by hydrolysis. The efficient weapon against the epoxides is the enzyme epoxide hydrolase, present in the endoplasmic reticulum of all tissues in the body, although in very variable amounts, and this will simply hydrolyze an epoxide to a diol. As always, there are many different kinds of epoxide hydrolases, with different substrate specificities, and some appear to be present in the cytoplasm also.

Cyclic diols formed by this procedure are always trans, as shown in Figure 5.24, indicating that the mechanism for the hydrolysis is that the water molecule that opens the epoxide is attacking from the opposite side of the epoxide ring (see Figure 5.24). The reaction with water as a nucleophile also takes place spontaneously, but without the enzyme it is much slower and much less selective. That is to say that without the enzyme epoxide hydrolase the epoxide would preferentially react with better nucleophiles, for example thiol groups in proteins and nitrogens in the nucleotides of the DNA. The enzyme picks up a molecule of water and a molecule of the epoxide, brings them together, and facilitates their reaction by creating conditions both acidic (to protonate the epoxide oxygen) and basic (to make the water oxygen more nucleophilic). The same reaction can be carried out in a chemical laboratory by adding acid or base as catalyst but never the two at the same time, and it is difficult to imitate the mild and efficient conditions from which enzymatic catalysis benefits.

For some compounds, for example, styrene which is used for the manufacture of polyfoam, it was suspected that the vinylic double bond is epoxidized by cytochrome P450, and synthetically prepared styrene oxide was shown to be genotoxic *in vitro*. However, it was not possible to obtain any conclusive results from *in vivo* experiments with respect to the carcinogenicity of styrene, although chronic exposure to styrene is well known to damage the CNS. The reason for the discrepancy of the genotoxicity is the existence of epoxide hydrolase. Styrene oxide is indeed formed *in vivo*, but it is so efficiently hydrolyzed by epoxide hydrolase (see Figure 5.25) that it has problems reaching the DNA to a significant extent. The diol obtained after the hydrolysis of styrene oxide is oxidized by alcohol dehydrogenase

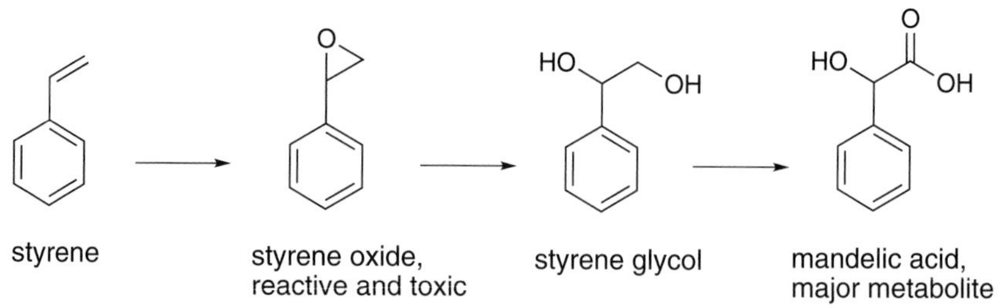

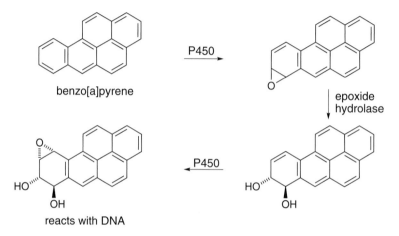

Figure 5.25 The metabolism of styrene.

Figure 5.26 Metabolic activation of benzo[a]pyrene.

to both the ketone and the aldehyde/carboxylic acid, and mandelic acid is the major urine excretion product of styrene.

A dangerous situation is at hand when an epoxide formed in a phase I oxidation is for some reason a bad substrate for the epoxide hydrolase. Such epoxides are not hydrolyzed, at least not efficiently, and they will be able to diffuse through the cell and react with critical targets. An example of this is benzo[a]pyrene, a chemical found in trace amounts in the exhaust gas from any combustion with a less than perfect balance between fuel and oxygen (car exhaust, cigarette smoke, etc.). Benzo[a]pyrene is genotoxic and carcinogenic, and is believed to be responsible for a significant proportion of the lung cancers caused by cigarette smoking. As a fairly large and lipophilic molecule it has no chance to be excreted without first being metabolized, and there the metabolism has little choice but to attack the aromatic system itself. Several different positions are oxidized, but the most important for the genotoxicity of benzo[a]pyrene is the C-7/C-8 epoxidation, as shown in Figure 5.26. The epoxide thus formed is a relatively stable, good substrate for epoxide hydrolase that will convert it to the *trans*-diol shown in Figure 5.26.

However, the diol will be epoxidized a second time, generating the diolepoxide, also shown in Figure 5.26, that also is relatively stable but for some reason not a suitable substrate for the epoxide hydrolase. This epoxide will be able to diffuse into the cell nucleus and react with the DNA.

The epoxide hydrolase present in the microsomes has been shown to be inducible, not by one specific compound but instead by a general need for 'epoxide hydrolysis'. Some compounds, for example, cyclohexeneoxide formed after the epoxidation of cyclohexene, have been shown to inhibit the enzyme, possibly by reacting with it. It is interesting in this context to imagine the possibility that endogenous compounds containing double bonds, of which there are many, could also be epoxidized. No systematic investigation has been carried out, but cholesterol has a carbon–carbon double bond that in principle could participate in such a conversion. The epoxide of cholesterol (prepared synthetically) is genotoxic to some simpler organisms but, because of its efficient hydrolysis, not to mammalian cells. This could be an indication that evolution has adapted our metabolism to the generation of reactive metabolites from endogenous compounds.

5.3.3.2 Hydrolysis of Esters and Amides

The need for our cells to hydrolyze ester and amide bonds is huge, and it is no surprise that we have a number of enzymes that carry out such hydrolyses. There are many examples of important tasks for esterases (hydrolyzing esters) and amidases (hydrolyzing amides), for example to hydrolyze fats to glycerol and free fatty acids, peptides to free amino acids, and some neurotransmitters to inactive compounds. Many of the esterases and amidases present in our cells are known to be nonspecific, and several esterases will even hydrolyze thioesters (and even some amides!). It is reasonable that exogenous esters and amides in most cases should be readily hydrolyzed. In addition, the products of the hydrolysis of an ester or an amide would be considerably more hydrophilic and good substrates for phase II reactions, as a ester yields a carboxylic acid and an alcohol while an amide yields a carboxylic acid and an amine (see Figure 5.27).

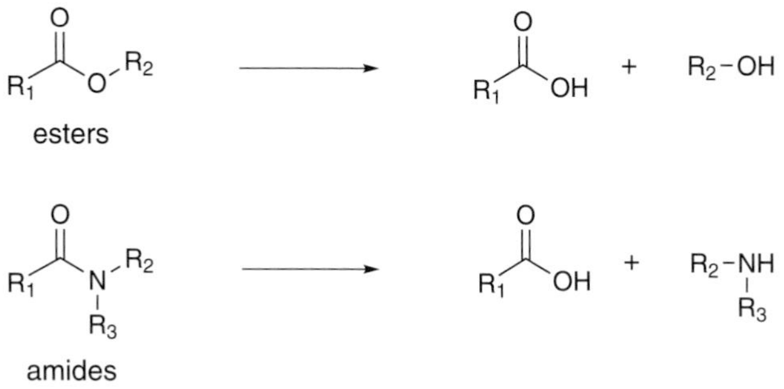

Figure 5.27 Hydrolysis of esters and amides.

The hydrolysis of an ester is normally much faster than that of an amide, partly because the amide bond is stronger (because the lone pair on the nitrogen will be more efficiently donated to the carbonyl group). However, this depends largely on the substituents on the oxygen/nitrogen. The esterases and amidases are both soluble and membrane bound, and besides hydrolyzing thioesters and some amides the esterases in the endoplasmic reticulum may also attack ureas, phosphoric esters, and thiophosphoric esters. Lately, it has been shown that esters may also be metabolized by α-hydroxylation by cytochrome P450.

5.4
Phase II Conversions (Conjugations)

The general purpose of phase II reactions in the metabolism of exogenous compounds can be said to be to increase the water solubility and to facilitate excretion, although some other important results are achieved in addition. It involves the reaction of a suitable compound, which may or may not have undergone phase I conversions, with different endogenous compounds which are said to be conjugated to the substrate. The reactions carried out are condensations (generating one molecule of water as a by-product), additions, and substitutions. Most conjugations are biosynthetic reactions and will consequently cost energy to perform. In general, this is achieved by activating the substrates prior to the reaction.

5.4.1
Conjugations with Sulfate

The attachment of a sulfate group to a compound that is to be excreted from the body is of course an attractive possibility, as this is a negatively charged group that will increase the water solubility of any compound considerably. The conversion of various compounds to their sulfate esters is an important phase II reaction, and the conjugates are excreted by the kidneys. The reaction takes place in the cytoplasm and is catalyzed by enzymes called sulfotransferases. They utilize a coenzyme with a complicated chemical structure (its name is abbreviated PAPS), which in principle is nothing but an activated sulfate group (see Figure 5.28). The origin of the sulfur atom in this sulfate group is originally cysteine, an essential amino acid that is used for many purposes in the body. However, the stock of PAPS is limited, because cysteine is valuable for the body, and the endurance of this rapid and very efficient phase II reaction is unfortunately not very good. For some potentially toxic substrates, conversion to the sulfate ester is competing with other phase II reactions, and in some cases it can be observed that low doses are harmless (because of the efficiency of the conjugation with sulfate) while higher doses are toxic (because the stock of PAPS is depleted and other, less efficient, phase II reactions take over and the time spent in the body is prolonged). The substrates for sulfate conjugations are primarily alcohols and phenols, both exogenous and

sulfate source:

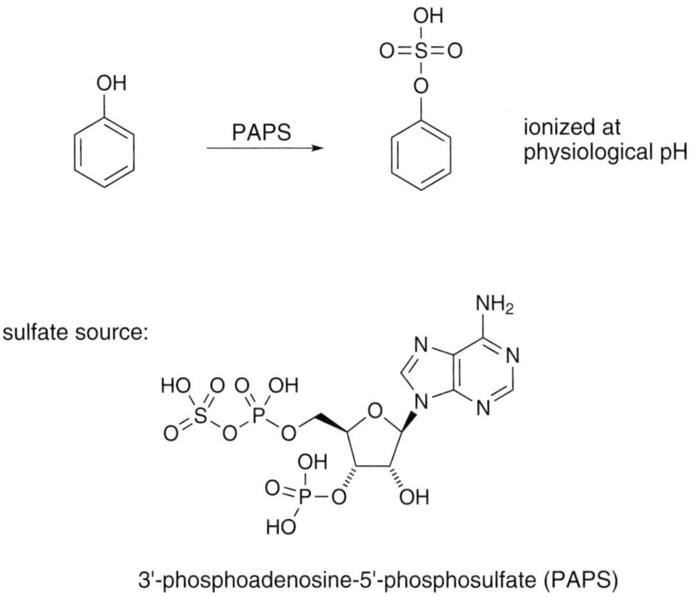

3'-phosphoadenosine-5'-phosphosulfate (PAPS)

Figure 5.28 Conjugation with sulfate.

endogenous, but also aromatic amines and aromatic hydroxylamines may use this phase II reaction.

Most sulfates formed are stable conjugates, in contrast to dialkylsulfates and alkylsulfonates, the monoesters of sulfuric acid which are not very electrophilic. However, when attached to strongly activated positions the sulfate ion may be substituted by a nucleophile. Sulfate esters of aromatic hydroxylamines especially have received a lot of attention. The sulfate group in such esters is not only benzylic but also attached to a heteroatom (the N–O bond is relatively weak). The benzylic position is also encountered in sulfates of benzyl alcohols. Examples are the sulfate groups of the hydroxylamine formed from 2-acetylaminofluorene (2-AAF), a carcinogenic chemical previously used in the dye industry, and safrole, a natural compound present in several spices (e.g., sassafras and black pepper) and associated with a weak carcinogenic activity (discussed also in Chapter 9). As indicated in Figure 5.29, both compounds react with DNA and are carcinogenic because the sulfate groups are poor but sufficiently good leaving groups in nucleophilic substitutions.

5.4.2
Conjugations with Glucuronic Acid

Glucuronic acid is derived from glucose, which by itself is water soluble. In addition to the hydroxyl groups, glucuronic acid also has a carboxylic acid

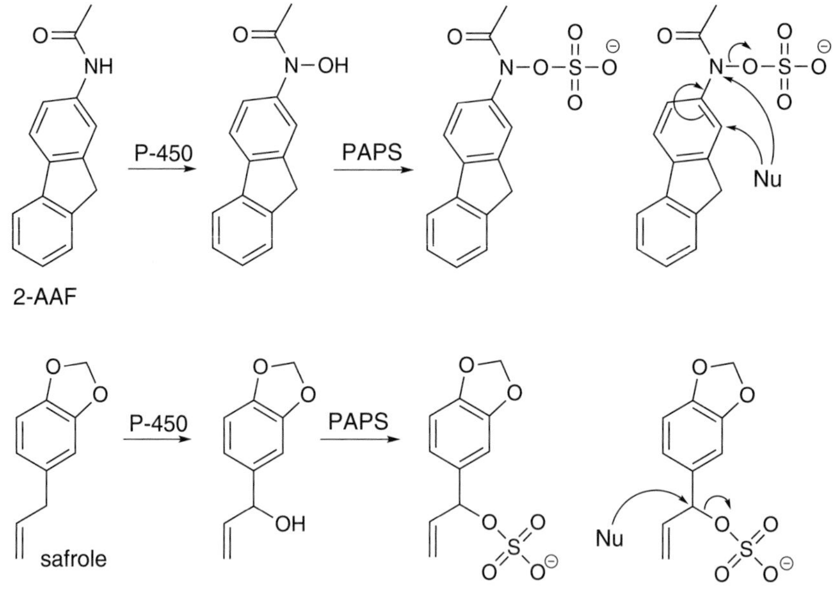

Figure 5.29 The sulfate group as a leaving group.

functionality that is ionized at physiological pH and that also may serve as a handle for specific excretion membrane transport systems. Unlike the other phase II reactions, the conjugation with glucuronic acid takes place in the endoplasmic reticulum, not in the cytoplasm, mainly in the liver cells. The enzyme that catalyzes the conjugation is called uridine diphosphate- (UDP)-glucuronosyl transferase, of which several forms (some that are inducible) have been identified. The enzyme uses UDP-glucuronic acid as a coenzyme, and again we notice (Figure 5.30) that although the coenzyme appears to be complicated it is simply an activated form of glucuronic acid. A special feature of the UDP-glucuronosyl transferases is that they are non-specific, and will conjugate almost anything with glucuronic acid. Examples of functional groups that are acceptable are alcohols, phenols, carboxylic acids, hydroxylamines, aromatic amines, and sulfonamides.

Because of the water solubility of the conjugates they can be excreted by the kidneys, but also by the liver as many of them have quite high molecular weights. The molar mass of glucuronic acid itself is 194 g/mol, and if the substrate is of the same size, excretion with the bile becomes significant. The glycoside bond between the substrate and glucuronic acid is relatively stable under normal conditions, but it can be hydrolyzed by certain enzymes and by acidic conditions. Enzymes called glucuronidases will efficiently hydrolyze *O*- and *S*-glycosides, and although they are not present to any large extent in our cells they are common in the bacteria of the intestines. It is therefore to be expected that glucuronic acid derivatives that are excreted by the liver will eventually be hydrolyzed again (see

glucuronic acid source:

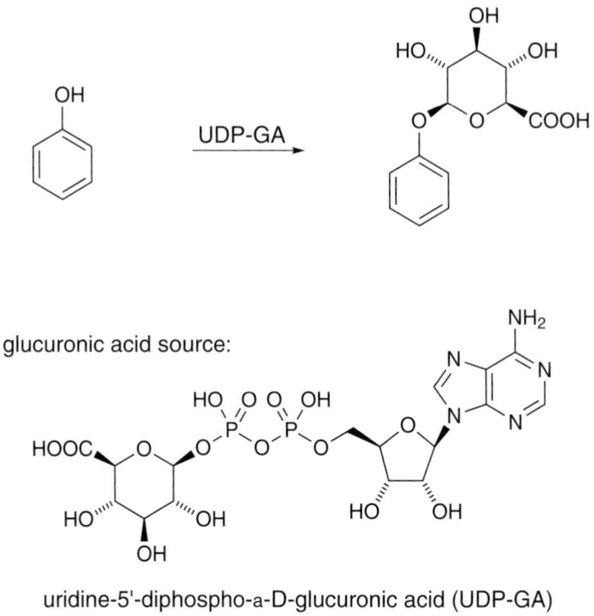

uridine-5'-diphospho-a-D-glucuronic acid (UDP-GA)

Figure 5.30 Conjugation with glucuronic acid.

Figure 5.31). If it as such is reasonably lipophilic it will be reabsorbed from the intestines, transported to the liver, conjugated with glucuronic acid once more, and be excreted with the bile. Even if it is not possible that this process, called the enterohepatic recirculation, is 100% efficient because of other metabolic conversions and other excretion routes, it will nevertheless increase the time that a toxic chemical spends in the body. Chemical hydrolysis catalyzed by acid (no enzymes involved) may take place in the urine bladder, as urine often has a lower pH compared to the rest of the body fluids. This has been suggested as the reason why the bladder is the target organ for aromatic amines. If they are excreted as the glucuronic acid conjugate of the corresponding hydroxylamine and hydrolyzed in the bladder, the bladder cells may absorb the hydroxylamines and possibly sulfate them to their electrophilic and genotoxic form (see Figure 5.31). In such a case one can say that the glucuronic acid masks the reactive functionality and delivers a toxic compound to its target.

5.4.3
Conjugations with Amino Acids

Carboxylic acids may be conjugated with the amino group of an amino acid to form an amide. In most cases this is carried out with the simplest amino acid, glycine, although in some cases others are used (e.g., glutamic acid and serine). The result of the conjugation is that the compound receives an extra amide

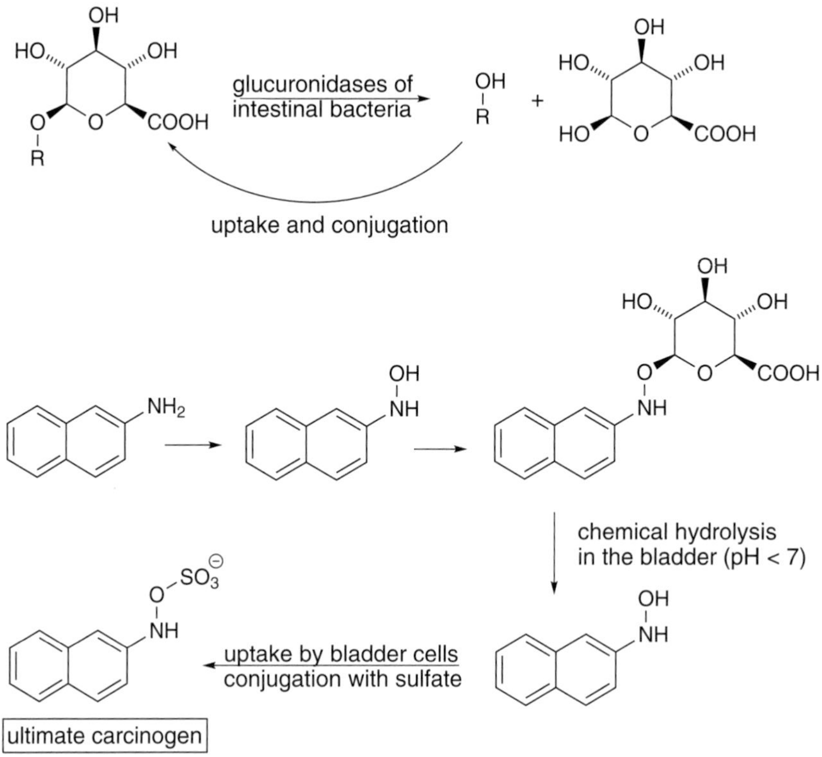

Figure 5.31 The hydrolysis of conjugates with glucuronic acid.

function, which increases the water solubility and serves as a handle for the specific excretion systems. The conjugation is carried out in two steps, first the substrate is activated as a coenzyme A thioester by the enzyme coenzyme A-ligase, after which the acyl transferases will catalyze the coupling of the activated acid and the amino acid. In Figure 5.32 the conjugation of benzoic acid with glycine is illustrated, and, as the end product (hippuric acid) was first isolated from the urine of a horse (Greek for horse is hippos), this phase II reaction is called 'hippuric acid synthesis'.

5.4.4
Conjugations with Glutathione

In several ways, the conjugation of exogenous compounds with glutathione is the most important phase II reaction, because it gives the body a direct possibility to take care of electrophilic compounds. Glutathione is a tripeptide composed of glycine, cysteine, and glutamic acid, and its nucleophilicity is of course due to the presence of a thiol group in the cysteine residue (see Figure 5.33). Cysteine itself is actually a better nucleophile than glutathione, because the sulfur has a slightly

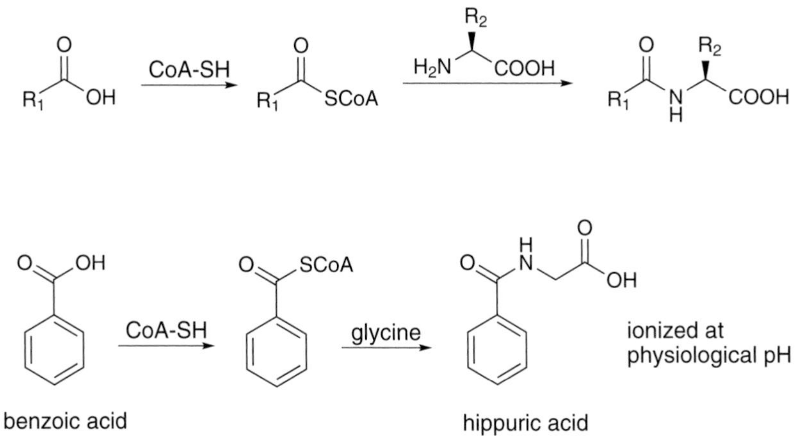

Figure 5.32 Conjugation with an amino acid.

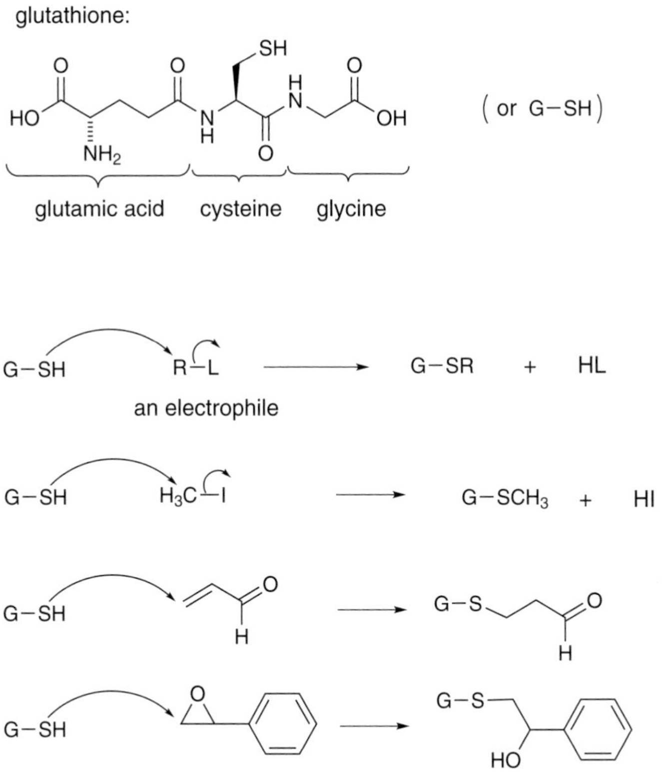

glutathione:

(or G−SH)

glutamic acid   cysteine   glycine

Figure 5.33 Conjugation of electrophiles with glutathione.

more ionic character in cysteine than in glutathione, and it is not evident why our cells go to the extra trouble of adding two other amino acids on each side of cysteine if it was only to be used in conjugations (which it is not!). The reaction between an electrophilic substrate and glutathione is catalyzed by enzymes, glutathione-S-transferases, and the trick is that a molecule of glutathione is a much more reactive nucleophile when it is in its position in the enzyme than when it is free. This is accomplished by enzymatic stabilization of the anion with a negative charge on the sulfur, and we have already seen how dramatically this will change the nucleophilicity. Most of the conjugations with glutathione take place in the cytoplasm, although some are carried out in the endoplasmic reticulum. In the cells of the liver, the concentration of glutathione is extremely high, 1–5 mM, and there are plenty of glutathione-S-transferases as well.

Note that the bond between cysteine and glutamic acid is not a normal peptide bond, and this makes glutathione less prone to degradation by peptidases (enzymes cleaving peptide bonds), although other enzymes (γ-glutamyl transferases) will be able to cleave this bond. After a conjugation with glutathione, glycine and glutamic acid may be hydrolyzed from the conjugate, and the remaining cysteine with the former electrophile attached to the sulfur atom can be further metabolized by two main routes. Either the glycine adduct is acetylated (see below) to a mercapturic acid derivative (this phase II reaction is sometimes called 'mercapturic acid synthesis'), or the bond between the sulfur and the rest of the cysteine is cleaved by a β-lyase, after which the resulting thiol may be methylated (see Figure 5.34). If an exogenous compound is conjugated with glutathione, indicating that it is electrophilic or is converted to electrophilic metabolites by the metabolism, one can expect to find mercapturic acids and thioethers in the urine (this is the case for example in the urine of cigarette smokers), and these can be detected by analytical procedures.

In almost all cases the product formed is less toxic than the original electrophile, but the reverse can of course happen. For example, acrolein, in spite of its general reactivity, has been found to be specifically toxic to the kidneys, and the toxic principle is believed to be formed from its mercapturic acid derivative. In the kidneys this can be oxidized to the sulfoxide, which is more prone to β-elimination and will regenerate acrolein (see Figure 5.35). Also 1,2-dichloroethane may form a potent electrophile when conjugated with glutathione, and this will be discussed in Chapter 9.

Another important role for glutathione is to assist during the conversion of hydrogen peroxide to water, as a coenzyme of glutathione peroxidase. In addition, it is believed to protect various membrane proteins from oxidation. Any situation causing the concentration of glutathione to drop would therefore be a potential threat, especially if it is accompanied by an exposure to electrophilic or pre-electrophilic chemicals. Extreme diets may result in the intake of insufficient amounts of cysteine, leading to a decrease in the glutathione levels in the liver. As a consequence, the toxicity of a number of chemicals (e.g., chloroform, carbon tetrachloride, acetaminophen, and thioacetamide) is aggravated. In order to strengthen the defense against reactive compounds, especially after intoxications,

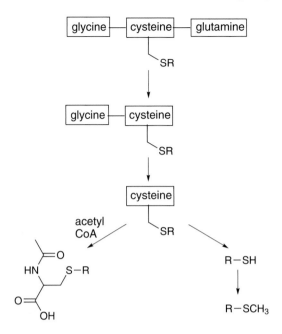

mercapturic acid derivative

Figure 5.34 Further processing of glutathione conjugates.

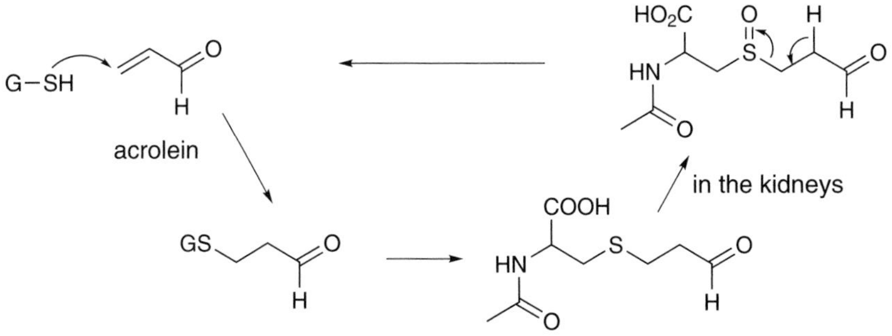

acrolein

in the kidneys

Figure 5.35 The conjugation of acrolein with glutathione.

it is possible to administer cysteine, normally as acetylcysteine, to the victim. Acetylcysteine is able to react as a nucleophile with the reactive compound, increase the supply of cysteine and glutathione, and also facilitate the production of PAPS. The protection offered by glutathione may, at least for some types of chemicals, have the result that no effects are observed up to a threshold concentration, above which glutathione is no longer able to take care of things. Besides all these protective functions, glutathione also acts as a reserve for cysteine, which when needed can easily be produced by the hydrolysis of glutathione.

5.4.5
Other Conjugations

In addition to the major phase II reactions discussed above, various acylations and methylations of heteroatoms also take place with exogenous compounds, of which the acetylation of nitrogen to form acetamides and the methylation of sulfur to form thioethers are especially interesting. However, acylations and methylations differ from the others as they decrease rather than increase the water solubility, and they are therefore of minor importance when it comes to converting exogenous compounds to an excretable form. The coenzyme for acetylations is acetyl-coenzyme A, which also plays a prominent role in energy production. The enzyme catalyzing the reaction is called *N*-acetyl transferase, and important substrates are aromatic amines and sulfonamides, and substituted hydrazines. The amide formed may of course be a substrate for amidases (see above). Acetylation is interesting because is has been found to be involved in the carcinogenicity of aromatic amines, since acetylated as well as free aromatic amines can be *N*-hydroxylated by cytochrome P450. An aromatic *N*-hydroxy-*N*-acetylamine may be converted by a special enzyme, an acyl transferase, to the corresponding aromatic *N*-acetoxyamine, which is an efficient electrophile because acetate is a very good leaving group (see Figure 5.36).

Methylation of heteroatoms is an important biochemical reaction that is used, for example, for the regulation of DNA and RNA. The coenzyme used in this reaction, S-adenosylmethionine (SAM for short), is unusual as an electrophilic endogenous compound. Interestingly, it has been shown that SAM will react with

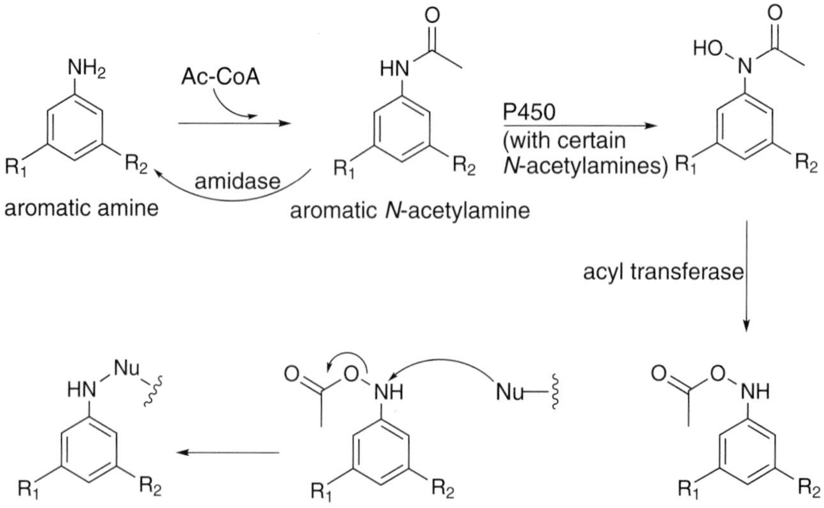

Figure 5.36 A possible way to activate aromatic amines.

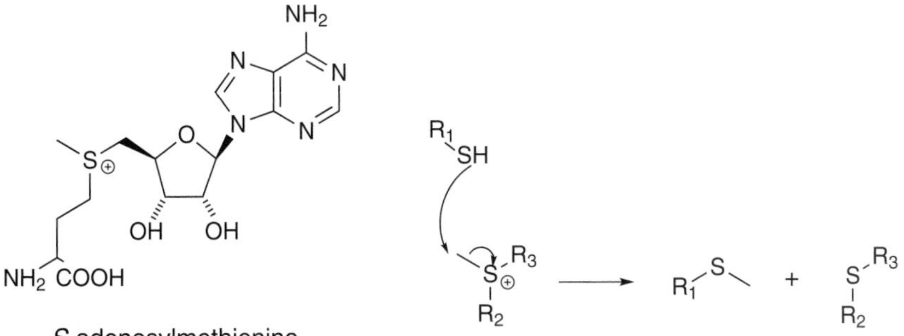

S-adenosylmethionine

Figure 5.37 Methylation of thiols with SAM.

and methylate DNA *in vitro*, although it is not known whether this reaction takes place to any significant degree *in vivo*, where in principle it is only used together with enzymes (methyl transferases). The methylation of exogenous compounds primarily takes place with thiols produced after glutathione conjugation (see above) as indicated in Figure 5.37.

5.5 Summary of the Major Phase I and Phase II Reactions

Oxidation	epoxidation of	alkenes
		alkynes
		aromatic systems
	hydroxylation of	allylic and benzylic C–H
		C–H α to heteroatoms
		allylic and benzylic N–H
		ω-1 position in hydrocarbons
	of single to double bonds	
	of heteroatoms	
	of alcohols/aldehydes	
Reduction	of aldehydes/ketones	
	of oxidized heteroatoms	
	reductive dehalogenation	
Hydrolysis	of epoxides	
	of esters	
	of amides	
Conjugation	with sulfate	
	with glucuronic acid	
	with amino acids	
	with glutathione	

5.6
Metabolism of Exogenous Compounds in Mammals

Section 5.6 gives a number of examples of how the metabolism of the major classes of exogenous compounds in mammals is carried out in practice, which is not necessarily exactly the same as the theoretical reactions discussed so far in Chapter 5. The most significant difference is that most 'real' compounds contain more than one chemical functionality that can be attacked by metabolic reactions, and it is not apparent in which order different conversions actually take place. The conclusions of Chapter 5 should be that although we cannot predict in detail which metabolites are formed from the metabolism of a certain compound, or their relative amounts, the general rules discussed in Chapter 5 still apply. With these in mind we can make 'intelligent guesses' about what to expect, and by anticipating the worst we can identify compounds that may be converted to toxic metabolites.

5.6.1
Metabolism of Hydrocarbons

Hydrocarbons are among the most common chemicals that we contact daily, as fuel but also in the working environment as, for example, solvents and pollutants, formed, for example, during the combustion of various materials. This section describes the metabolism of some saturated, unsaturated, and aromatic hydrocarbons, and also some other compounds which are metabolized by conversions involving carbon–hydrogen or carbon–carbon bonds.

Among the alkanes, the smallest (methane to pentane) are very volatile and are not likely to stay long enough in the body to be metabolized at all, unless we are continuously exposed to high concentrations. With hexane (boiling point 69 °C) the volatility starts to become limited. Hexane is a cheap and excellent solvent for various purposes and has been used in enormous quantities by the chemical industry. However, it has been noted that chronic exposure to hexane may affect the peripheral nervous system, causing loss of sensation in the hands and feet. In addition it may damage the testicular tissues. The cause of these injuries is a metabolite of hexane, formed from the cytochrome P450-mediated hydroxylation of both ω-1 carbons according to Figure 5.38.

The toxic metabolite is 2,5-hexanedione, which is believed to react irreversibly with primary amino groups of the amino acid lysine in proteins in the peripheral nerve cells. The product is an *N*-substituted 2,5-dimethylpyrrole derivative (for organic chemists this reaction is known as the Paal–Knorr pyrrole synthesis), which is aromatic and thereby stabilized, and alkylpyrroles are susceptible to autoxidative dimerization, which also is believed to be an important step in the neurotoxicity of 2,5-hexanedione (see Figure 5.38). This has led to the replacement of hexane in many industrial processes by, for example, heptane, which does not give the same toxic effects. Heptane is oxidized in the corresponding way, but the 2,6-heptanedione formed cannot form an aromatic (and

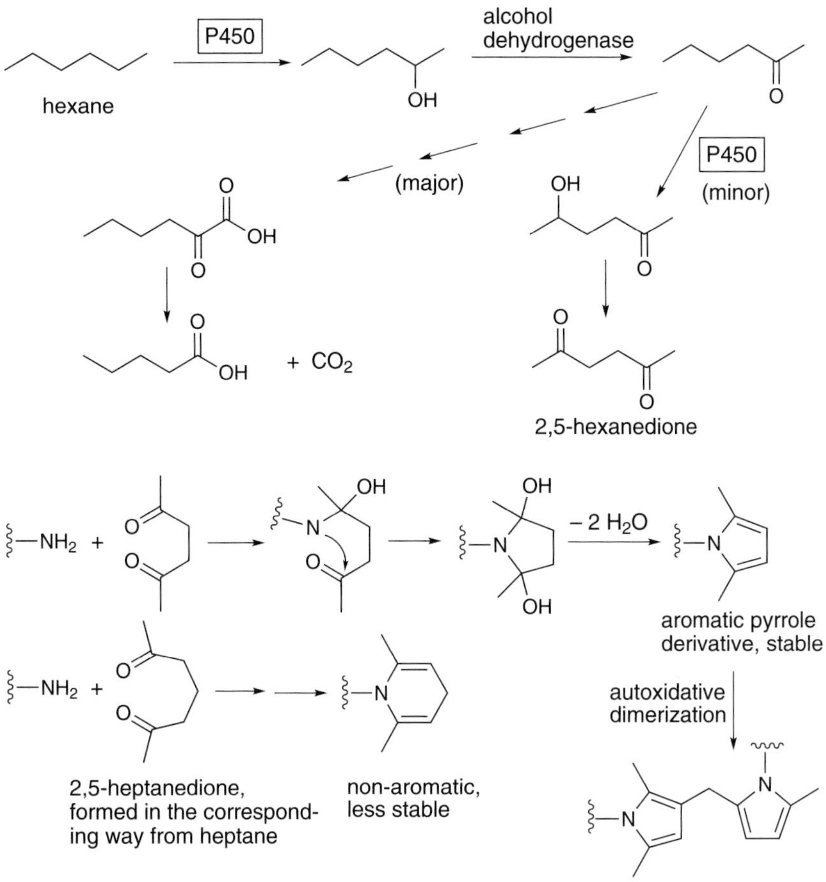

Figure 5.38 The conversion of hexane.

stabilized) end product after reaction with an amine, which makes this reaction reversible.

Alkanes are hydroxylated in various positions, although the ω-1 position is considered to be slightly promoted by being substituted but not too sterically crowded. However, hydroxylations are much easier in positions that are activated, for example, by the presence of an aromatic system or a double bond next to the carbon to be hydroxylated. An example is the natural products of the safrole type, which have been shown to be weak carcinogens and also hepatotoxic in animal experiments. This will be discussed in Chapter 9.

Hydroxylation is the most common conversion that saturated hydrocarbons are subject to, but carbon–carbon single bonds can also be oxidized to double bonds. We have already encountered the example of valproic acid, and another is urethane, or ethyl carbamate. Urethane has been used for many applications, but has been found to be carcinogenic and should be avoided. It may react with genetic

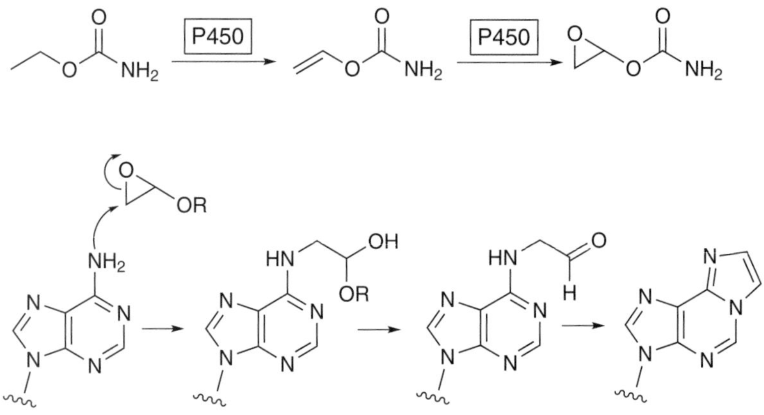

Figure 5.39 Bioactivation of the carcinogenic compound urethane to an epoxide, and its reaction with adenosine.

material, for example, the DNA base adenosine according to Figure 5.39, after oxidation to the alkene and epoxidation of the double bond by cytochrome P450.

Besides the oxidation of the carbon–carbon single bond, urethane will be oxidized by hydroxylation of both the nitrogen and the methyl group, although these reactions are not considered important for toxicity.

Alkenes can be both epoxidized and hydroxylated, preferably in an allylic position, by cytochrome P450, and although it is not evident which route will be chosen, epoxidation seems to more common. Cyclohexene is epoxidized and the epoxide is hydrolyzed to the diol by epoxide hydrolase (note the formation of the *trans*-diol). Small amounts of the epoxide will also react with glutathione (see Figure 5.40). In 4-ethenylcyclohexene (4-vinylcyclohexene), with a double bond in both a ring and a chain, it is almost exclusively the ring that is epoxidized (see Figure 5.40). On the other hand, in styrene, where the choice stands between epoxidation of the vinyl group and the aromatic ring, it is the side chain that is converted to give styrene oxide (see Figure 5.25 for a complete picture of the metabolism of styrene).

Aromatic compounds can be epoxidized in the ring, resulting in unstable epoxides that readily rearrange to phenols. The simplest aromatic compound, benzene, is a relatively cheap organic solvent that was used in large quantities until it was realized that exposure to benzene over years increased the risk of leukemia. Benzene is still used in industry, although its handling is controlled by restrictions, and it is still a component of gasoline, which means that most people actually come into contact with the compound in small amounts. There are consequently a number of reasons for investigating how benzene gives rise to its carcinogenic effect, but in spite of the many man-years of research into this matter we still do not have a clear picture of its metabolic fate. It is obvious that most is converted to phenol, and the fact that some amounts of the corresponding 1,2-dihydro-1,2-diol and the glutathione conjugate are formed suggests that the phenol formation

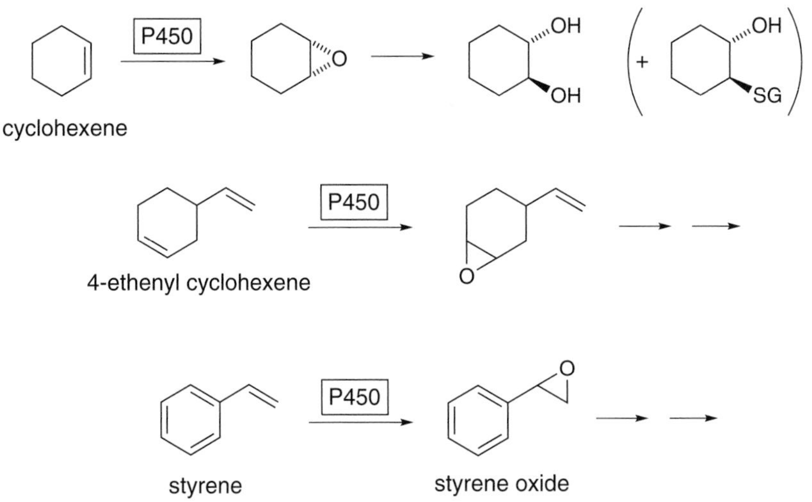

cyclohexene

4-ethenyl cyclohexene

styrene   styrene oxide

Figure 5.40 Epoxidation of various alkenes.

proceeds via the epoxide. However, phenol itself does not give the same effects as benzene (phenol is also a carcinogen but acts by a different mechanism). This difference could be due to the fact that phenol formed from the oxidative metabolism of benzene is directly present in the endoplasmic reticulum and can more efficiently be further converted to the final carcinogenic metabolite, or that this is formed by another metabolic route, not via phenol.

The epoxide was long considered to be responsible for the toxicity of benzene. It survives long enough to be hydrolyzed by epoxide hydrolase and conjugated with glutathione, but these conversions only take place with small amounts (a few percent) of the epoxide. The diol formed can spontaneously eliminate water and be transformed to phenol, it can be oxidized by alcohol dehydrogenase to the keto alcohol, which spontaneously turns into the corresponding enolic phenol, or it can be further oxidized to open the ring. The detection of both Z,Z-muconic acid and the more stable E,E-muconic acid (with both carbon–carbon double bonds in the cis configuration, see Figure 5.41) as metabolites of benzene indicates that the unsaturated dialdehyde Z,Z-muconaldehyde is formed after oxidative opening of the ring (either via dihydrocatechol or via another route, and an alternative metabolic route to Z,Z-mucoaldehyde via benzene oxide and its isomer oxepin is shown in Figure 5.41). Although the dialdehyde itself has not been isolated and the evidence for its formation is only circumstantial, it is nevertheless likely that this very reactive compound is formed and is responsible for at least part of the toxicity of benzene. As described previously, the epoxide of benzene spontaneously yields phenol, and the half-life of the epoxide in water at 37 °C and pH 7 is only approximately 2 min. Most of the phenol formed is conjugated with sulfate and glucuronic acid, but a portion is further oxidized by epoxidation/rearrangement to form catechol and hydroquinone (the ortho and para positions of phenol are actually

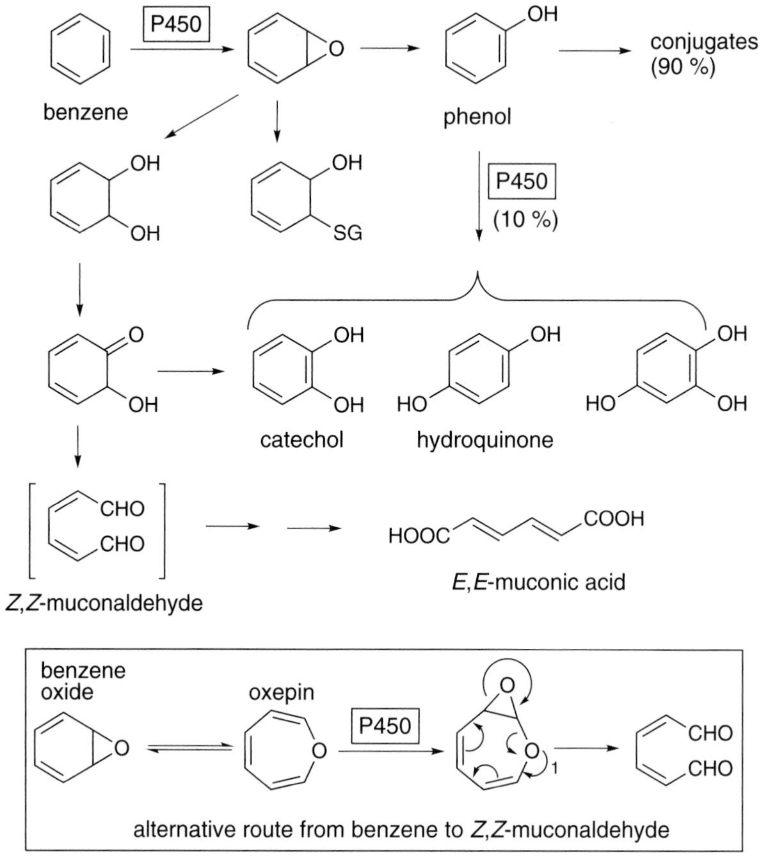

Figure 5.41 The conversion of benzene.

activated for further oxidation by the electron-donating hydroxyl group already present), and even to trihydroxybenzenes. Catechol and hydroquinone can be oxidized to the corresponding quinones, which are electrophilic and toxic, and hydroquinone is actually autoxidized by the molecular oxygen present in the cells (no enzymes required). Besides the formation of quinone, which with its unsaturated keto functionality is electrophilic and mutagenic and can react with DNA bases, this autoxidation also generates superoxide, which adds to the toxic effect (see Figure 5.42).

However, the oxidation of hydroquinone is faster in the presence of peroxidases, which normally have higher activity in tissues where the monooxygenase activity is low, and this may be a piece in the puzzle to explain the specific carcinogenic effect of benzene on the blood cells.

Toluene has replaced benzene as a solvent: it has similar chemical properties but it is considerably less toxic. The reason for this difference is that the methyl

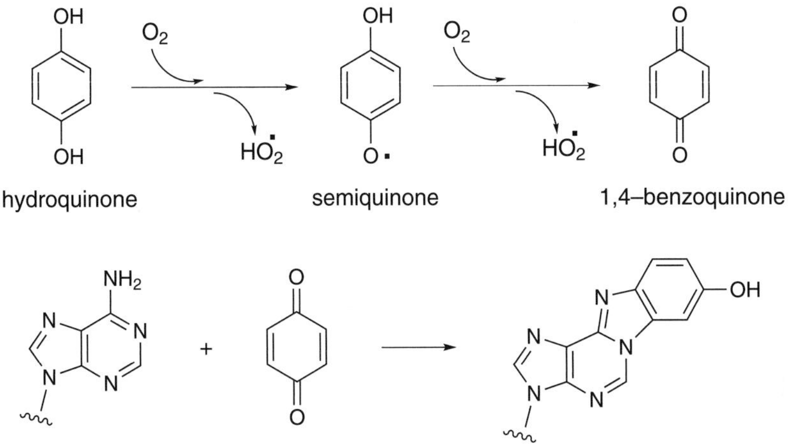

Figure 5.42 Autoxidation of hydroquinone yields reactive quinone.

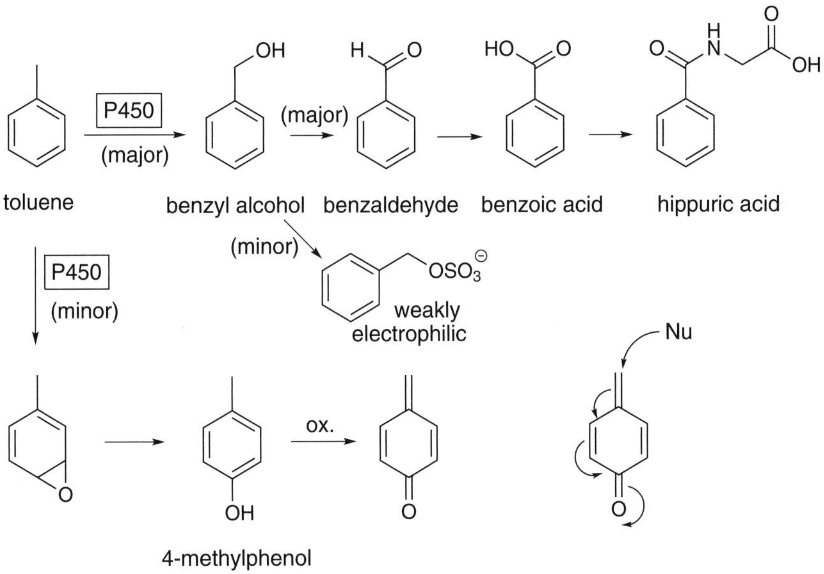

Figure 5.43 The conversion of toluene.

group in toluene is more readily oxidized than the aromatic ring, and this metabolic path leads to less toxic metabolites (see Figure 5.43). Benzylic hydroxylation leads eventually to hippuric acid after conjugation of the carboxylic acid with glycine, which is efficiently eliminated with the urine.

However, epoxidation of the aromatic ring of toluene takes place to some extent, generating, for example, 4-methylphenol, and this minor route is believed to be responsible for the toxic effects that toluene will give after chronic exposure. 4-Methylphenol, as well as any 4-alkylphenol, will be oxidized by peroxidases to a quinone methide that is electrophilic and harmful (see Figure 5.43). In addition, it has been shown that small amounts of the benzyl alcohol formed are conjugated with sulfate, yielding a slightly electrophilic product having a sulfate group in a benzylic position (similar to the toxic metabolite of safrole, see above). The hazards associated with toluene and the xylenes (dimethylbenzenes) have therefore been re-evaluated during the last decade.

Polyaromatic hydrocarbons will also be epoxidized, and the epoxides formed may be hydrolyzed or conjugated, or may spontaneously rearrange to phenols which are conjugated and eventually may be oxidized to quinones. Differences between the rates of the various reactions are observed, which will make the polyaromatic hydrocarbons differ in toxicity. We have already discussed benzo[a]pyrene (Section 5.3.3), which is one of the most toxic members of this class (a potent carcinogen), the reason for this being the relative stability of the epoxide formed, which is a poor substrate for epoxide hydrolase and is therefore given time to react with DNA. In addition, we will discuss the simplest PAH, naphthalene, in Chapter 9.

5.6.2
Metabolism of Compounds Containing Nitrogen

The use of simple, aliphatic amines in industry is increasing, and some toxic effects have been noticed (besides the general irritating effect that all amines have because of their basicity). Amines in which the amino group is attached to a saturated carbon having at least one hydrogen bound to it are oxidized by cytochrome P450 or by amine oxidases to aldehydes. When cytochrome P450 is responsible for the oxidation it proceeds via an α-hydroxylation and a dealkylation, as described previously, while the amine oxidases will oxidize the amine to an imine which spontaneously will be hydrolyzed to the aldehyde. Methylamine (aminomethane), for example, is oxidized by an amine oxidase to the imine that is hydrolyzed to formaldehyde (see Figure 5.44). Allylamine (3-aminopropene), a highly irritating and toxic chemical used, for example, in the pharmaceutical industry, gives a specific and strong toxic effect on the muscle cells of the blood vessels of the heart, even after a single exposure. The effect resembles that seen in arteriosclerosis.

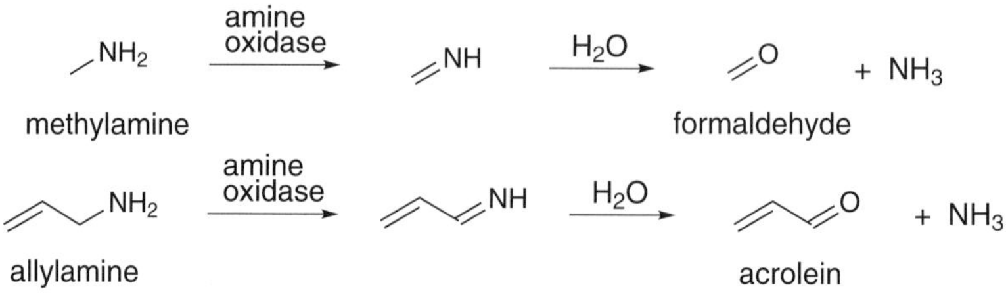

Figure 5.44 The oxidation of amines to aldehydes by amine oxidases.

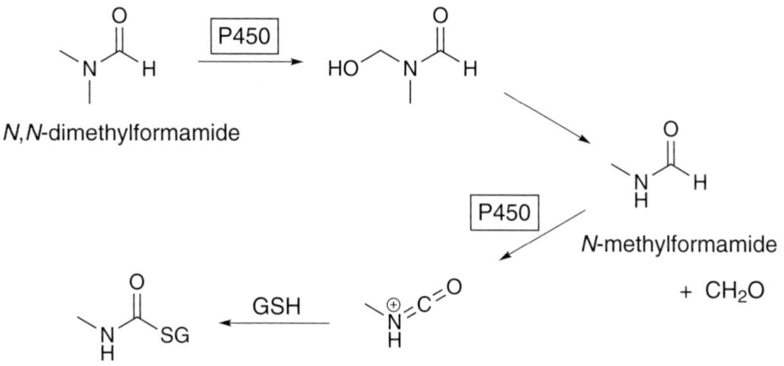

Figure 5.45 The oxidation of *N*-alkylformamides yields an electrophile.

Allylamine is oxidized by an amine oxidase, which is present in relatively large amounts in the cells of the damaged heart tissue, to acrolein (see Figure 5.44) which is responsible for the effect.

N-Alkylformamides, for example *N*-methylformamide and *N*,*N*-dimethylformamide (see Figure 5.45), are important solvents used both in laboratories and in industrial processes. They have been found to be toxic to the liver and in some cases also to be teratogenic. *N*,*N*-Dimethylformamide undergoes oxidative dealkylation (α-hydroxylation) catalyzed by cytochrome P450 in the liver to *N*-methylformamide (and formaldehyde). Further oxidation by cytochrome P450, at the formyl carbon, yields a reactive intermediate that is believed to be protonated methyl isocyanate (see Figure 5.45), that will rapidly react with nucleophiles [e.g., glutathione to form *S*–(*N*-methylcarbamoyl)glutathione].

The aromatic amines have attracted considerably more attention, because they are toxic to the blood in general and are in many cases carcinogenic, especially to the bladder. The aromatic amides, in which the amino function has been transformed to an amide by condensation with a carboxylic acid, are normally discussed together with their corresponding amines as they are oxidized in the same way in our bodies and are easily converted to amines (by amidases). They were among the first chemicals to be used by a growing chemical industry in the 19th century, for example, for the production of pigments and polymers. Aromatic amines can be metabolized in different ways, and Figure 5.46 summarizes the most important routes leading to toxic metabolites.

Hydroxylation of the nitrogen followed by conjugation with sulfate creates a reasonably good electrophile that may be attacked by nucleophiles either directly on the nitrogen but also in conjugated positions in the aromatic ring. The hydroxylamine is actually electrophilic as it is, because the nitrogen–oxygen bond is relatively weak, but conjugation increases its reactivity. In addition, the hydroxylamine may be oxidized by peroxidases to an oxygen radical, which will initiate radical reactions in the cells. Finally, the epoxidation of the aromatic nucleus, a less likely conversion that still will take place, will produce aminophenols that will be

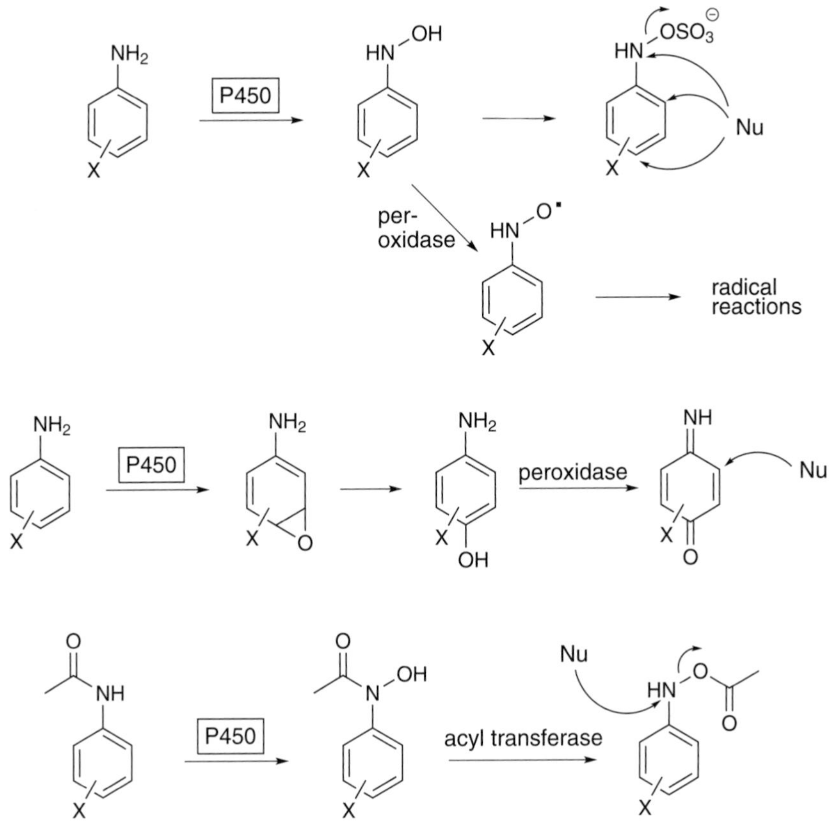

Figure 5.46 Principal metabolic activations of aromatic amines. (X denotes any substituent in any position.)

oxidized by peroxidases to reactive benzoquinone imines. Aromatic amides will also be N-hydroxylated, and a special enzyme (an acyl transferase) can move the acetyl group of the hydroxyamide from the nitrogen to the oxygen resulting in an aromatic N-acetoxyamine. This is very electrophilic, because an acetoxy group in this position is an excellent leaving group in nucleophilic substitutions.

Aniline itself and acetanilide (the N-acetyl derivative of aniline) have not been proven to be carcinogenic, although this is suspected, but o-toluidine (aniline with a methyl group in the *ortho* position) is clearly carcinogenic. Considerably more potent are, for example, 2-aminonaphthalene and N-acetyl-2-aminofluorene (both used in large quantities in the early days of the chemical industry), for which it has been possible to clearly establish that they cause tumors in exposed workers. Interestingly, the analogs having the amino (or acetylamino) function in position 1 instead of 2 are only very weak carcinogens, while both isomers of the hydroxy-lamines (or N-acetyl-N-hydroxylamines) prepared synthetically and assayed in animal experiments are equally potent (see Figure 5.47).

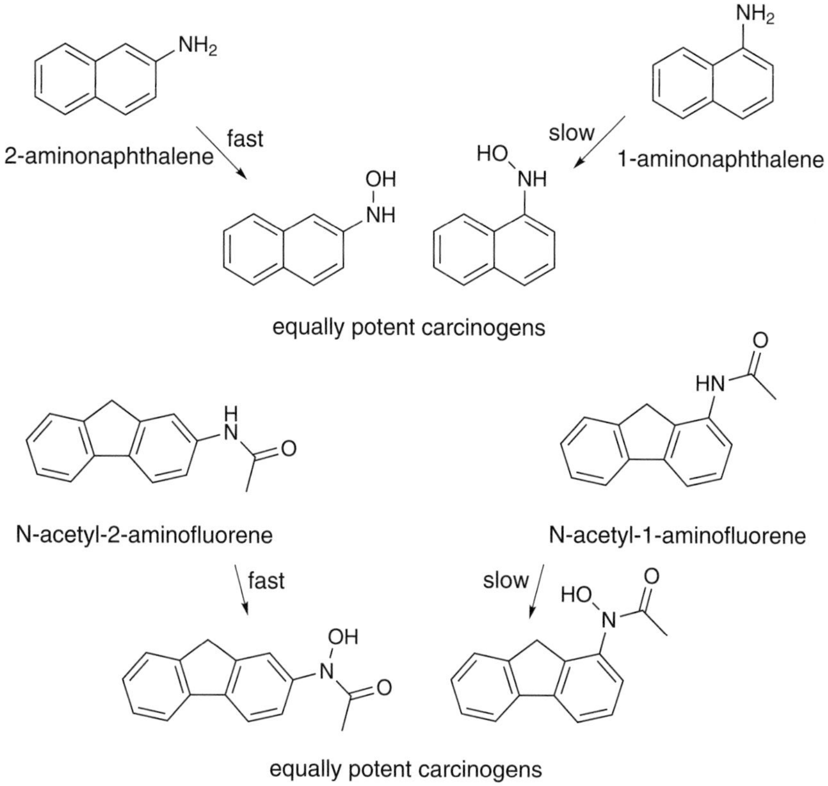

Figure 5.47 The hydroxylation rate differs between various aromatic amines (amides).

The observed difference is caused by an apparent sensitivity to steric hindrance of the hydroxylation of the nitrogen atom by cytochrome P450. The 1-hydroxylamine (and 1-N-acetyl-N-hydroxylamine) are therefore formed much more slowly, but once formed they are equally potent. Besides N-hydroxylation, both 1- and 2-aminonaphthalene are also oxidized in the ring to yield 1-amino-2-hydroxynaphthalene and 2-amino-1-hydroxynaphthalene, which after peroxidase oxidation to the corresponding quinone imines will add to the toxic effects. In addition, conjugation will take place in parallel to oxidation, both of the original amine and the oxidation products; the conjugates are normally less toxic, but we have seen that the reverse may also be true.

In this context it should be noted that the toxic effects of aromatics containing nitro groups resemble those of the corresponding aromatic amines, simply because the nitro group can be reduced to the hydroxylamine (via the nitroso derivative). As noted when the carcinogenic activity of 4-aminoquinoline-N-oxide (which is inactive) was compared to that of the corresponding nitro derivative (which is a potent carcinogen) (see Figure 5.48), there are no steric factors impeding the

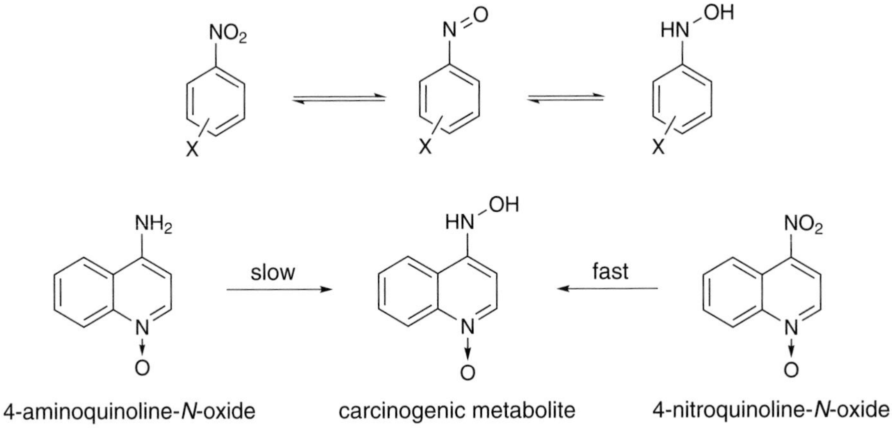

4-aminoquinoline-*N*-oxide   carcinogenic metabolite   4-nitroquinoline-*N*-oxide

Figure 5.48 Aromatic nitro compounds can be reduced to the corresponding hydroxylamines and are often toxic as the aromatic amines.

reduction of a nitro group to a hydroxylamine, in contrast to the oxidation of an amino group to a hydroxylamine.

An interesting aromatic amide that is used in large amounts as a medicament is acetaminophen, which will be discussed in more detail in Chapter 9.

As well as the amines that we contact, for example, in the working environment, there are also a number of endogenous amines that are not necessarily completely harmless. The DNA base adenine has attracted special attention, and studies have shown that even adenine can be oxidized by cytochrome P450 to the corresponding *N*-hydroxyadenine, which turns out to have genotoxic properties *in vitro*. However, it has so far not been possible to determine whether this activation takes place also *in vivo*, and, if so, whether it has any biological significance.

The formation of secondary *N*-nitrosoamines from secondary amines and nitric acid in the stomach has already been discussed in Chapter 4. In addition they are formed in lubricating oils in which sodium nitrite is used as an anti-bacterial additive, during the vulcanization of rubber if nitrite is a component, and in some foods (e.g., some beers). Natural tobacco contains both amines and nitrite, and the tobacco used for smoking contains several *N*-nitrosoamines (a cigarette smoker inhales approximately 1 µg per cigarette). Some secondary *N*-nitrosoamines, for example, dimethyl-*N*-nitrosoamine, are potent carcinogens after metabolic activation via a cytochrome P450 hydroxylation of the carbon α to the *N*-nitrosoamine group (see Figure 5.49), and they give tumors in various tissues.

Dimethyl-*N*-nitrosoamine is converted to diazomethane, a highly toxic and volatile chemical (that in addition is highly explosive), which in its protonated form is a powerful electrophile and will methylate nucleophiles. The hydroxylation of secondary *N*-nitrosoamines is most efficient with methyl groups,

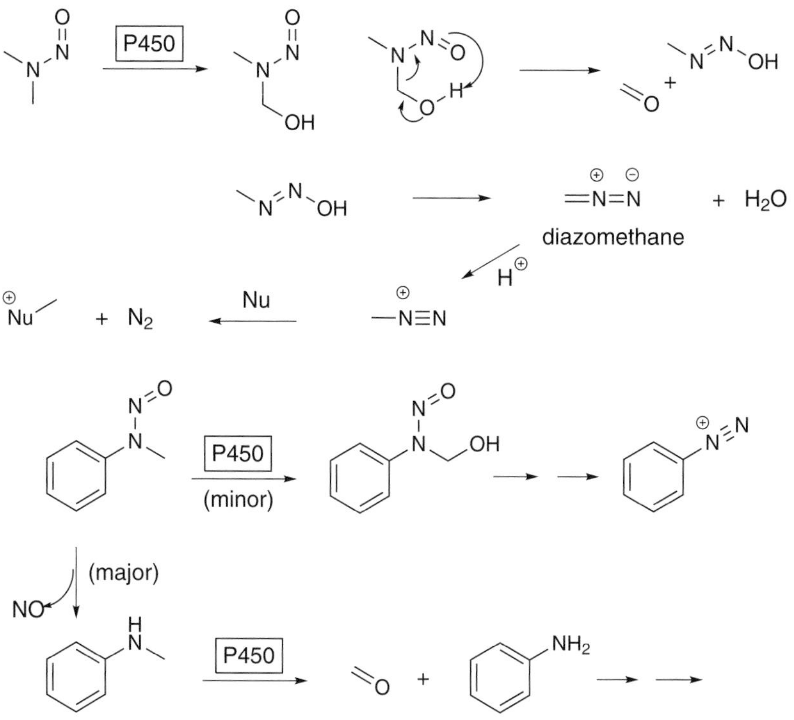

Figure 5.49 The metabolic activation of secondary N-nitrosoamines.

where there is as little steric hindrance as possible, and asymmetric compounds will preferentially yield the electrophilic diazonium ion of the bigger alkyl group. Another possibility for N-nitrosoamines is a reduction to nitric oxide and the corresponding secondary amine, and for N-nitroso-N-methylaniline this is actually the most important route (see Figure 5.49). However, in spite of the limited amounts of the electrophilic benzene diazonium ion formed from N-nitroso-N-methylaniline, it is still carcinogenic and causes cancer of the esophagus.

The cyanide functionality in nitriles is peculiar in that the carbon atom is not really part of the carbon skeleton of the molecule. Instead, the cyanide group can be regarded as a substituent similar to a hydroxyl group or a halogen and is even a (poor) leaving group in nucleophilic substitutions. Nitriles may consequently be subjected to oxidative dealkylation via an α-hydroxylation, yielding an α-hydroxynitrile that will spontaneously eliminate hydrogen cyanide. As this product is a highly toxic compound, it is at least in theory possible that nitriles may be hazardous because of the hydrogen cyanide they generate as a metabolite. However, as will be discussed in detail in Chapter 7, hydrogen cyanide is not very long-lived in our bodies, and it is difficult for the cytochrome P450

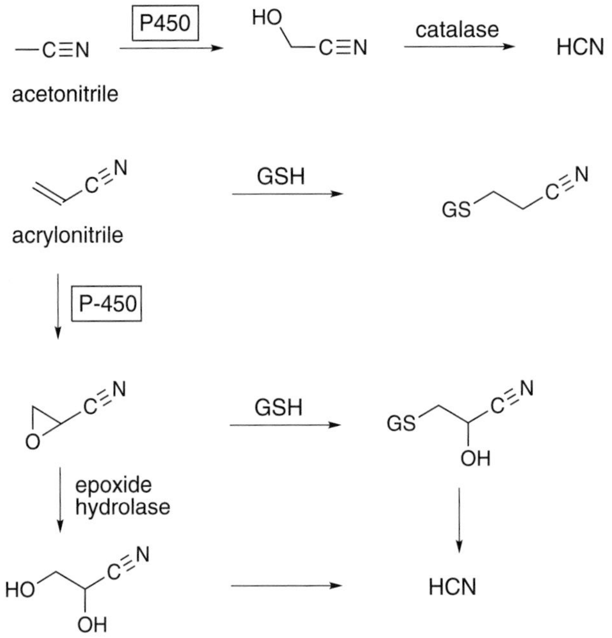

Figure 5.50 The metabolism of acetonitrile and acrylonitrile. (GSH = glutathione.)

α-hydroxylation to generate life-threatening concentrations of this compound. Acetonitrile is a much used solvent that many working in chemical laboratories are exposed to. It has been shown to be oxidized by cytochrome P450, although the intermediate hydroxyacetonitrile appears to produce hydrogen cyanide after oxidation by catalase and the expected metabolite formaldehyde (or formic acid) has not been detected (see Figure 5.50). Because of its high water solubility, a substantial part is actually excreted directly with the urine without being metabolized. In bacteria, acetonitrile may be converted to acetamide by addition of water to the triple bond, but this metabolite is not formed in mammals. Acrylonitrile, used for the preparation of various polymers, is an electrophile because the double bond is activated by the electron-withdrawing cyanide group and may consequently react with glutathione (or other nucleophiles) directly. However, the double bond may also be epoxidized by cytochrome P450, and after hydrolysis/conjugation of the epoxide a labile α-hydroxynitrile-generating hydrogen cyanide is formed (see Figure 5.50).

Methyl isocyanate, the chemical that killed more than 3000 people after an accident in a pesticide manufacturing plant in India in 1984 (discussed further in Chapter 9), is a highly reactive compound that in high concentrations will destroy any tissue. In lower amounts it can be taken up by the body and react with glutathione in a reversible way, and systemic effects can appear some time after an exposure (see Figure 5.51).

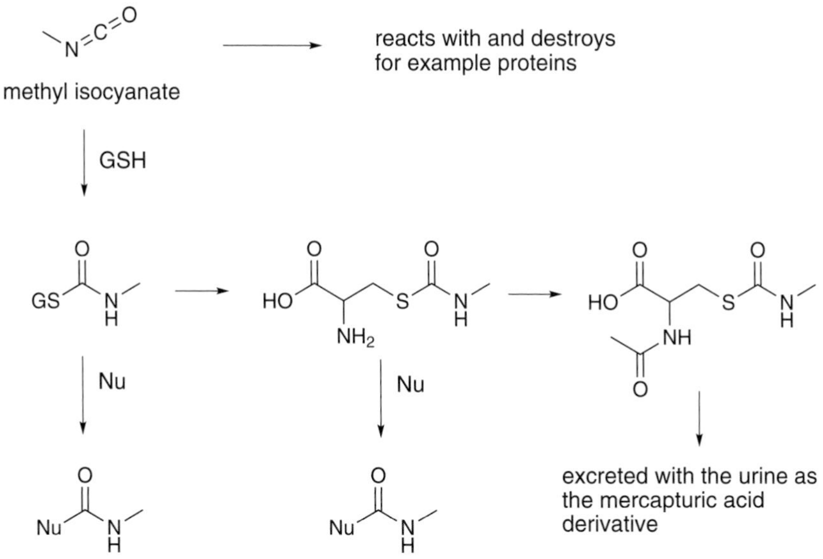

Figure 5.51 The metabolism of methyl isocyanate. (GSH = glutathione.)

5.6.3
Metabolism of Compounds Containing Halogen

The principal oxidations and conjugations of halogenated organic compounds have been discussed previously in Chapter 5. Many chlorinated compounds were formerly used as solvents in huge quantities, but have since been replaced when it turned out that they were in fact hazardous. Carbon tetrachloride (tetrachloromethane), for example, is a very good solvent indeed, but rather quickly turned out to be carcinogenic and toxic to the liver. It is to some extent metabolized by reductive dehalogenation (Section 5.3.2) to the trichloromethyl radical, which is able to initiate radical reactions and is responsible for the toxic effects observed. The dimer of this radical, hexachloroethane, is also observed as a metabolite, proving that the radical actually is formed (see Figure 5.52). Carbon tetrachloride was replaced by chloroform (trichloromethane), which is less toxic, but after having been used for some time turned out to have similar effects on the liver. It appears to be mainly oxidized to the electrophile phosgene, although a reduction generating radicals has also been suggested. Chloroform was in turn replaced by methylene chloride (dichloromethane). This solvent is less toxic than carbon tetrachloride or chloroform, but there are indications that it is genotoxic and possibly carcinogenic. Methylene chloride may react slowly as a weak electrophile with, for example, DNA, and the conjugate formed with glutathione, S-(chloromethyl) glutathione (see Figure 5.52), would be an excellent electrophile. As well as being toxic, some chlorinated solvents have the ability to affect the ozone layer in the stratosphere, and measures are today being taken to decrease the use of this class

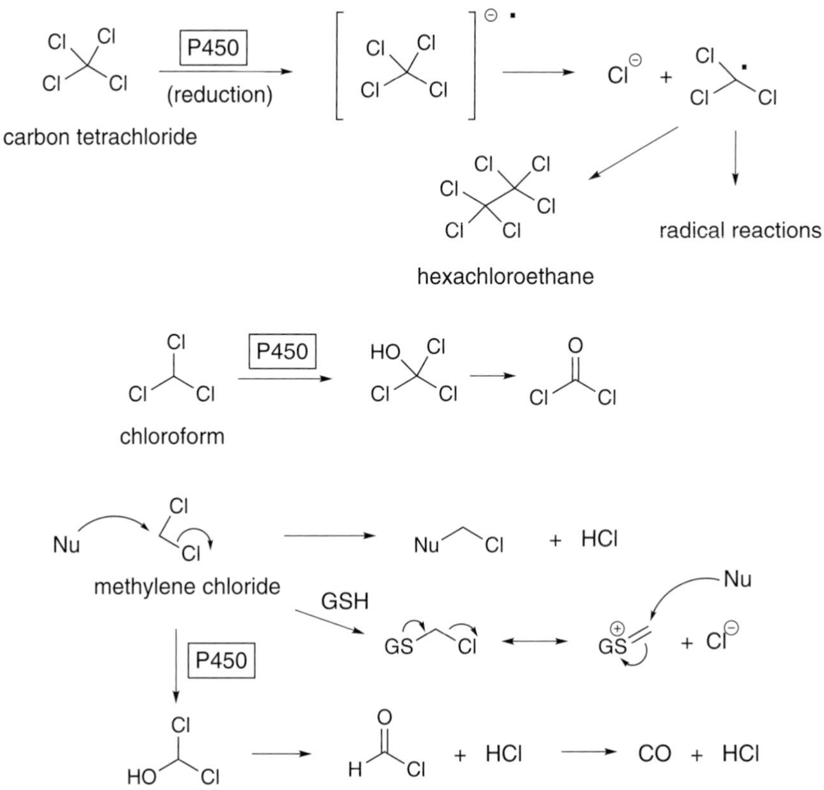

Figure 5.52 Metabolism of carbon tetrachloride, chloroform and methylene chloride. (GSH = glutathione.)

of solvents. Methylene chloride can possibly be replaced by methyl *tert*-butyl ether (MTBE), a volatile compound that is used in large amounts as an octane booster in gasoline. MTBE lacks chlorine altogether and is considered to be quite safe from a toxicological point of view, but its properties as a solvent are different from those of methylene chloride.

Methylene chloride is also metabolized by oxidative dehydrohalogenation, generating the acid chloride of formic acid (see Figure 5.52), which should be a highly reactive and toxic product. However, it is so unstable that it is simply transformed to carbon monoxide and hydrogen chloride the moment it is formed. This is probably the reason why methylene chloride is less toxic to the liver than carbon tetrachloride and chloroform, although large doses of the compound will generate toxic (and possibly fatal) amounts of carbon monoxide in the body.

Halothane (2-bromo-2-chloro-1,1,1-trifluoroethane), used as an inhalation anesthetic, has low toxicity but has in some cases been shown to provoke an allergic

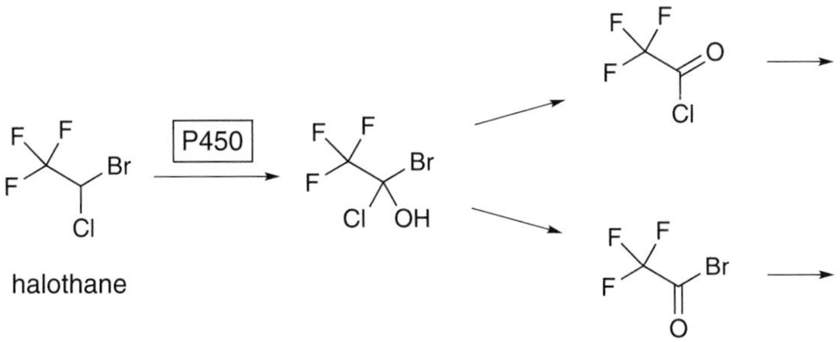

Figure 5.53 Halothane may be converted to reactive acid halides.

response and may also be toxic to liver cells. Also, there has been some concern especially for persons exposed to halothane daily in their work. It may be oxidized by cytochrome P450 to form trifluoroacetic acid chloride (or bromide) via oxidative dehydrohalogenation (see Figure 5.53), and will react with, for example, proteins in the liver cells. There is also evidence that halothane can be reduced by cytochrome P450 to form reactive radicals.

Trichloroethylene, a solvent that also has been used for general anesthesia, is considered less toxic than vinyl chloride. It will also undergo epoxidation by cytochrome P450, but the epoxide is so unstable that it will more or less instantly be rearranged (in a variant of the pinacol rearrangement!) to trichloroacetaldehyde (see Figure 5.54). This reaction is so fast that no epoxide hydrolase product and no conjugate formed from the reaction with glutathione can be observed. Trichloroacetaldehyde, or chloral, is a well-known and popular sedative and hypnotic which is still used. Chloral can be oxidized to the corresponding acid (trichloroacetic acid) or reduced to 2,2,2-trichloroethanol, which is conjugated with glucuronic acid and excreted. Trichloroethylene can also react with cysteine and glutathione to form various S-(dichlorovinyl) derivatives that are especially toxic to the kidneys because these ionic compounds are concentrated there. This reaction is quite slow and probably of little importance *in vivo* but has been shown to be important when trichloroethylene is used for the extraction of fat from soybean oil, as the corresponding S-(dichlorovinyl)-cysteine formed during the extraction is toxic to animals that are fed the residues after the extraction. (Chemically the reaction is not a nucleophilic substitution but rather an addition of the thiol to the double bond followed by an elimination.) The enethiol ether is bioactivated via the S-(dichlorovinyl)-cysteine by the enzyme β-lyase, which cleaves the bond between the sulfur atom and the rest of the cysteine. The enethiols are transformed to the corresponding thiocarbonyl compounds, or by elimination of hydrochloric acid to a chlorothioketene (see Figure 5.54), which are all reactive compounds (compared with the corresponding oxygen derivatives).

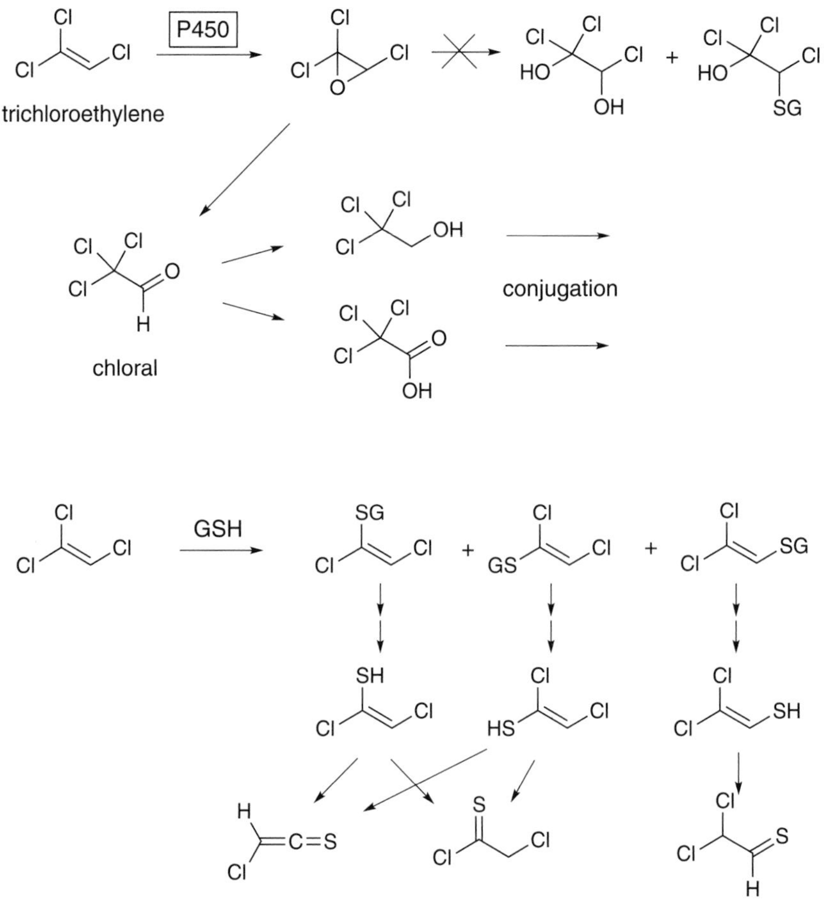

Figure 5.54 Conversions of trichloroethylene and its glutathione derivative. (GSH = glutathione.)

5.6.4
Metabolism of Compounds Containing Sulfur

Dimethyl sulfoxide (DMSO) is an unusual solvent that readily dissolves both polar and nonpolar compounds. It is characterized by being absorbed relatively easily through the skin and is known to rapidly induce a distinct taste (of onions) in the mouth if applied or spilled on the skin. This is due to the reduction of small amounts of dimethyl sulfoxide to dimethyl sulfide (see Figure 5.55), a compound that has an extremely strong (and characteristic!) smell even in very low concentrations, in our bodies. Most DMSO is oxidized to dimethyl sulfone. Both dimethyl sulfoxide and dimethyl sulfone are water soluble and can be eliminated with the urine.

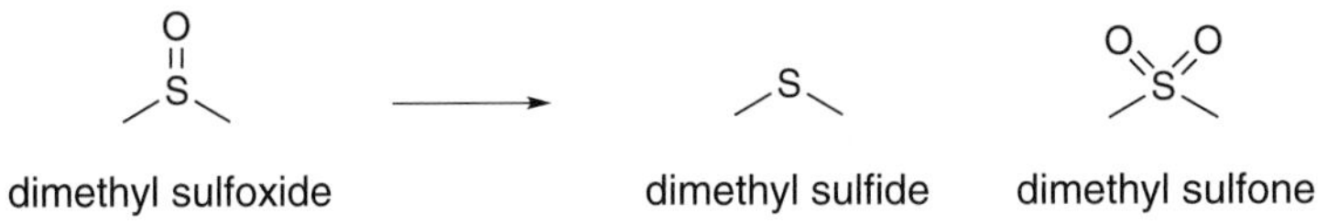

dimethyl sulfoxide   dimethyl sulfide   dimethyl sulfone

Figure 5.55   Dimethyl sulfoxide is both reduced and oxidized.

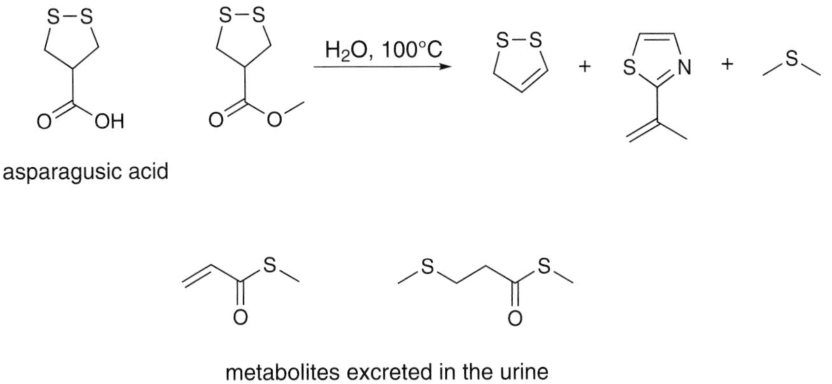

asparagusic acid

metabolites excreted in the urine

Figure 5.56   Transformations and conversion of asparagusic acid.

A final example of a compound that has little importance for the environment but that some may have noticed the existence of is asparagusic acid, present together with its methyl ester in one of the most delicious comestibles available; asparagus. After boiling 1,2-dithia-3-cyclopentene, dimethyl sulfide, and 2-acetylthiazol are the major components in the aroma, while the thioesters shown in Figure 5.56 are the metabolites responsible for the characteristic odor of the urine after the delight of an asparagus meal.

6
Conversions and Transformations in the Environment

In the environment (= outside the body of humans) chemicals are subjected to conversions/transformations (degradation) by reactions with other chemicals present in the same medium, by photochemical reactions initiated by sunlight, and by enzymatic conversions in organisms. Such conversions/transformations will change the structure and thereby the toxicity of a chemical, and it is consequently of interest to understand what may be formed under various circumstances. Organic compounds will eventually be mineralized, meaning that they are completely degraded to inorganic chemicals (CO_2, H_2O, etc.), although this may take a long time and take place via more hazardous intermediates. Decisive for the fate of compounds in the environment is their distribution in the different media. For example, the probability that volatile compounds will undergo photochemical reactions is of course high compared to nonvolatile compounds. Some compounds, even if they contain chemical functionalities that appear to be easily converted/transformed, have such limited water solubility and volatility that they are simply sorbed onto macroscopic surfaces and withdrawn from circulation. Such compounds may survive for very long times in various compartments, for example in sediments, and will not pose any hazard until they for any reason are mobilized again.

6.1
Enzymatic Conversions

Biotic conversion of chemicals in the environment is especially important for the degradation of organic compounds in general and pollutants in particular, because they include reactions that would hardly be accomplished, at least not on a reasonably short time scale, in abiotic ways. Enzymes may facilitate a chemical reaction that is feasible but slow without enzymatic catalysis by lowering the activation energy or by the input of energy (from cofactors produced by the primary metabolism) that promotes the reaction, and we have already seen several examples of conversions carried out by mammalian enzyme systems in Chapter 5. As was indicated in Chapter 2 by far the most common organisms are the microorganisms, and they will consequently be responsible for most of the metabolic conver-

Chemistry, Health, and Environment. Olov Sterner
© 2010 WILEY-VCH Verlag GmbH & Co. KGaA, Weinheim
ISBN: 978-3-527-32582-5

sions of organic compounds in the environment (some examples of conversions of toxic metal ion are given in Chapter 10). There are many different types of microorganisms (e.g., some algae, bacteria, fungi, and protozoans) possessing different metabolizing capacities, and microorganisms often have an incredible ability to adapt to new conditions (e.g., the presence of a new compound in their environment). However, although several of the basic metabolic conversions are similar to those of mammals (Chapter 5), there are many more, and it is not possible to summarize the microbial metabolism of organic compounds in this chapter. In addition, it has turned out that some microorganisms carry out a sequence of metabolic steps in cooperation, and that the rate of the overall conversion depends on the presence of all partners at the same place. A major factor that influences the metabolic fate in different organisms is the amount of molecular oxygen present. In mammals, which normally have relatively high concentrations of molecular oxygen, oxidative conversions predominate, while organisms living in a environment containing only a little molecular oxygen may have a completely different metabolic profile. Many organic compounds, for example, hydrocarbons, are more efficiently metabolized by oxidation, but halogenated compounds are often converted more rapidly under reducing conditions. The metabolites that are eventually formed will be substrates for other metabolic conversions as well as for chemical/photochemical transformations, which further complicates the picture.

The ability of some microorganisms to perform unexpected metabolic conversions at a high rate has proven to be useful for the purge of contaminated soil and water. The uncontrolled handling of hazardous chemicals during the 20th century left a large number of industrial and military sites where the soil was heavily contaminated, and to avoid the leaking of such contaminants into the subsoil water and to make such areas exploitable again the soil has to be decontaminated. This can be achieved in several (expensive) ways, and lately biodegradation with microorganisms has become popular. By screening large numbers of microbial species it is often possible to find at least one that can be used for the degradation of a certain organic compound. The conversion of organic compounds by microorganisms can be divided into 4 principal classes:

1) The compound is relatively non-toxic to the microorganism and readily metabolized to components of the primary metabolism. In this case the microorganism may use the compound as nutrient, multiply, and mineralize the compound completely.

2) The toxicity of the compound is limited and, even if it is not mineralized it is metabolized to new compounds that are as toxic or less toxic compared with the original compound. The products may then be further metabolized by other organisms or transformed by chemical/photochemical reactions.

3) The toxicity of the compound is limited but the metabolites formed are more toxic compared with the original compound. This poses a threat not only to the microorganism itself, which will have problems to multiply, but also to

other organisms. The products may be further metabolized by other organisms or transformed by chemical/photochemical reactions. The conversion rate will be low, and as a biodegradation this means low efficiency.

4) The compound is toxic to microorganisms, hampering its multiplication, and probably also to other organisms. Only low concentrations of the compound in the soil/water will be tolerated. If the compound is toxic because it is chemically reactive, a chemical or physical method for degrading the pollutant may be preferable.

In addition, the mass transport from the sorbed state (in soil) to the microorganism is also a critical factor, as is the concentration of oxygen. Completely different metabolites are produced under aerobic compared to anaerobic conditions, and in the case where biodegradation is to be performed in a controlled way there are many parameters to optimize. It is important to remember that biodegradation is not simply a question of getting rid of an organic contaminant in soil or water; it is also necessary to be aware of the possibility that even more hazardous compounds are formed. Most attempts with biodegradation have been made with aromatic compounds, because many members of this class are hazardous, used in very large amounts, and released into the environment. Consequently, most of our knowledge about biodegradation of organic compounds concerns aromatic compounds.

6.1.1
Biodegradation of Saturated Hydrocarbons

Hydrocarbons in crude oil from natural oil leaks or from spills resulting from oil production and oil refining are the most common pollutants in both aquatic and terrestrial ecosystems. Even though hydrocarbons lacking functional groups and unsaturations are not easily converted by the metabolic reactions, many microorganisms will nevertheless metabolize hydrocarbons so efficiently that they are of practical use for biodegradation of oil discharges. We have seen in Chapter 5 that two metabolic pathways for the degradation of saturated hydrocarbons are available in mammals, hydroxylation (the insertion of an oxygen between a carbon and a hydrogen) and oxidation of a carbon–carbon single bond to a double bond. Microorganisms will primarily use hydroxylation, catalyzed by cytochrome P450 monooxygenases or related enzyme systems, and the most important conversions are summarized in Figure 6.1. Hydroxylation may occur either in a terminal position, that is, of a methyl group, or in a chain or a ring, that is, of a methylene or a methine group.

The hydroxylation of a straight-chained hydrocarbon in either of its methyl groups will produce a primary alcohol that will be further oxidized (alcohol dehydrogenases and aldehyde dehydrogenases) via the aldehyde to the carboxylic acid, which will be degraded by the β-oxidation of the primary metabolism. It is not unusual that both ends of an unbranched alkane are oxidized simultaneously, forming diacids that also are further oxidized by β-oxidation.

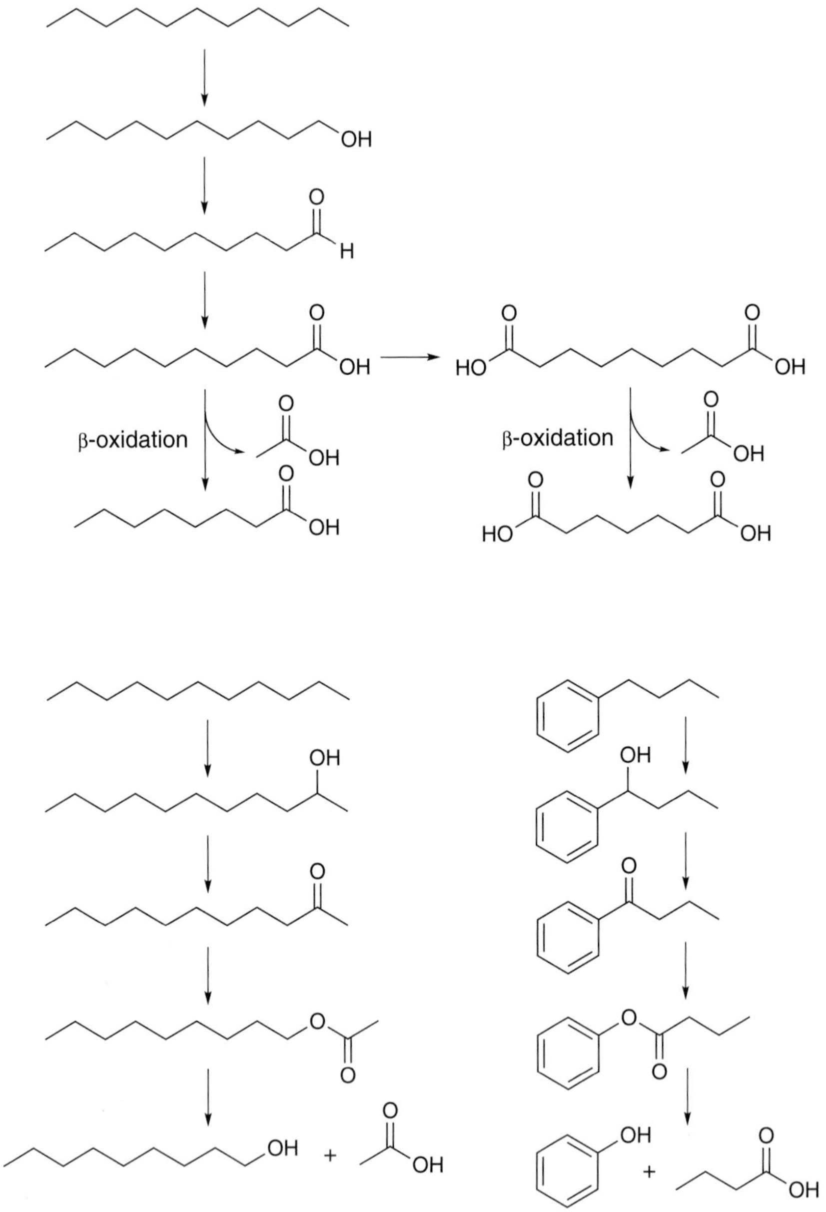

Figure 6.1 Major biodegradation pathways for saturated hydrocarbons by microorganisms.

Methylene groups that are hydroxylated to secondary alcohols are further oxidized to ketones by alcohol dehydrogenases. Monooxygenases may then incorporate an additional oxygen next to the keto function, in a conversion that is analogous to the chemical transformation called the Baeyer–Villiger reaction, to produce an ester. The ester can be hydrolyzed, and the resulting carboxylic acid and alcohol will be further oxidized. Tertiary alcohols, if formed by the hydroxylation of methine groups, are not oxidized further.

6.1.2
Biodegradation of Benzene

In general, different organisms will attack the aromatic hydrocarbons in different ways, and while, for example, plants and fungi predominantly use monooxygenases and phase II reactions, bacteria will use dioxygenases, which produce metabolites that are easier to mineralize. The major biodegradation routes for benzene proceed via catechol, formed after the oxidation of benzene by a dioxygenase followed by a dehydrogenase. The intermediate product, benzene dihydrodiol, (note that the two hydroxyl groups are cis!) is not too stable and may eliminate a molecule of water to form phenol (and thereby regain its aromaticity). Nevertheless, catechol appears to be the major product and is further metabolized by two principal pathways (see Figure 6.2). Additional oxidation of the diol moiety by the enzyme catechol 1,2-dioxygenase will, via ring fission, produce muconic acid via the so called ortho-cleavage route. Muconic acid will spontaneously cyclize to muconolactone, which will be isomerized to the enol lactone of 2-oxoadipic acid, as shown in Figure 6.2. This will yield acetic acid and succinic acid, which are completely degradable. The other pathway, called the meta-cleavage route, involves further oxidation of the carbon–carbon bond adjacent to the diol moiety by catechol 2,3-dioxygenase and produces 2-hydroxymuconic acid semialdehyde as its initial product. Its keto form is oxidized and decarboxylated to 2-oxo-4-pentenoic acid, which after the addition of a molecule of water to the carbon–carbon double bond (to 2-oxo-4-hydroxypentanoic acid) is converted to pyruvic acid and acetaldehyde.

6.1.3
Biodegradation of Alkylbenzenes

Substituted benzenes are metabolized by the same routes, although variations are of course observed. Alkylbenzenes, for example, toluene, are also oxidized in the alkyl group, and a methyl group after oxidation to a carboxylic acid functionality may be split off as carbon dioxide. Dimethylbenzenes (xylenes), for example, are oxidized to methylbenzoic acids, which are decarboxylated during the subsequent oxidation to methylcatechols (see Figure 6.3 for 1,3-dimethylbenzene).

6.1.4
Biodegradation of Fused-Ring Aromatic Compounds

Some organisms are also able to degrade fused-ring aromatic compounds (compounds with two or more benzene rings joined side by side). Of special interest are of course the polycyclic aromatic hydrocarbons (PAHs), because some of

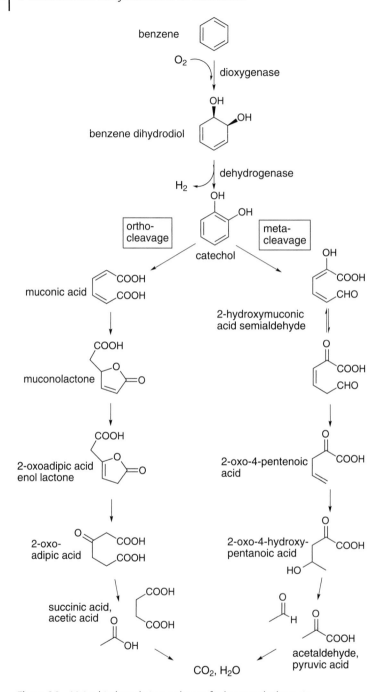

Figure 6.2 Major biodegradation pathways for benzene by bacteria.

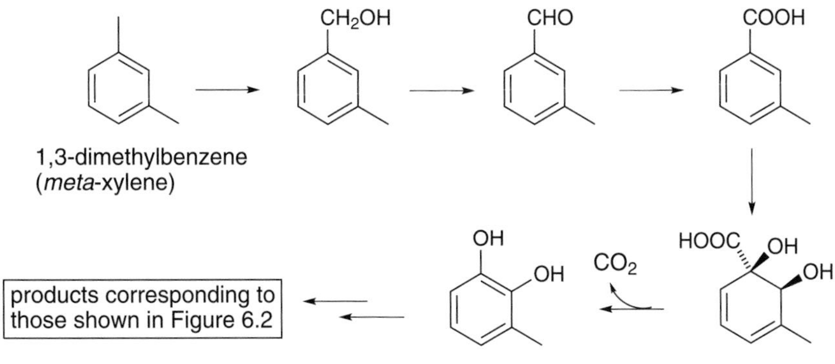

Figure 6.3 Bacterial degradation of 1,3-dimethylbenzene (*meta*-xylene).

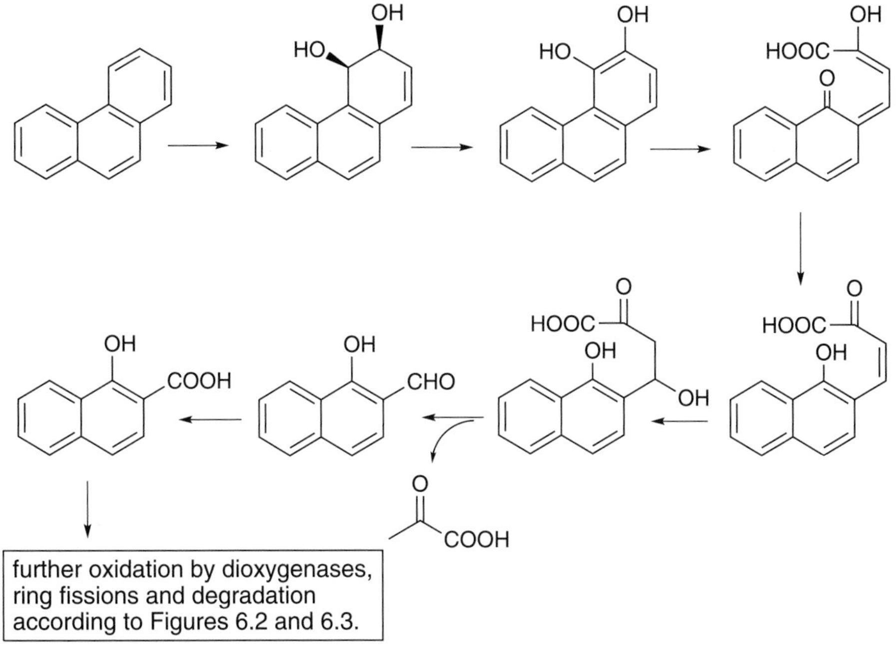

Figure 6.4 Biodegradation of phenanthrene by bacteria.

these are highly toxic and carcinogenic. Figure 6.4 shows how phenanthrene is believed to be metabolized by certain bacteria similarly to the conversions discussed above.

However, biodegradation will generate all kinds of metabolites, including toxic ones. Figure 6.5 shows an example of the products identified from the degradation of pyrene by an *Aspergillus* species, and we can see that this metabolism is more similar to that of mammals. The formation of reactive and potentially hazardous

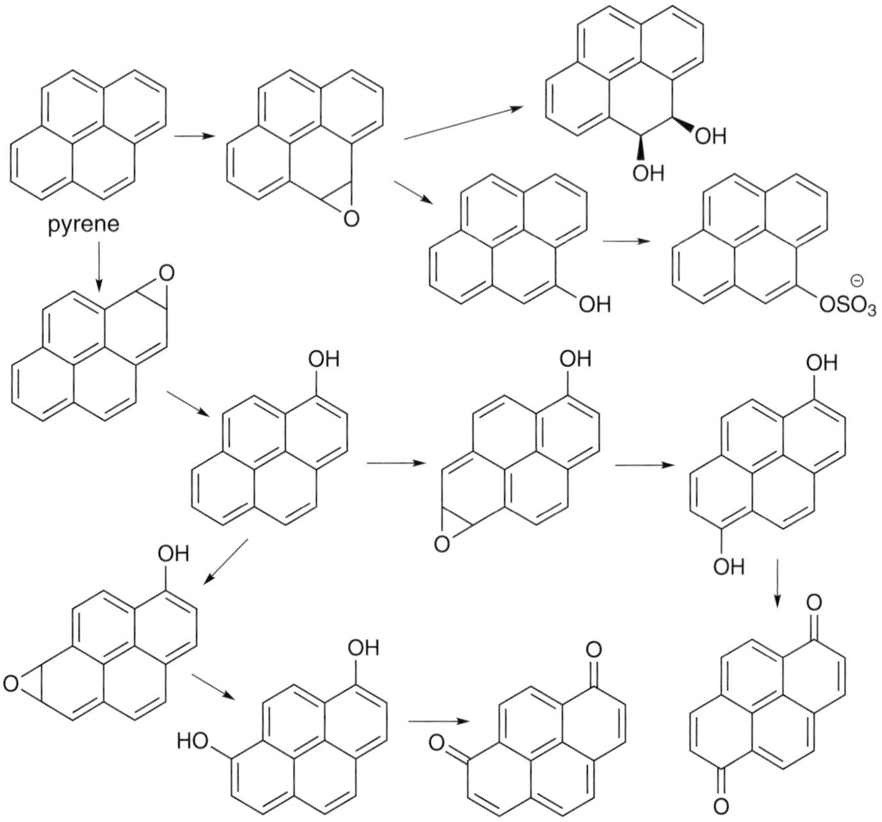

Figure 6.5 The degradation of pyrene by an *Aspergillus* species.

quinones is always a possibility, and we have seen that oxidation of some hydroquinones to the corresponding benzoquinones may take place spontaneously
under aerobic conditions.

6.1.5
Biodegradation of Halogenated Aromatic Compounds

As indicated in the previous section, substituted benzenes may be metabolized by
microorganisms in the same way as benzene itself, via the ortho- or meta-cleavage
route. In addition, halogenated aromatic compounds may also undergo oxidative
dehalogenation (see Figure 6.6), where the halogen is lost during oxygenation of
the ring, a sequence that also generates catechols. Another possibility is hydrolytic
halogenation, where a halogen is simply replaced by a hydroxyl group without any
oxidation or reduction taking place. Finally, under anaerobic conditions halogenated aromatic compounds may undergo reductive dehalogenation, resulting in the
replacement of a halogen with a hydrogen.

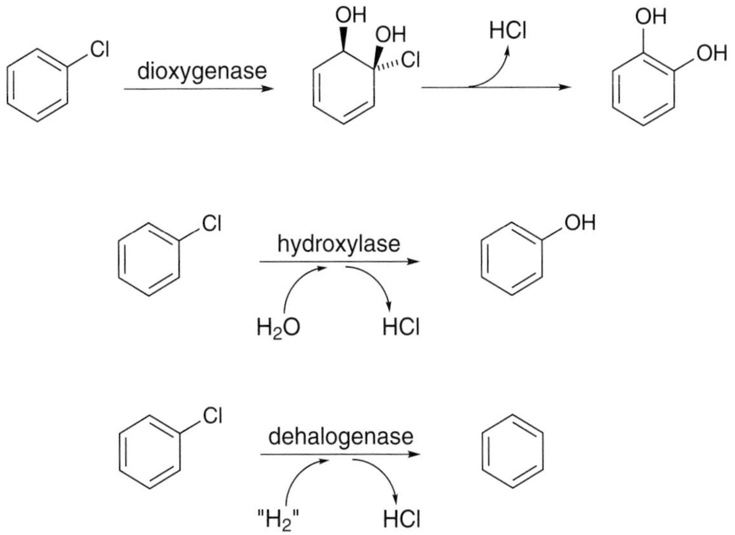

Figure 6.6 Principal dehalogenations of halogenated aromatic compounds.

Again, one should not be surprised to find completely different products. Figure 6.7 shows what a fungus managed to make of 3,4-dichloroaniline, a microbial product from the biodegradation of several chlorinated phenylcarbamate herbicides. Instead of the 3,4-dichloroaniline being degraded, it was condensed to dimers, trimers, and tetramers, and it is likely that such products can be incorporated into, for example, the humic substances.

6.2
Chemical Transformations

Chemical or abiotic transformations of chemicals do not involve the action of enzymes, although this does not necessarily mean that they are not catalyzed. They may take place in the air, in water, or at solid surfaces, and can be divided into the major categories oxidations/reductions, hydrolyses, and substitutions/additions/eliminations. In principle, all existing reactions described in the chemical literature (as well as those not yet described!) take place, although some reactions are more important. As well as catalysis, temperature is an important factor that promotes chemical reactions. So far we have ignored it, as the temperature in mammals is constant, but in the rest of the world temperature differences may be substantial. At the surface of earth, heat can be generated by, for example, geothermic processes and sunlight. The effect of increased temperature is to increase the energy of the molecules that react, enabling them to reach the activation energy for the reaction.

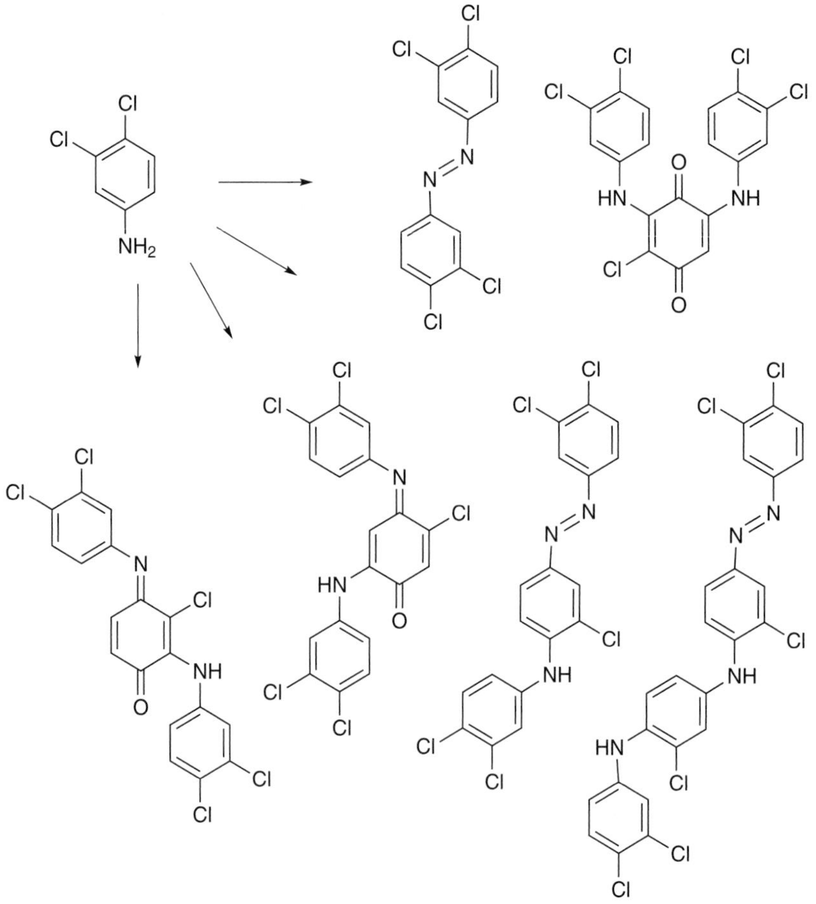

Figure 6.7 The conversion of 3,4-dichloroaniline by a fungus.

6.2.1
Oxidations

Oxidations (and also reductions) require a transfer of electrons to the oxidation agent from whatever is being oxidized, and at normal temperatures most types of oxidations have to be catalyzed. However, the presence of large amounts of oxygen in our atmosphere of course makes it oxidative in general, and easily oxidized materials will be oxidized at ambient temperatures. We have already seen examples of such so-called autoxidations, for example, of hydroquinones to benzoquinones via the radical semiquinone (a process that also generates superoxide), and the oxidative reactions initiated by the reactive oxygen species (which occur naturally in low concentrations). Other examples of spontaneous oxidations with molecular oxygen are the oxidation of certain aldehydes to carboxylic acids, of thiols to dithioethers, of mercaptans to sulfoxides and sulfones, and of anilines to hydroxylamines and nitrobenzenes. Fortunately, molecular oxygen in its ground state is not very reactive and will only oxidize certain compounds, although the energy released by the oxidation of organic compounds should promote autoxidations. However, as we shall see in Section 10.4, molecular oxygen may be activated to more reactive forms by radiation.

6.2.2
Hydrolyses

We have already encountered various hydrolyses, with and without catalysts, and it is evidently an important reaction type both in biochemical conversions and chemical transformations. The most common functionality to be hydrolyzed is the carboxylic acid ester, a reaction which, like hydrolyses in general, can be catalyzed by enzymes (esterases) as well as by acids and bases. In the absence of a catalyst, the hydrolysis of a carboxylic acid ester depends on the nature of the groups bound to the carbonyl carbon and the oxygen, respectively. As indicated by the examples given in Figure 6.8, electron-withdrawing substituents on the carbon bound to the carbonyl group will increase the rate of hydrolysis in pure water simply by making the carbonyl carbon more positively charged and more prone to be attacked by water. Electron-withdrawing substituents on the carbon bound to the oxygen in the ester functionality will also increase the rate, especially if they can stabilize the corresponding alkoxy (or phenoxy) ions (e.g., in 2,4-dinitrophenyl acetate).

In addition to being hydrolyzed by water according to the classical mechanism involving an attack of the water molecule on the carbonyl carbon, the acylate ion (R-COO⁻) of a carboxylic acid ester may also be a leaving group in a nucleophilic substitution or elimination (see below). Amides are less prone to hydrolysis, as we have already noted, because amines are in general poorer leaving groups than alcohols and because the nitrogen in an amide has a greater tendency to donate

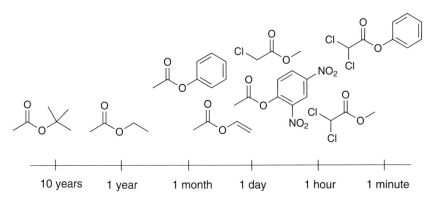

Figure 6.8 The half-lives of various carboxylic acid esters in water at 25 °C and pH 7.

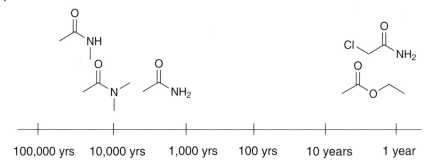

Figure 6.9 The half-lives of various carboxylic acid amides in water at 25 °C and pH 7.

electrons to the carbonyl group, giving the CO–N bond a partial double bond character. The half-lives of a few simple amides are compared to that of ethyl acetate in Figure 6.9, and again we see that electron-withdrawing substituents on the carbon bound to the carbonyl group increase the rate of hydrolysis.

Carbamates are widely used as insecticides and herbicides, and we will encounter an example (carbaryl) in Chapter 9. The carbamate functionality has both an oxygen and a nitrogen bound to the same carbonyl carbon, making it an ester–amide of carbonic acid. Hydrolysis can take place of both the ester moiety (yielding an alcohol) and the amide (yielding an amine), and as we can guess that the ester will normally be hydrolyzed more rapidly than the amide. In some cases both are hydrolyzed, yielding an alcohol, an amine, and carbonic acid. The usual rules apply (see Figure 6.10), but there is a remarkable difference in rate of hydrolysis between carbamates that have one alkyl group attached to the nitrogen and those that have two, especially if the free alcohol is relatively acidic (e.g. 4-nitrophenol). The reason for this is that two different mechanisms operate. The hydrolysis may be initiated not only by an attack by the water molecule on the carbonyl carbon, but also by an elimination of the alcohol to form an isocyanate (see Figure 6.10). The isocyanate will then react fast with water (an addition) to form the hydrolysis product. Isocyanate formation is facilitated by the acidity of the nitrogen proton, both being attached to a heteroatom and situated α to a carbonyl group.

Other compounds that are frequently employed as pesticides are derivatives of phosphoric and thiophosphoric acid esters. Such esters will also be hydrolyzed, either by the attack of a water molecule on the phosphorus atom followed by the departure of an alcohol moiety or by reactions corresponding to those discussed above for carboxylic acid esters (nucleophilic substitution and β-elimination). Some examples of phosphoric and thiophosphoric acid esters that are hydrolyzed and their half-lives in water at 25 °C, are given in Figure 6.11.

A number of other functionalities are also subject to hydrolysis in contact with water, for example epoxides and aziridines, lactones and lactams, anhydrides and imides, carboxylic acid chlorides, carbonates and ureas, phosphonic and thiophosphonic acid esters, phosphinic and thiophosphinic acid esters, and sulfuric and sulfonic acid esters and amides.

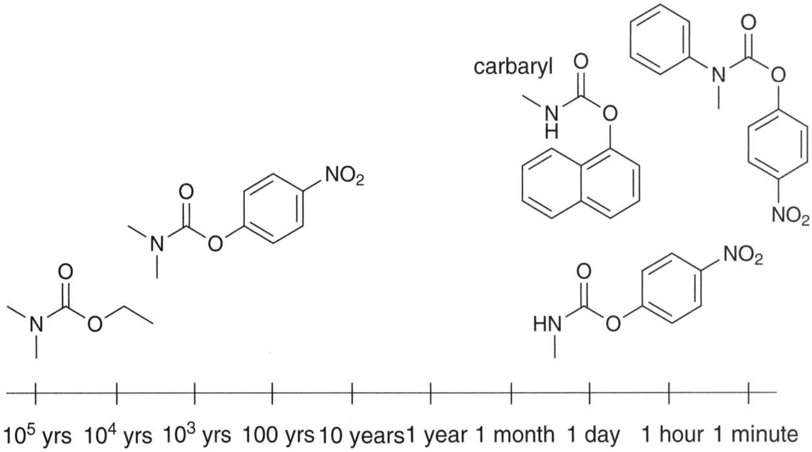

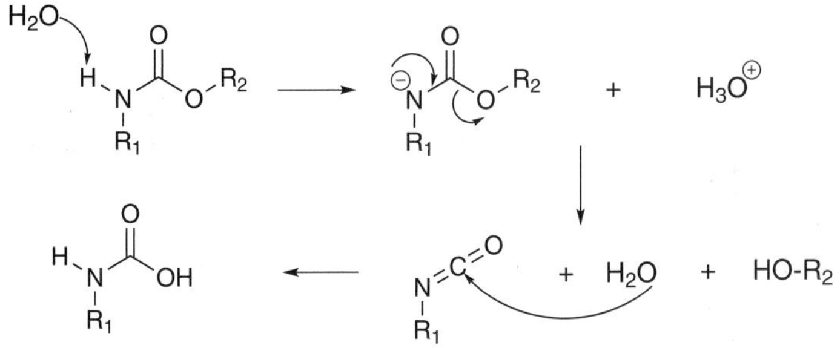

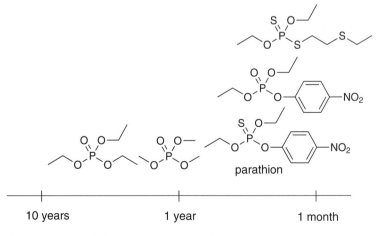

Figure 6.10 The half-lives of various carbamates in water at 25 °C and pH 7. The mechanism for hydrolysis, involving an elimination and an addition, is shown.

Figure 6.11 The half-lives of phosphoric and thiophosphoric acid esters in water at 25 °C and pH 7.

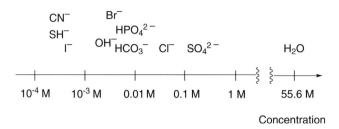

Figure 6.12 Concentrations of various inorganic nucleophiles in water that will compete with water itself for the reaction with bromomethane (see text for details).

6.2.3
Substitutions, Eliminations, and Additions

Soil and natural waters contain, besides water itself, a number of nucleophiles that may react with electrophiles (either present as contaminations or produced by organisms) in nucleophilic substitutions. Figure 6.12 compares the nucleophilicity of the most common ones with that of water, and shows which approximate concentration (in M, i.e. $mol\,L^{-1}$) of a nucleophile X^- in water (the concentration of water in water is $1000\,g\,L^{-1}$, i.e., 55.6 M) will result in a 50:50 reaction of an electrophile like bromomethane with the two nucleophiles water and X^- (yielding 50% CH_3–OH and 50% CH_3–X).

Uncontaminated fresh water normally contains low concentrations of the nucleophiles shown in Figure 6.12, leaving water itself as the major nucleophile, while seawater contains, for example, approximately 0.5 M Cl^-, and the reaction between water and various electrophiles is of minor importance. In contaminated waters the concentrations of nucleophiles (and electrophiles!) may of course be very different. As can be seen, the hydroxide ion will not react significantly in water with a pH below 10. The cyanide ion is normally present in very low concentrations and may actually be hydrolyzed itself, as discussed in Chapter 7, but hydrogen sulfide (and hence the H_2S/HS^- couple) can be formed from sulfate ion by sulfate-reducing bacteria. The reaction between H_2S/HS^- and electrophiles will produce mercaptans (RSH/RS^-), which are even better nucleophiles. The reactivity of the electrophile will, of course, also influence the reaction rate, and in Figure 6.13 the half-lives of various halogenated hydrocarbons in pure water at 25 °C are shown.

An alternative reaction for some electrophiles, for example, 1,2-dibromoethane and 1,1,1-trichloroethane (Figure 6.13), is β-elimination (see Figure 6.14). This reaction will be more important if the nucleophile has basic properties and if the electrophile is sterically hindered, and some examples are shown in Figure 6.14.

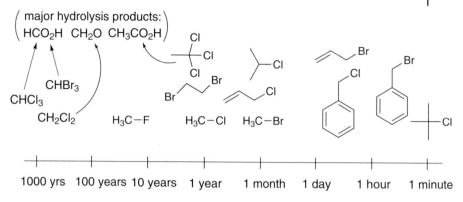

Figure 6.13 The half-lives of some halogenated hydrocarbons in water at 25 °C and pH 7.

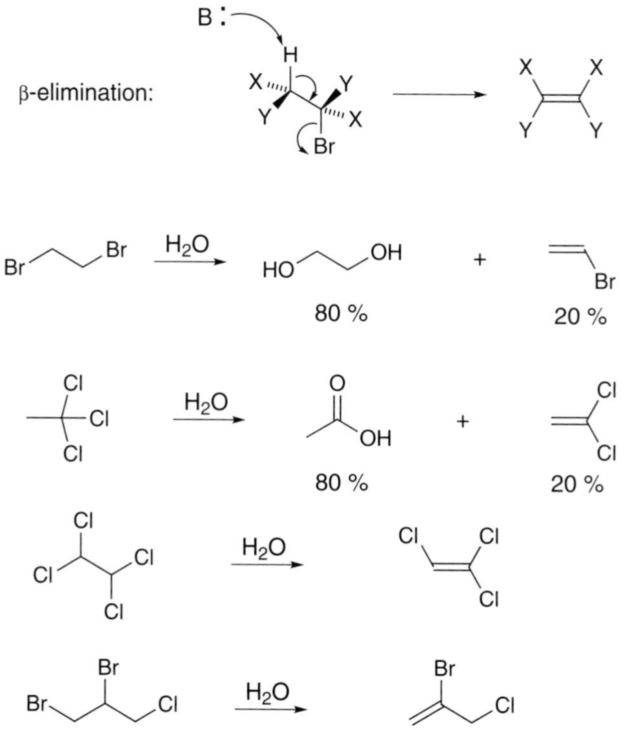

β-elimination:

Figure 6.14 Examples of eliminations of halogenated hydrocarbons in pure water.

The alkenes produced are in general more volatile, and in some cases (e.g., the pesticide 1,2-dibromo-3-chloropropane) more reactive.

Additions are the reverse of eliminations, when HX (X being for example OH, Cl, Br, SH, etc.) adds to a double bond. The combination of an elimination and an addition (in either order) will result in, for example, a substitution or hydrolysis.

6.3
Transformations Induced by Radiation

The earth is constantly irradiated from sources in space, and the various types of radiation may affect the reactivity of the chemicals that are hit. Most important is of course the electromagnetic radiation coming from the sun, flooding the atmosphere and the surface of the hydrosphere and pedosphere with incredible amounts of energy. Actually, the yearly energy consumption by human activities is equivalent to the energy reaching the earth from the sun every 20 minutes! The electromagnetic radiation from the sun has a maximum in the visible region (wavelength 400–760 nm), but substantial parts are also in the infrared region (wavelength over 760 nm, containing less energy) and in the more harmful ultraviolet region (wavelength 100–400 nm). Fortunately, most of the ultraviolet radiation is absorbed in the atmosphere (for example, by molecular oxygen, which is split to two atoms of oxygen) and never reaches the ground, and as this kind of photochemistry is important for the formation and degradation of ozone it will be further discussed in Chapter 10. Radiation is the activation factor in several important photochemical reactions, and a summary is given in Figure 6.15.

6.3.1
Photolysis

The energy of electromagnetic radiation is inversely proportional to the wavelength:

$$\text{Energy} = \text{constant}/\text{wavelength}$$

The energy of radiation with a certain wavelength can be calculated and compared to the energy of a bond between two atoms (the dissociation energy). If the wavelength of the radiation is so low that the energy exceeds that of the bond, the radiation will split the bond, and this will happen with molecular oxygen if the wavelength of the radiation is less than 243 nm. The oxygen atoms formed will react with molecular oxygen and form ozone, and we can notice this reaction (by smelling ozone) close to sources of ultraviolet radiation (e.g., ultraviolet lamps and photocopiers). While the photolysis of molecular oxygen requires radiation with relatively high energy, other molecules are more easily photolyzed. Chlorine (Cl_2), for example, will be dissociated if irradiated with radiation with a wavelength of 492 nm (blue–green light) or less. Among organic molecules, ketones, peroxy ethers and aliphatic azo compounds are easily cleaved by photolysis (Figure 6.16).

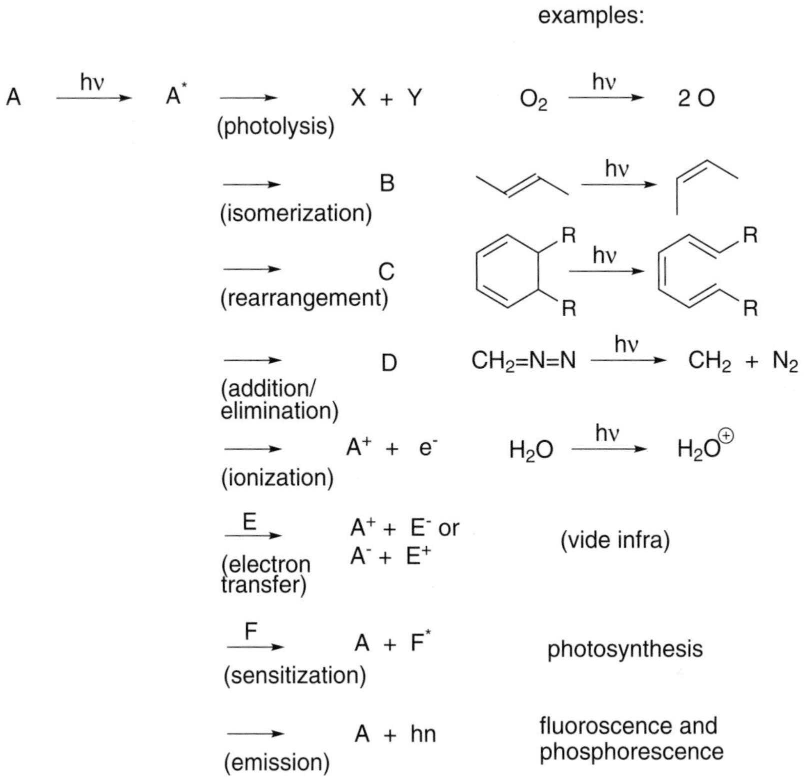

examples:

$$A \xrightarrow{h\nu} A^* \longrightarrow X + Y \qquad O_2 \xrightarrow{h\nu} 2\,O$$

(photolysis)

$$\longrightarrow B$$

(isomerization)

$$\longrightarrow C$$

(rearrangement)

$$\longrightarrow D$$

(addition/
elimination)

$$CH_2{=}N{=}N \xrightarrow{h\nu} CH_2 + N_2$$

$$\longrightarrow A^+ + e^-$$

(ionization)

$$H_2O \xrightarrow{h\nu} H_2O^{\oplus}$$

$$\xrightarrow{E} A^+ + E^- \text{ or} \atop A^- + E^+$$

(electron
transfer)

(vide infra)

$$\xrightarrow{F} A + F^*$$

(sensitization)

photosynthesis

$$\longrightarrow A + h\nu$$

(emission)

fluoroscence and
phosphorescence

Figure 6.15 Important photochemical reactions.

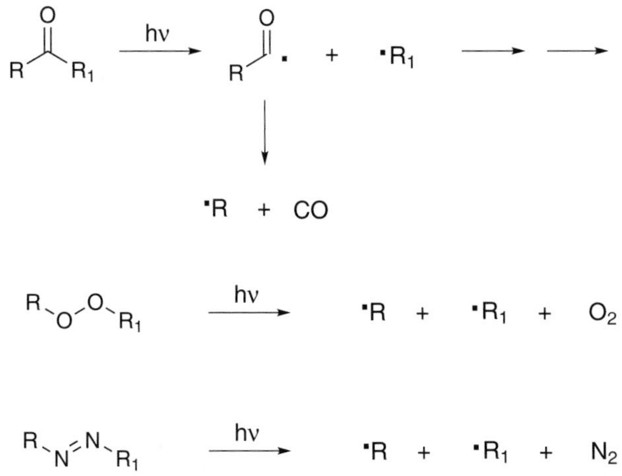

Figure 6.16 Examples of photolyses.

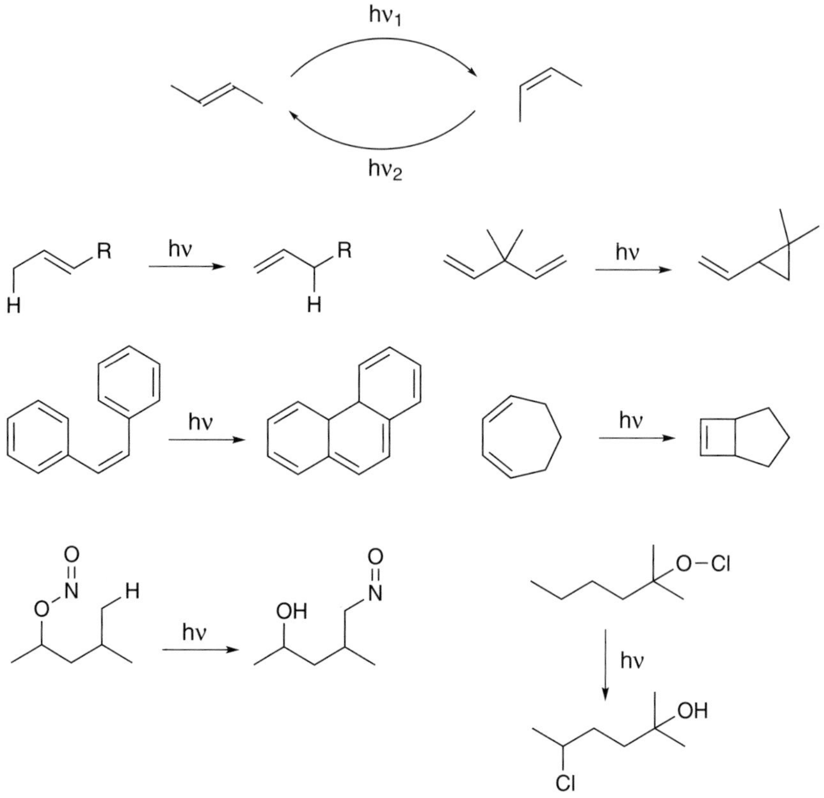

Figure 6.17 Some examples of rearrangements.

6.3.2
Isomerization/Rearrangement

The isomerization of double bonds is a common photochemical reaction that has great practical importance in that the direction of the isomerization can be decided by the frequency of the light used (see Figure 6.17). Many different reactions of this type are known, and a few examples are given in Figure 6.17.

6.3.3
Additions/Eliminations

Figure 6.15 shows the formation of a carbene from a diazo compound by the elimination of N_2 induced by light, this being an example of an important and useful type of chemical reaction. The carbenes can be said to be diradicals in which the two unpaired electrons are paired, so that there are no unpaired electrons but also no complete octet around the carbon. Carbenes are therefore highly

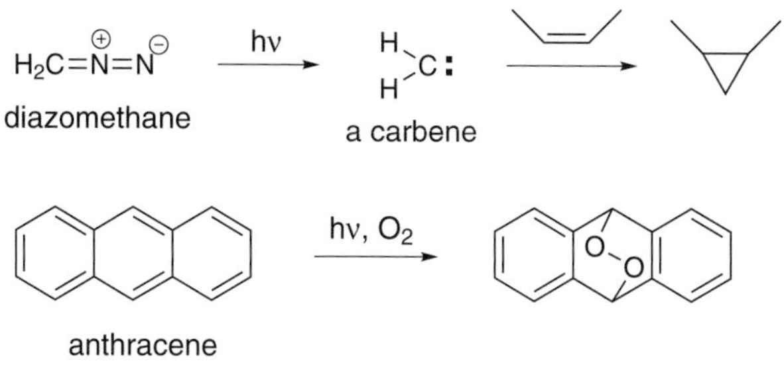

Figure 6.18 Examples of light-induced additions/eliminations.

electrophilic and will rapidly react with compounds that can provide an electron pair, for example, olefins, and a product from the reaction between a carbene and a olefin is a cyclopropane derivative. An example of an addition is shown in Figure 6.18, where a conjugated diene reacts with oxygen under the influence of light to form an internal peroxide.

6.3.4
Photoionization

We have already mentioned the effect of ionizing radiation, which can be electromagnetic radiation with a very short wavelength (approximately 100 nm to ionize molecular oxygen and water). However, it is obvious that ionizing radiation will be hazardous, and we do not need to enter more deeply into that here.

6.3.5
Electron Transfer

Simple aldehydes and ketones will (at approximately 280–320 nm) be converted into an excited state which has a diradical character (see Figure 6.19). Besides adding to carbon–carbon double bonds forming an oxetane, it can also take part in a photooxidation reaction with a hydrogen atom donor (e.g., cyclohexane in Figure 6.19) or accept an electron from another compound with a nitrogen (or sulfur) atom. The latter reaction, an electron transfer, will result in a radical anion/radical cation pair, which can subsequently react to form other products.

6.3.6
Sensitization

Photosensitization is a process in which one excited molecule returns to its ground state by exciting another, which subsequently takes part in a chemical

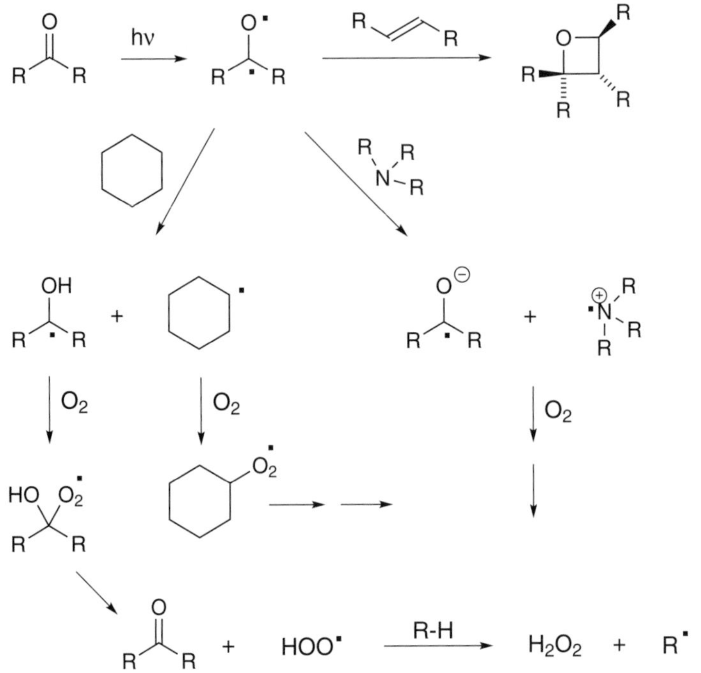

Figure 6.19 Photochemically excited aldehydes and ketones may induce an array of reactions, including electron transfers.

reaction. In the photosynthesis of plants, the energy of sunlight is converted into chemical energy via the excited state of chlorophyll (which is consequently a photosensitizer). A common reaction is that a molecule of oxygen in its ground state is excited by a photosensitizer to form singlet oxygen. The singlet state is for most molecules the ground state, but oxygen is a rare exception, singlet oxygen being an excited state (in which the electron configuration actually can be regarded as normal with an oxygen–oxygen double bond and two unshared electron pairs on each oxygen atom). This is highly reactive, the half-life of singlet oxygen in water being approximately $2\,\mu s$. Ground-state oxygen can be excited by a number of photosensitizers (e.g., dyes like eosin), and in its excited form will react with, for example, dienes in Diels–Alder reactions or olefins to form hydroperoxides (see Figure 6.20).

Note that the allylic hydroperoxidation with singlet oxygen proceeds via a different mechanism from that of autoxidation with ground state oxygen, something that is evident when the stereochemical outcome of the reactions is analyzed; a probable mechanism is shown in Figure 6.20. A carbon–carbon double bond may also be attacked if it has an electron-donating atom (e.g., nitrogen) attached; the

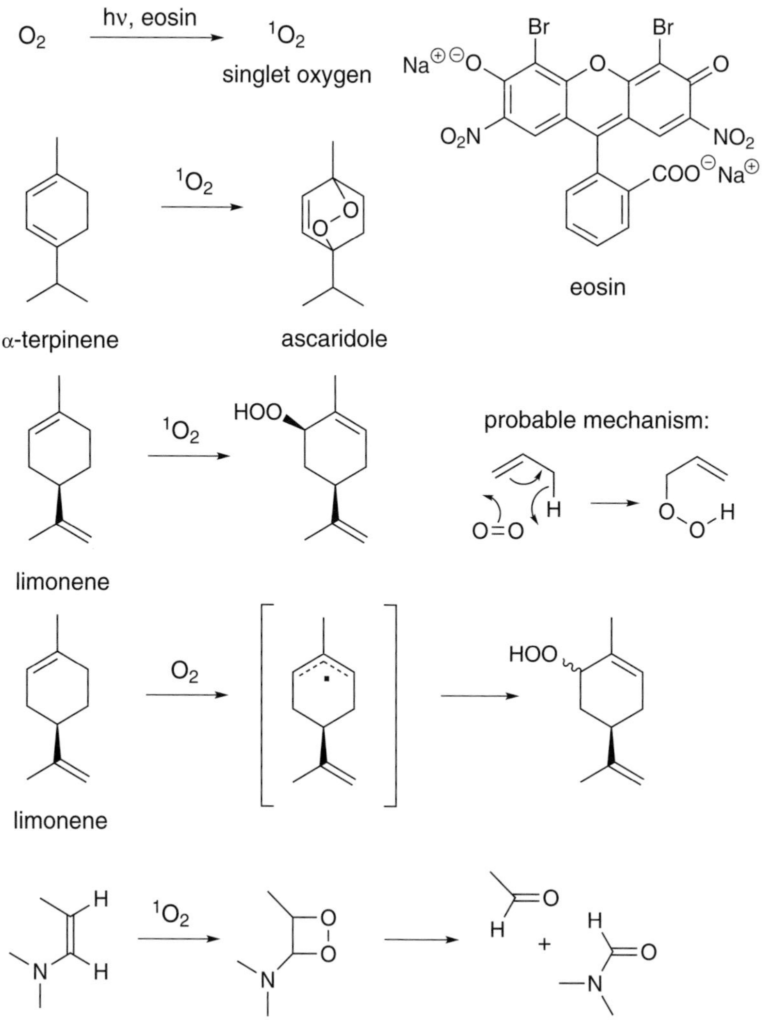

Figure 6.20 Reactions with singlet oxygen produced by photosensitization.

product, a dioxetane, is unstable and generates two carbonyl compounds spontaneously. Singlet oxygen can be formed in the body if photosensitizers are present, and we will return to this topic when we discuss the effects of the psoralenes (Section 7.3.6).

7
Toxic Effects of Chemicals

Chemicals exert systemic toxic effects essentially by disturbing the normal course of biochemical reactions, and in principle it should be possible to describe the mechanisms on a molecular level for any systemic toxicity. However, we still do not have a complete understanding of how most biochemical processes function, and it is actually only in a few cases that we fully understand what is happening. It should not be forgotten that a toxic effect observed with a chemical is often the result of several toxicities, specific and nonspecific, and that it may be difficult to determine their relationships. In this chapter we will discuss the molecular mechanisms underlying some common toxic effects of chemicals, and the effects have been divided into 'general toxic effects' and 'organ-specific toxic effects'. However, it is obvious that such a division is not straightforward. Although general toxic effects will influence all tissues in the body, one organ is always more sensitive than another (for various reasons) and will show symptoms of poisoning first. On the other hand, what we see as organ-specific effects are often accompanied by effects on other organs, and general toxic effects will in some cases reappear in an organ-specific way. Remember also that the general intention is to link chemical properties and functionalities with toxic effects, not to give a comprehensive toxicological description of toxic effects. To begin with, it is necessary to define the toxicological concepts that we need to use in the discussion.

7.1
Toxicological Concepts

A general definition of the term toxicology is 'the study of adverse effects of chemicals on living organisms', and the organism in focus is of course the human body. Toxicologists around the world are constantly generating new knowledge about chemicals, making users more aware of the associated potential hazards and thereby enabling them to reduce the risk to themselves as well as to the environment. The 'adverse effects' studied can be anything from irritation to sudden death, and the chemicals can be anything from minerals to proteins, making toxicology a truly interdisciplinary science. In this chapter we will not go into toxicol-

Chemistry, Health, and Environment. Olov Sterner
© 2010 WILEY-VCH Verlag GmbH & Co. KGaA, Weinheim
ISBN: 978-3-527-32582-5

ogy in any depth, but will just define some concepts that we will need to use in later chapters and will try to give an idea of how difficult it is to determine how hazardous any given chemical is.

7.1.1
The Toxicity of Different Chemicals

In Table 7.1 the toxicity of 13 compounds is compared, toxicity being expressed as the LD_{50} value, which is a measure of the short-term fatal toxicity and indicates the dose, in mg per kg body weight of the test organism (rats in this case), that would kill approximately 50% of the individuals exposed. The compounds chosen have completely different chemical structures and properties; some are familiar and present in most homes while others are classical poisons.

Ascorbic acid is a compound that we cannot live without, a daily intake of approximately 50 mg being recommended, although it is commonly used in prophylaxis against common colds and other diseases (we will return to ascorbic acid later in the chapter). 2-Propanol is a very common solvent present in many household products (e.g., windscreen washer liquid). N-Acetyl-4-aminophenol and acetylsalicylic acid are common medicaments used as antipyretics, to treat inflammations, and to alleviate headaches, and show low acute toxicity if not abused. 2,4,5-Trichlorophenoxyacetic acid was a popular herbicide (weed-killer) in the 1960s and 70s until it (and similar phenoxycarboxylic acids) came under suspicion for causing various illnesses. Acrylamide is a high volume industrial chemical, used for various purposes. 1,1,1-Trichloro-2,2-bis(4-chlorophenyl)ethane, or DDT, is the famous insecticide mentioned in Chapter 1 that has been banned in most parts of the world. Nicotine is the stimulating principle of tobacco, but also quite toxic. Arsenic oxide and mercury chloride are examples of toxic inorganic chemi-

Table 7.1 Approximate LD_{50} values (rats, oral intake) for 13 chemicals.

Chemical	LD50 (mg/kg)
Ascorbic acid (vitamin C)	12 000
2-Propanol (isopropanol)	6 000
N-Acetyl-4-aminophenol (acetaminophen)	2 400
Acetylsalicylic acid (aspirin)	1 500
2,4,5-Trichlorophenoxyacetic acid [(2,4,5)-T]	300
Acrylamide	120
1,1,1-Trichloro-2,2-bis(4-chlorophenyl)ethane (DDT)	110
Nicotine	50
Arsenic(III) oxide	40
Aflatoxin B1	5
Mercury(II) chloride	1.4
Strychnine	0.24
2,3,7,8-Tetrachlorodibenzodioxin (TCDD)	0.022

cals that are widespread in the environment in large quantities. Aflatoxin B₁ is a mycotoxin formed by a mold that may attack food, and is a potent carcinogen. Strychnine is an alkaloid isolated from a plant, and a classical poison much featured in detective stories. Finally, 2,3,7,8-tetrachlorodibenzodioxin (also known simply as dioxin or TCDD) is an extremely toxic chemical formed as a by-product during various chemical procedures.

Even though Table 7.1 does not include the least toxic (e.g., water with an approximate LD_{50} value of 500000 mg/kg) or the most toxic compounds (e.g., the botulin toxins which are neurotoxic proteins capable of killing a grown man at doses of around 1 μg), the differences are still striking, the acute toxicity of TCDD being approximately 500000 times higher than that of ascorbic acid. As can be seen in Figure 7.1, the chemical structures of the 13 compounds differ greatly, giving different chemical properties (e.g., solubility and reactivity) and thereby

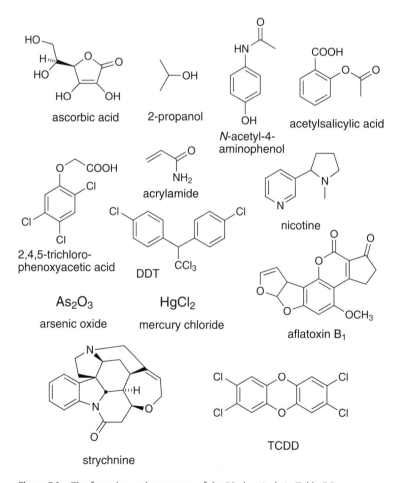

Figure 7.1 The formulae and structures of the 13 chemicals in Table 7.1.

different abilities to interfere with the normal chemical processes that keep an organism alive. More details of how these interferences create toxic effects on the molecular level for most of these 13 compounds will be discussed later. However, as well as such intrinsic properties of a chemical, there are a number of other factors related to exposure that will modulate the toxic effect of a chemical, and these, somewhat surprisingly, may vary substantially. Before we explore these factors, we need to define a number of toxicological concepts.

7.1.2
Toxicological Concepts

7.1.2.1 Acute and Chronic Effects

Acute effects of a chemical are observed immediately or shortly after an acute exposure, that is to say the victim has received the dose on one single occasion or as multiple doses within 24 h. When volatile chemicals are tested, animals are normally caused to inhale air containing a certain concentration of the compound for 4 h. Chronic effects are the result of chronic exposure, which goes on continuously for a long time (at least 3 months). Between these two, when the exposure goes on for more than 24 h but less than three months, the terms subacute (multiple dose, short duration, less than 1 month) and subchronic (multiple dose, long duration, 1–3 months) are used. Note that a toxic compound can give both acute and chronic effects by different mechanisms. An example is benzene, which after a single exposure will give an acute effect on the central nervous system (CNS), as most organic solvents do, while it may give leukemia after chronic exposure. Most acute toxic effects are immediate, developing rapidly after an exposure, but some are not perceived immediately and are referred to as delayed. An example is a tumor, which in principle can be caused by an acute exposure but may have a latency period of many years.

Chronic effects can be caused, for example, if a compound is taken up faster than it is excreted, resulting in an accumulation of the compound in the body that eventually will produce the effect. Heavy metals, for example, mercury and cadmium, are not easily excreted, and very small daily intakes can lead to accumulated amounts in the liver and kidneys that can affect their normal function. However, several other mechanisms for chronic effects also exist, and it can be a question of a decreasing ability to withstand the effects of a compound (e.g., as a result of poisoning, aging, or sickness).

7.1.2.2 Reversible and Irreversible Effects

Reversible effects will disappear if the exposure stops. For example, inhalation of solvent vapors may result in dizziness which will pass away if the solvent is removed and the inhaled amounts are allowed to be eliminated. Irreversible effects, on the other hand, never disappear, and the nervous tissues are especially sensitive as they not are regenerated if damaged. Carcinogenic and teratogenic effects are other examples of irreversible effects. Note that one compound may give both reversible and irreversible effects at the same time.

7.1.2.3 Local and Systemic Effects – Target Organs

Local effects of a chemical are observed on the part of the body that first came in contact with it. Examples of compounds that give local effects are strong acids and bases that rapidly burn the skin, for example, and other highly reactive chemicals. Systemic effects on the other hand require that the chemical is taken up by the body, distributed, and eventually reaches the organ (the target organ for that chemical) that is affected. The target organ may be injured because, for various reasons, it receives most of the compound or because it is more efficiently converted to its final toxic form in this organ. However, it may also be especially sensitive, and then the amounts are not critical. An example is DDT (see Figure 7.1), which, because of its lipophilicity, is distributed mainly to the adipose tissue (the body fat) although it is a nerve poison. A primary injury on one organ may also lead to a situation where other organs are affected by a secondary mechanism. For example, if the kidneys are injured and malfunctioning, this will change the liquid balance in the body in a way that puts extra strain on the heart and lungs, and the effect first noticed may come from this. The CNS is the most commonly affected target organ, followed by the circulatory system, the liver (where most of the metabolism of exogenous compounds takes place), the kidneys, the lungs, and the skin. The muscles and the bones are seldom affected by chemicals, though this can occasionally happen. Metabolic activation is often required for a chemical to be toxic, indicating that it is the metabolized form that is toxic, while metabolic inactivation indicates that the compound is rendered less toxic by the metabolic conversions (see Chapter 5).

7.1.2.4 Independent, Additive, Synergistic, and Antagonistic Effects

Normally, at workplaces where chemicals are handled, several chemicals are used at the same time, and in practice people are exposed to mixtures. However, the total effect from a mixture of chemicals, though important, is difficult to estimate. Although almost nothing is known about this, we can say that two compounds given simultaneously act together to give a toxic effect in one of the following ways:

1) Independently (independent effect), if the compounds exert their own toxic effects (both target organ and potency) independently of each other.

2) Additively (additive effect), if the compounds each give the same type of effect and together give an effect that is the sum of the individual effects. Additive effects are commonly observed for compounds with similar chemical structure, an example being toluene and *p*-xylene. Both compounds affect the CNS, as most organic solvents do, and the effect of 100 mg of any of them is similar to the effect of a 100 mg mixture containing 50 mg of each.

3) Synergistically (synergistic effect), if the two compounds give a combined effect that is stronger than the additive effect. Ethanol and carbon tetrachloride both produce liver damage, although with different potencies, but together the effect is much stronger. This is caused by the induction by ethanol of the enzymes converting carbon tetrachloride to liver-damaging metabolites. The influence of one compound on the metabolism of another is a very common mechanism for synergistic effects. As tobacco smoke contains

several compounds that have this ability, smokers may be especially sensitive to synergistic effects.

4) Potentiating (potentiating effect), if one of the compounds by itself does not give a toxic effect but will enhance the effect of another. 2-Propanol is not known to be toxic to the liver, but just like ethanol (and by the same mechanism) it will enhance the liver toxicity of carbon tetrachloride.

5) Antagonistically (antagonistic effect), when one compound opposes the effects of another. An antidote taken to counteract the effects of a poison is an example of antagonistic effect, which is very common in the pharmaceutical sciences. An antagonistic effect may be functional, when the two compounds give neutralizing effects (e.g., one compound raises the blood pressure while another lowers it) resulting in no observable net effect. It can also be chemical, when the two compounds react with each other and are transformed to something less toxic. Chemical antagonism is used to treat poisonings by metal ions (for example arsenic and mercury) by chelating and masking them with 2,3-dimercapto-1-propanol (see also Chapter 10). Dispositional antagonism occurs when one compound decreases the concentration and/or duration of another in the target organ, for instance by influencing its uptake, metabolism, or excretion. The parallel case leading to a synergistic effect has already has been mentioned, and further examples were given in Chapter 4. Finally, receptor antagonism is observed when two compounds bind to the same receptor and the response is weaker when the effect of the two together are compared to the sum of their individual effects. A receptor antagonist (blocker) is a compound that binds to a receptor without stimulating it but blocks it against other active compounds, and we shall come back to this mechanism later in this chapter.

7.1.2.5 Tolerance

The development of tolerance to the toxic effects of a chemical means that previous exposure to the chemical has made the individual more resistant to it. Tolerance is more rare than generally thought, and definitely not to be counted on. It can in principle develop as a result of the induction (increased production) of detoxifying proteins or of the depletion of enzymes that converts the compound to a more toxic form. Cadmium will induce the production of metallothioneins (discussed in Chapter 5), proteins which complex cadmium and take it out of circulation, while the reactive metabolites of carbon tetrachloride will destroy the enzymes that activate it to a reactive form, thereby preventing it from being metabolized so efficiently.

7.1.3
Factors Modulating Toxicity

7.1.3.1 The Purity and Physical State of the Toxicant

The presence of impurities in a chemical is in principle not relevant if we want to discuss the toxicity of only one compound. However, completely pure chemicals do not exist. As small amounts of a very toxic impurity will change the outcome of a toxicity test significantly, the purity of a compound being evaluated is an important and relevant property that has to be assessed. A number of inconsistent examples in the toxicological literature, for example, the carcinogenic effects of various aromatic amines (see Chapter 5), have later been explained by the fact that compounds of different purity had been used. The physical state of a compound will influence the principal route of exposure, which in turn will determine how efficiently the compound is taken up by a human (discussed in Chapter 4). In workplaces, the most important route for uptake of chemicals into the body is via the lungs, and any compound that can exist as a vapor or an aerosol will normally be more dangerous in this state. Obvious examples are volatile compounds, but also nonvolatile solids that can exist as either heavy crystals, a granulate, or finely ground like flour. The latter will raise dust during handling and be more easily inhaled than the same compound in a nondusty solid form. If swallowed, solids that are not easily dissolved in the intestines will be better absorbed if they are finely ground (having a large surface area) than they would be in a granulated form. Arsenic oxide is an example, and the LD_{50} value given in Table 7.1 actually depends a lot on the form given orally to the rats. Some compounds will be safer and easier to handle as solutions in a suitable solvent, although, as was also discussed in Chapter 4, some solvents will facilitate the uptake of a solute through the skin and thereby enhance its potential ability to give rise to toxic effects.

7.1.3.2 The Exposure

The route of exposure is an important modulator, and the same amount of a toxicant administered to the same victim by different routes may give completely different effects. In workplaces the principle routes are via the lungs (by inhalation), via the skin, and via the intestines (after accidental swallowing), but when the toxicity of chemicals is assessed during animal experiments other routes of exposure are also used. As much of the toxicological data available in the literature are based on animal experiments, it is important to know about the techniques used, whatever one's feelings toward animal experiments may be.

ivn   intravenous administration means that the compound to be tested is injected into a vein of the animal. It enters directly into the circulation and will rapidly be spread throughout the whole body.

ihl   inhalation of the chemical to be tested, which is volatile.

ipr   intraperitoneal administration is when a compound is injected into the peritoneal cavity (into the belly but beside the intestines and not into a blood vessel). From there, the compound is taken up by the blood and distributed to the rest of the body. Although administration by ivn and ipr does not have any apparent relevance to the exposure encountered in workplaces and in the environment, these routes have been used frequently, and there are a large amounts of ivn and ipr data in the literature.

scu   subcutaneous administration means that the compound was injected into the skin, between epidermis and dermis (Section 4.2.1).

orl   oral administration is used when the chemical to be tested is given via the mouth of the animal, together with food or simply forced down with a syringe. The compound will pass through the stomach and will eventually end up in the intestines where it can be taken up by the blood and delivered into the body.

d (skn) dermal (skin) application administration is used if the uptake through the skin is to be investigated. The compound to be tested is simply applied to the outside of the skin and has to penetrate through the epidermis down into the dermis where there are blood vessels. How fast this takes place depends greatly on the structure of the chemical (discussed in Chapter 4).

These are the most common routes for the administration of chemicals tested in animal experiments, given in order of efficacy. The natural ways to absorb chemicals (natural = without needles) are by inhaling air containing pollutants, by swallowing them, or by spilling them on the skin, and ihl, or, and d correspond to the routes that we see in workplaces. Which one is actually used in an experiment depends on what kind of information is sought, and also on practical considerations. If, for example, the same amount of the same compound was given to three animals by ipr, ivn, and or administration, the blood concentrations in the three animals as a function of time would look something like the graphs shown in Figure 7.2.

If a certain blood concentration is required to trigger a toxic response, this may be reached by using ivn administration but not by the two other routes. In general, ivn administration is the most efficient, followed by inh, ipr, scu, or, and d. In

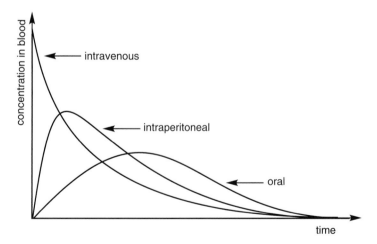

Figure 7.2 Different routes for the administration will produce different effects.

addition to affecting the rate of absorption, the route of exposure also influences the possibilities of the victim to convert the compound by the metabolic processes (discussed in Chapter 5).

7.1.3.3 The Victim

The major reason for choosing animals like rats and mice for assessing the toxicity of chemicals is that they are mammals like humans and results from animal experiments should in principle be transferable to us. This is true in many cases, and chemicals that are toxic to rats are probably also toxic to man and vice versa. However, there are numerous exceptions, and we need to be aware of the reasons for these.

Species-Related Differences. Even though most mammalian species are very similar in function and generally believed to be approximately equally sensitive to toxic chemicals, there are nevertheless a number of cases known where this is not true. A well-known example of how man is more sensitive than other animals to a chemical is that of methanol, or wood alcohol. Approximately a teacupful of methanol may be fatal to a man, methanol being much more toxic than the apparently very similar compound ethanol (this will be further discussed later in this chapter). However, in mice and rabbits, for example, there is no great difference in the toxicity of the two compounds, this being due to the fact that these animals are able to convert both compounds to carbon dioxide and water, while we can only do this with ethanol. Differences in the metabolism of exogenous compounds are the major reason for species-related differences in toxicity, although there are others. Another example of species-related differences is shown in Table 7.2.

TCDD (dioxin) is an environmental toxicant which has been of great concern during the last decade or so. Its presence as a by-product in various circumstances will be discussed in Chapter 10. TCDD is extremely toxic to some species (see Table 7.2), while others are less sensitive. In addition, it is carcinogenic. For the layman there may not be any striking differences between a guinea pig and a hamster, but the former is nevertheless more than 1000 times more sensitive to

Table 7.2 The toxicity of TCDD toward various mammalian species.

Species	LD50 (mg/kg, oral intake)
Guinea pig	0.0005
Dog	0.001
Monkey	0.002
Rat (male)	0.022
Rat (female)	0.045
Mouse	0.11
Frog	1
Hamster	1.2

TCDD. It is believed that also TCDD is metabolized differently in various species, but it is not yet known in detail how TCDD is toxic at the molecular level. Neither is it known in what position the human species should be placed in Table 7.2, because no fatal poisoning with TCDD has been reported for man, but we probably do not belong among the most sensitive species. Species-related differences are more a rule than an exception, although differences as large as those with TCDD are unusual.

Sex-Related Differences. Small differences are normally observed between males and females of one species. A major reason is that the need for sex hormones differs and males and females therefore differ slightly in their secondary metabolism.

Age-Related Differences. In general, infants and old people are more sensitive to toxic chemicals. In infants, some metabolic systems as well as biological barriers (see Chapter 4) are poorly developed, and old people may be more sensitive as, for instance, their immune systems are becoming less efficient.

Individual Differences. Individuals of the same species, same sex, and same age vary in health and nutritional status, and this will modulate the sensitivity to toxic chemicals. In addition, as individuals we all look slightly different and we all have our own variants of the human metabolic system with small but significant differences in, for example, the balances between competing metabolic pathways. As a result, a larger group of individuals of the same species, sex, and age exposed to a toxic chemical by the same route will not be affected in the same way by exactly the same dose. The individual variations become apparent if the fraction of the exposed group afflicted is plotted against the dose, as in Figure 7.3. As the dose increases from zero a few sensitive individuals are affected long before the

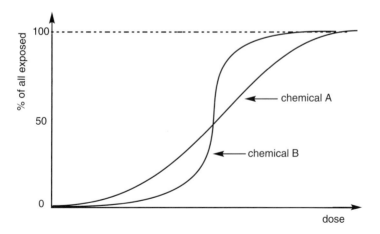

Figure 7.3 The effects of different doses of two chemicals on a group of individuals.

majority, while other individuals withstand surprisingly high doses. Such graphs are normally S-shaped, although the shape of the S can vary a lot.

However, for some effects, for example, chemical allergy, or in cases of chemical idiosyncrasy (when a hereditary abnormality increases the sensitivity for a chemical), it may look completely different.

7.1.4
Acute Toxicity

7.1.4.1 Measures for Acute Toxicity

From such dose–effect relationships (see Figure 7.3), one can extract values for the dose that will give rise to a certain effect in X % of the exposed individuals (same species, sex, age, and exposure route). The most common approach is to enter the diagram at 50% affected and read the dose for that, but any percentage of the exposed group can of course be chosen. If the effect studied is death, then one can read the Lethal Dose on the y-axis, and the value is called the LD_X-value (X being the number in %). The median lethal dose LD_{50} is therefore the dose that will kill 50% of the individuals in the exposed group, while LD_{10} indicates the dose that will kill the 10% most sensitive. As species and exposure route are the two most important modulators of lethal doses, it is normal to state at least these two facts with the value, for example, $LD_{50} = 25$ mg/kg (rat). If it is interesting to assay volatile compounds and exposure via the lungs, the toxicity is expressed instead as LC values (Lethal Concentration) measured in mg/m^3 in the inhaled air. For natural reasons, few (although nevertheless shockingly many!) LD_{50} or LC_{50} values are known for man, most having been obtained with rats and mice. Besides LD and LC values, one also encounters TD and TC values for acute toxicity in the literature. T stands for Toxic, and TD_X (TC_X) values give the dose (concentration) of a compound that will give a toxic effect (that has to be defined) in X % of the exposed group. In addition, for many compounds an LD_{lo} (or LD_{low}) value is reported in the literature, and such values state the lowest known dose that has caused death in an individual. LD_{lo} values are often reported for man, as a result of a fatal accident when the dose involved could be estimated, and are therefore important.

Measuring acute toxicity with LD_{50} values obtained from animal experiments (involving something in the order of 100 animals) has been criticized as unethical and cruel, and this is certainly the case. In addition, animal experiments are always expensive and the information obtained is actually rather limited unless a detailed (and very expensive) *post mortem* investigation is carried out with the dead animals. It is therefore possible to replace animal experiments for LD_{50} determinations with an array of simpler, more ethical, and less expensive *in vitro* assays based on mammalian cells, at least if a new standard for preliminary testing of acute toxicity was to be established. However, animal experiments for testing more specific toxic effects (e.g., tumors in certain organs) are not easily replaced, and will continue to be used for some time. Whole animals, with their metabolic and excretion systems intact, do provide more information about uptake, distribution, metabolism, excretion, recovery, and so forth, which so far is difficult to simulate in *in vitro* studies. In spite of its various disadvantages, the LD_{50} value is still used when the acute toxicities of chemicals are compared, simply because there are so many LD_{50} data in the literature.

7.1.4.2 Specific and Nonspecific Toxicity

Toxic effects can be highly specific because of the interaction of the toxicant with only certain proteins or certain stretches of the DNA helix. Examples are compounds that bind to receptors in, for instance, the CNS, or that inhibit enzymes, and such compounds are in many cases very toxic and give dramatic effects.

The recently discovered antitumor antibiotic calicheamicin γ_1^{Br} (see Figure 7.4) is also an example of a compound with a specific toxic effect. It will become associated with DNA through hydrogen bonding between the carbohydrate-based tail

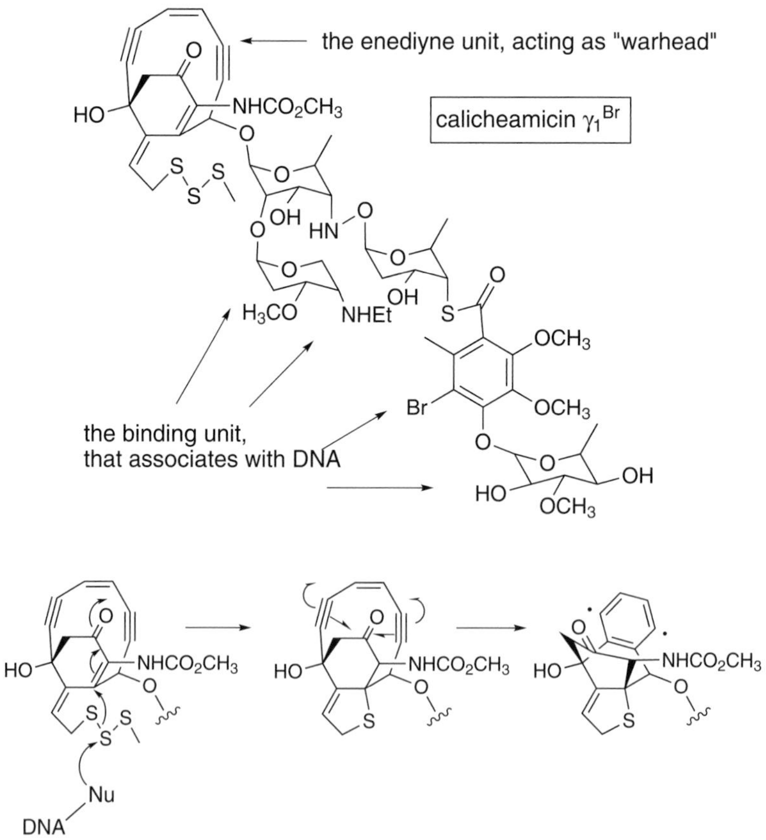

Figure 7.4 The antitumor antibiotic calicheamicin γ_1^{Br} and its interaction with DNA.

and the minor groove of DNA in a way that positions the sulfur bridge so that it can be attacked by a nucleophilic group in DNA (see Figure 7.4). This results in a conformational change in the enediyne ring that promotes its closure to an aromatic system (relieving the substantial ring strain in the enediyne ring). This change leaves two unpaired electrons; that is, the product formed is a diradical which is highly reactive and will attack both strands of the DNA polymer and cleave it. This is a severe injury that in most cases will kill the cell, although it may lead to chromosome mutations (see Chapter 8).

Nonspecific toxic effects, on the other hand, do not rely on the combination between toxic chemicals and specific targets; instead, the effects are caused by a reversible lipophilic association with the membranes of the cell or cell organelles. Compounds giving nonspecific effects are therefore less dependent on their reactivity and more on their lipophilicity, although the same compound can of course give both specific and nonspecific effects. A typical nonspecific effect is the dizziness caused by organic solvents because they are gathered in the cell membranes in neurones in the CNS and disturb the normal function of the receptors in the nerve cell membranes. It has been suggested that there are receptors that mediate this effect, but so far no such receptors have been identified. Instead it is believed that the lipophilic molecules are collected inside the membrane, which will expand and become more fluid. This will disturb the proteins associated with the membrane and even change the conformation and function of integral proteins. Besides effects on the nervous system, the cardiac function may also be affected, and despite the reversible nature of the interactions even nonspecific toxicity may be fatal.

In general, there is a correlation between lipophilicity (measured, for example, as the log P value) and nonspecific toxicity (see Figure 7.5).

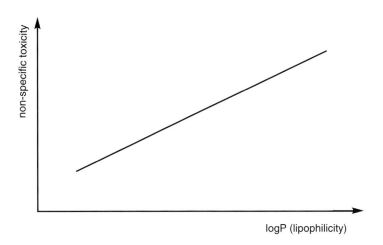

Figure 7.5 The general correlation between lipophilicity and nonspecific toxicity.

However, extremely poorly water soluble compounds will have problems reaching their targets (e.g., CNS), at least within a limited time period, and will not show the acute toxicity expected by their lipophilicity.

7.1.4.3 Selective and Nonselective Toxicity

A chemical that shows selective toxicity is more toxic to one organism than to another, or to one organ in an organism than to another organ in the same organism. Nonselective chemicals are consequently equitoxic to both organisms (organs, cells). The mechanism for selective toxicity is either that the compound is accumulated in one organism (organ, cell) due to differences in absorption, distribution, metabolism and/or excretion, or that it interferes specifically with cellular or biochemical components that are only present or of importance in one organism (organ). Selective toxicity is, of course, of great importance when antibiotics, pesticides, and herbicides are developed. Examples are the penicillins, which kill bacteria without affecting the cells of the person infected because they block the construction of the bacterial cell wall (which mammalian cells do not have). DDT is an efficient insecticide with relatively low toxicity for humans that handle it because it is poorly absorbed through the skin, but it is highly toxic if injected. Norbormide (see Figure 7.6) has found use as a rodenticide because of its selective toxicity towards rats. Its LD_{50} value (oral intake) is 3.8 mg/kg for rats, while the corresponding value is 2200 mg/kg for mice and 1000 mg/kg for monkeys, dogs, cats, and rabbits. Norbormide acts irreversibly on the smooth muscle of peripheral blood vessels, leading to contraction and subsequent necrosis, but the molecular target is not known, and the reason for the exceptional sensitivity of rats is not understood. Glyphosate (see Figure 7.6) is both selective, killing plants but not affecting microorganisms or mammals (LD_{50} for rats is 4900 mg/kg orally), and nonselective, killing all plants except conifers. It apparently interferes with the biosynthesis of aromatic amino acids, although the exact mechanism for its action has not yet been elucidated, and we will return to this product in Chapter 9.

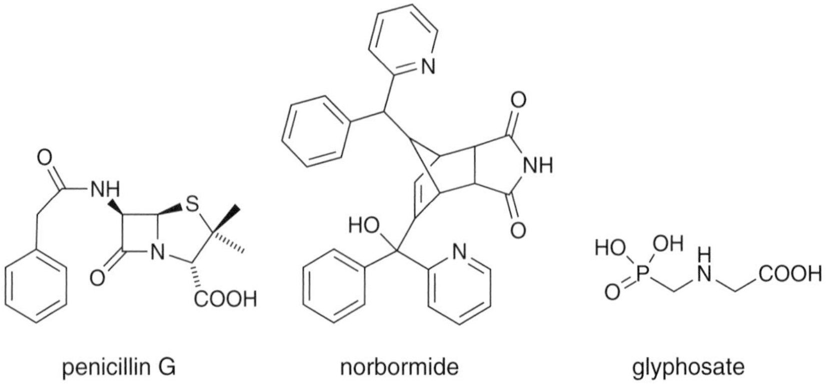

Figure 7.6 Examples of chemicals exhibiting selective toxicity.

7.1.5
Interpretation of Results from Toxicological Testing

In principle, all toxicity testing is carried out to obtain knowledge that can be used to protect man from exposure to hazardous compounds, or to assess the risks connected with exposure. However, the assessment of any hazard or risk is difficult; there is no sharp dividing line between hazardous and nonhazardous chemicals, and a judgment will not be based on objective information alone (see also Chapter 1). For example, many industrial chemicals are of great economical importance, and such factors are normally considered when their use is being regulated. The adverse effects of cigarettes and alcoholic beverages are well documented, but society has chosen not to ban such products because they are considered, at least by many, to be means of enjoyment, and the risks are (again by many but not all) acceptable. Also, the risks involved with handling a hazardous chemical are difficult to estimate, for example, which level of exposure should be regarded as acceptable in a workplace. Hazard (qualitative) and risk (quantitative) assessments of chemicals often to some degree have an element of economical/political decisions. Humans are very seldom the test organisms during toxicological testing, although the results should be applicable to them. Therefore, the results have to be translated from whatever test organism was used, and this is of course not trivial. We have already discussed how different species and even individuals may be differently sensitive to toxic effects, and in cases where results from assays other than animal experiments are to be evaluated the estimations will be even more precarious. This is normally accounted for by a safety margin if, for example, an exposure limit should be determined from data from an animal experiment. If a volatile chemical is considered to be harmless to rats breathing air containing $1000 \, mg/m^3$, a human would be considered to be able to resist 10, 100 or 1000 times less (depending on the character of the toxic effect – 10 times for nonselective toxicity and 1000 times for, for example, carcinogenicity). However, the safety margins are no guarantee that humans will not be affected, something that we have learned the hard way several times over the years.

The problems with translating data from animal experiments to humans are also linked to the way the experiments are normally designed. It is important to obtain information about two things: whether the chemical assayed possesses a certain toxicity; and how potent this is. To obtain the necessary correlation between dose and response it is necessary to test several doses, one being $0 \, mg/kg$ (the so-called negative control). For economic and ethical reasons, as few animals and as few doses as possible are used, but in order to obtain unambiguous effects the dose should be as high as possible. The situation may look like the graph in Figure 7.7, in which the incidence of tumors in animals exposed to 0, 50, and $100 \, mg/kg/day$ of a chemical (which, for example, is being considered as a new preservative in food) is plotted against the dose. The results clearly suggest that the chemical is carcinogenic.

However, humans are unlikely to be exposed to the same massive doses. Instead the expected human daily intake is perhaps less than $1 \, mg/kg$. If we look at that

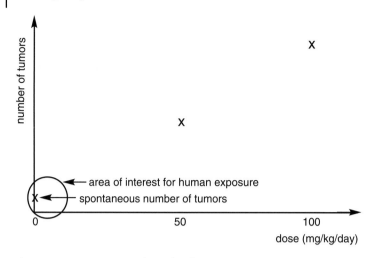

Figure 7.7 Dose-response relationship from a carcinogenicity assay.

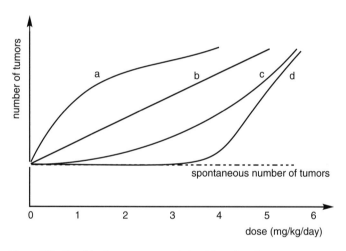

Figure 7.8 Possible dose–response relationship at low doses.

part of the diagram, there are no data to suggest whether low doses are carcinogenic. It can, of course, be assumed that the apparent linear dose–response relationship indicated in Figure 7.7 can be extrapolated all the way to zero (case b in Figure 7.8), and that low doses actually are slightly carcinogenic. However, we do not know if this is the case from the data shown in Figure 7.7. Other possibilities are that low doses are more carcinogenic than expected (case a in Figure 7.8), or less (case c or d in Figure 7.8).

Many carcinogenic compounds appear to behave as indicated by graphs c and d in Figure 7.8 because of the endogenous defensive systems that to some extent can protect us from reactive compounds (further discussed in Chapter 5). Such systems have a limited capacity and will only be able to neutralize small amounts of reactive chemicals (the threshold dose). If we want to know for certain about the low-dose toxicity, it would be very expensive and consume large numbers of animals. Say that the preservative discussed above in low doses (1 mg/kg/day) gives cancers in 0.1% of the exposed people, that would mean that 1 individual among 1000 exposed would be affected. That is an unacceptable incidence and the preservative should be banned. However, at this dose it is not a potent carcinogen, and in order to statistically establish the effect in an animal experiment several thousands of animals would have to be used. As this is not possible in practice, unrealistically high doses are tested instead, with the result discussed above. We shall return to this topic.

7.2
General Toxic Effects

7.2.1
Nonspecific Effects of Lipophilic and Amphipathic Compounds

The nonspecific effects that relatively large amounts of lipophilic compounds have on cell membranes and the toxic effects this may cause have already been discussed briefly, and we will return to this. Amphipathic compounds will also interact with membranes and may disturb any biochemical activity that takes place in them. Aliphatic alcohols, which have one polar and one nonpolar end, will in high concentrations lyze cells because of this property, and ethanol especially is commonly used as an antiseptic, for example to kill bacteria on the skin. When phenol was introduced as an antiseptic agent in hospitals it dramatically decreased the spread of infections, killing bacteria because it too is amphipathic. It is not known exactly how nonspecific toxicants in very high doses cause death, because no pathologic marks are produced by this type of toxicity, but is probably due to the depression of either the CNS or the heart muscle.

7.2.2
Lipid Peroxidation

Lipid peroxidation is caused by the oxidation, initiated by radicals and propagated by molecular oxygen, of membrane components containing unsaturated fatty acids (see Figure 7.9). The radicals are formed, for example, during the metabolism of exogenous compounds, and this was discussed in the last chapter. The initial relatively stable products formed between unsaturated fatty acid radicals and molecular oxygen are hydroperoxides, which explains the name 'lipid peroxidation'. Identical reactions will also take place in other circumstances and are responsible, for example, for butter and other foods going rancid when stored improperly. They are in general stimulated by temperature, oxygen, and radical

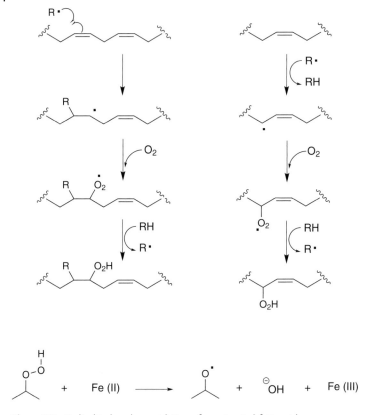

Figure 7.9 Radical-induced peroxidation of unsaturated fatty acids.

initiators (e.g., radicals or light). The bad smell of rancid butter is caused by the formation of shorter saturated fatty acids, for example, hexanoic acid, as the carbon–carbon double bonds are cleaved by the oxidative processes. Another example (also mentioned in Chapter 3) of spontaneous oxidation of unsaturated fatty acids is the use of paints based on linseed oil (containing large amounts of linolenic acid) which will 'dry' because they are oxidized, and a piece of cloth containing linseed oil may even autoignite because of the heat of reaction developed when it oxidizes. The fact that radicals are so short-lived and difficult to study has delayed the recognition of the importance of lipid peroxidation, and it is only during the last decades that it has received proper attention. A radical may attack an unsaturated fatty acid either by abstracting a hydrogen atom adjacent to a double bond or by adding directly to the double bond (see Figure 7.9). Remember that the reaction between a radical and a nonradical (or a diradical like molecular oxygen) always yields a new radical, and as a result radical reactions are chain reactions.

The initially formed radical may react in a number of ways, but it is not unlikely that it will react with molecular oxygen, which is always available and always willing to participate in a radical reaction. This would result in peroxy radicals, which readily abstract hydrogens from, for example, other unsaturated fatty acids to become hydroperoxides. These may, as hydrogen peroxide (H_2O_2), be reduced enzymatically by glutathione peroxidase, but are also oxidizing agents prone to participate in reactions that generate new radicals. The corresponding hydroxy radical may, for example, be formed by reduction with Fe(II), which is oxidized to Fe(III). So, in addition to being a continuous process constantly generating one radical for each one consumed (until the chain is terminated by, for instance, the combination of two radicals to give a product with paired electrons), lipid peroxidation also generates hydroperoxides that may produce additional radicals. The final product formed from the peroxidized unsaturated fatty acid is a complex mixture (which to some extent depends on the unsaturated fatty acid itself) of hydrocarbons, alcohols, aldehydes, and carboxylic acids formed by chain cleavage and further oxidation (see Figure 7.10). Several of the products are toxic themselves, for example, the α,β-unsaturated carbonyl compounds, which are electrophilic (see Figure 7.10). Especially malonaldehyde has received attention, as it is known to be genotoxic, probably because it, as the enol tautomer, is a good electrophile and may react with the DNA bases (see Figure 7.10). 4-Hydroxy-2-nonenal (R = pentyl in Figure 7.10) is another product that is relatively reactive because of the stabilization of the product as a γ-lactol (precluding the reversal of the Michael addition sometimes observed with β-substituted carbonyl compounds, see Figure 5.35 in Section 5.4.4 for an example). It can also react with primary amines to form pyrroles (see Figure 7.10).

Radicals that can initiate lipid peroxidation are continuously formed, for example, by the additional loop of cytochrome P450 oxidations discussed in Chapter 5. However, such formation of superoxide and hydrogen peroxide is normally kept very low by the protective enzymes, and very few radicals escape. If a person is exposed to exogenous compounds that are metabolized by cytochrome P450, more superoxide is produced, and eventually the protective systems may be overcharged, resulting in increasing concentrations of reactive oxygen species. In addition, as we have seen exemplified in Chapter 5, several types of exogenous compounds generate radicals directly when they are metabolized by cytochrome P450 and other oxidative enzymes. Another source of radicals is ionizing radiation, to which we are constantly subjected from space. Ionizing radiation injures cells mainly because it will transform water molecules to the extremely reactive hydroxyl radical by knocking off an electron, and the hydroxyl radical will cause the damage observed (see Figure 7.11). (To a minor extent, ionizing radiation will also hit important proteins and DNA and thus directly kill a cell.)

As well as the enzymes superoxide dismutase, catalase, and glutathione peroxidase, which will take care of superoxide and hydrogen peroxide, we are also protected against radicals by endogenous radical scavengers. Examples are α-tocopherol (vitamin E), a lipophilic but also amphipathic molecule situated in the membranes with the hydroxyl group directed toward the water solution, and the hydrophilic ascorbic acid (vitamin C). Both are essential chemicals with several important functions, and it appears as if they collaborate in their efforts to eliminate radicals

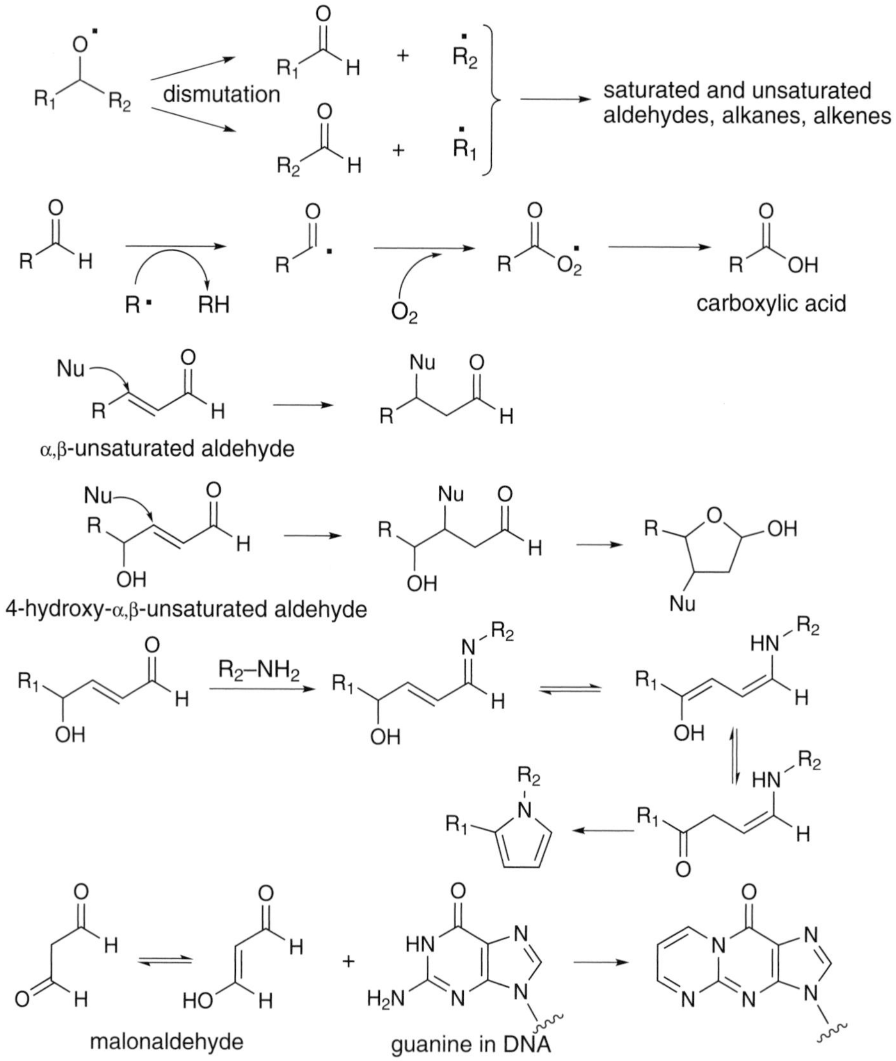

Figure 7.10 The formation of electrophilic compounds from lipid peroxidation.

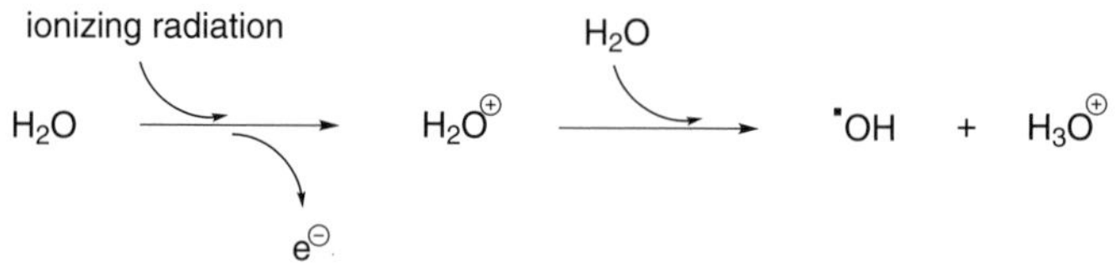

Figure 7.11 The generation of hydroxyl radicals by ionizing radiation.

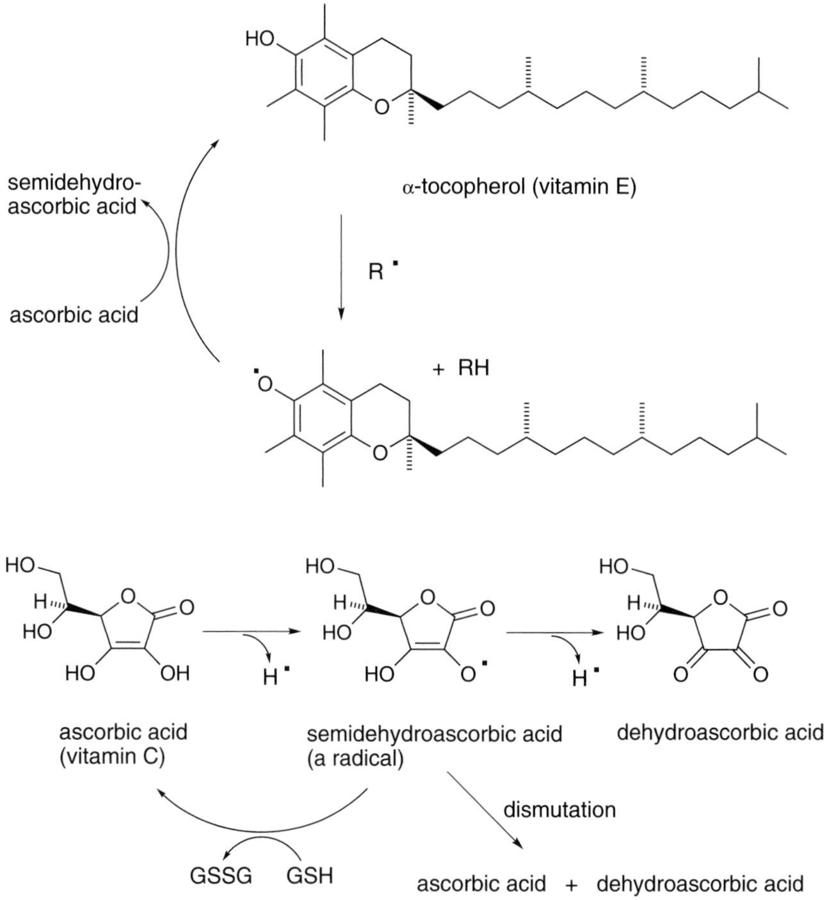

Figure 7.12 The scavenging of radicals by vitamins C and E.

in the way that α-tocopherol (and other similar tocopherols) will scavenge radicals formed in the membranes, while ascorbic acid will regenerate oxidized α-tocopherol. Ascorbic acid will in its oxidized form (semidehydroascorbic acid, see Figure 7.12) either be reduced back to ascorbic acid by a reductase utilizing GSH as coenzyme, or be dismutated to ascorbic acid and dehydroascorbic acid. Glutathione obviously plays an important role in the protection against lipid peroxidation, and glutathione depletion (by, for example, electrophilic compounds that are conjugated with it) will promote lipid peroxidation.

Ascorbic acid is a relatively strong reducing agent and will take care of any radical including superoxide directly, as well as of singlet oxygen. However, it has been shown that it may also reduce Fe(III) to Fe(II), and as the latter is involved in the generation of hydroxyl radicals from hydrogen peroxide it is not unanimously considered that high doses of ascorbic acid are completely harmless.

In conclusion, lipid peroxidation can produce toxic effects on a cellular level in three principal ways:

1) The damage to the membranes caused by the peroxidation of its unsaturated fatty acids will, in general, be difficult for any cell. Cell membranes, internal as well as external, are designed to perform a number of important tasks that may be hampered if the structure of the membrane is changed.

2) The radicals formed during lipid peroxidation may react not only with the unsaturated fatty acids but also with the components of membrane-bound enzyme systems. This will affect, for example, the protein synthesis, and we will later in this chapter see how this may result in a toxic effect on the liver.

3) Some of the products formed from the peroxidation of unsaturated fatty acids are nonradical but still reactive as electrophiles and may diffuse from the membrane where they are formed to, for example, the nucleus of the cell and react with the DNA causing genotoxic effects.

7.2.3
Acidosis

Acidosis is a condition that is caused by the body fluids having too low a pH, while alkalosis is the opposite. The pH of any cell in the body is extremely important; no cell or organism will survive an internal pH that is significantly different from the normal value. Even a small change of the pH would have a considerable impact on the protonation of, for example, proteins, which in turn determines their three-dimensional form and thereby their function. In the blood of a human being, the pH is strictly regulated between 7.35 and 7.40 (the blood going to the lungs is more acidic because of the presence of higher concentrations of carbonic acid formed from carbon dioxide and water), and anything above or below these limits will be attended to by regulatory systems. However, the pH of the urine is more variable, and is normally below 7. In the kidneys, the equilibrium between water plus carbon dioxide and carbonic acid is regulated by the enzyme carbonanhydrase (see Figure 7.13), which performs an important task in resorbing sodium ions from the primary urine.

Sodium ions are exchanged for oxonium ions formed by the action of carboanhydrase, and the oxonium ions thus excreted give the urine an acidic pH (which may catalyze the transformations discussed in Chapter 5). The pH of the blood is regulated in several ways, for example, by the buffering capacity of the blood, which contains HCO_3^- ions that can neutralize acid by forming carbonic acid. The carbonic acid is excreted via the lungs as carbon dioxide, and labored breathing

$$2\,H_2O \;+\; CO_2 \xrightarrow{\text{carboanhydrase}} H_2CO_3 \;+\; H_2O \rightleftharpoons HCO_3^{\ominus} \;+\; H_3O^{\oplus}$$

Figure 7.13 Carboanhydrase catalyzes the addition of water to carbon dioxide to form carbonic acid.

will speed up the excretion of carbon dioxide, thereby eliminating acid in the blood at the expense of HCO_3^- ions. Acid may of course also be eliminated by direct excretion with the urine, as oxonium ions are water soluble. The pH of the blood is surveyed by a biological pH meter in the brain, which will respond to a decreasing pH value by increasing the breathing frequency. Note that the different mechanisms to eliminate acid work on different time scales. While the buffering of HCO_3^- ions (and also other inorganic ions as well as proteins) acts immediately, the excretion of carbon dioxide by the lungs or oxonium ions by the kidneys will take minutes to hours. However, the capacity of the body to regulate its pH is limited, and it can only neutralize relative small amounts of acid. Acidosis will result if the capacity is overloaded, and a pH of the blood below 7.0 (or above 8.0 during alkalosis) is immediately life-threatening.

Acids may in theory gain access to the body by being absorbed by the routes discussed in Chapter 4, but it is evident that acids (and bases) that are sufficiently strong to give a serious effect on the pH of the blood are ionized and therefore not readily absorbed. Instead, they give rise to local effects. Weaker acids (e.g., salicylic acid) may be absorbed, but can only cause acidosis if the amounts taken are very large (multigram doses). However, stronger organic acids may be absorbed in a masked form that itself is not acidic but can be converted/transformed to acid inside the body (see Figure 7.14). This is in principle true for esters of, for example, formic acid (pKa 3.75). A carboxylic acid functionality masked as an ester is lipophilic, and esterases are relatively efficient in hydrolyzing esters to acids. However, the esters of formic acid, for example, methyl formate, are actually hydrolyzed so easily that they are highly irritating locally if inhaled (as vapor), swallowed, or spilled on the skin. Carboxylic halides, for example, acetyl chloride, and anhydrides, for example, acetic anhydride, are more reactive and more easily hydrolyzed than the esters, and will give severe local effects due to their spon-

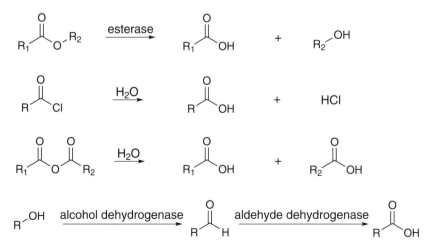

Figure 7.14 Ways in which acidic carboxylic acids can be formed inside the body.

taneous hydrolysis upon contact with any tissue containing water. Both acetyl chloride and acetic anhydride are corrosive, especially in concentrated form, and besides generating acetic acid (and hydrochloric acid with acetyl chloride!) the hydrolysis reaction generates heat that will aggravate the effect of the acid. As the eye is often the most sensitive part of the body that may be subjected to splashes, any handling of halides or anhydrides of carboxylic acids must be carried out with eye protection.

Another possibility for the generation of carboxylic acids inside the body is, of course, the oxidation of primary alcohols via the corresponding aldehyde, and this is the cause of most cases of acidosis caused by an exogenous chemical. Methanol is the simplest of the alcohols and is used in enormous amounts, mainly as an organic solvent, and many people are exposed to methanol at their workplaces every day. However, when methanol causes acidosis it is generally because it is mistaken for ethanol and consumed with the intention of getting drunk. It is fairly toxic to humans (an adult may die after drinking approximately 100 mL), and at workplaces one should be aware of its high volatility and excellent ability to be absorbed through the skin. Methanol is only slowly converted in the body, and approximately 30% of a dose is still present unchanged after 48 hours. Because of its volatility and hydrophilicity, more than half of an absorbed dose will actually be excreted unchanged by the lungs and the kidneys, while the rest is oxidized to formaldehyde and formic acid. The acute effects of methanol are irritation of the mucous tissues of the stomach, an unpleasant intoxication due to the nonspecific effect on the CNS, reversible and irreversible effects on the optic nerve, and acidosis. After a typical accident when a number of persons have been consuming methanol in the belief that it is ethanol, one third get away with becoming very drunk, one third lose their eyesight temporarily or permanently, while one third die as a result of acidosis. The effects on the optic nerve, which are unique to higher mammals such as humans and monkeys, are believed to be caused by formaldehyde, possibly due to the high concentrations of alcohol dehydrogenase, which has an important role to play in the biochemistry of vision. In rats, methanol is oxidized mainly by the enzyme catalase and will not affect the optical nerve of these animals. The acidosis is caused by formic acid, the principal end-product of the oxidation of methanol in man. A lethal dose of formic acid if injected directly into the blood would be approximately 3 g for an adult, corresponding to 2 g of methanol, reflecting how inefficient the conversion of methanol to formic acid actually is in the body. In other species, for example, rats and rabbits, formic acid can be converted to carbon dioxide and water, and methanol is consequently considerably less toxic in such species.

Ethanol is oxidized by exactly the same enzymes in man, via acetaldehyde to acetic acid. Although acetic acid is a slightly weaker acid (pKa 4.75) than formic acid, it is still in principle sufficiently acidic to cause acidosis. However, acetic acid is the unit used by the cells for the generation of energy (see below), and our cells will have no problems to oxidize it to carbon dioxide and water. Acetaldehyde is certainly a toxic intermediate, but during normal conditions the acetaldehyde formed is immediately converted to acetic acid (several times faster than ethanol is oxidized to acetaldehyde) and is only present in low concentrations. However, individuals that for example suffer from an aldehyde dehydrogenase deficiency will not be able to convert acetaldehyde, at least not as efficiently, and will suffer badly from the effects of acetaldehyde if exposed to ethanol. The conversion rates of methanol and ethanol are quite different. Ethanol, being more lipophilic and less volatile, is oxidized more rapidly (approximately 10 g per hour in an adult), and only a fraction (approximately 1%) of the absorbed amounts are excreted unchanged (by the lungs). If both alcohols are available, the alcohol dehydrogenase prefers to oxidize ethanol and leaves the methanol unchanged, and this is taken advantage of when methanol poisonings are being treated. Small amounts of ethanol are continuously administered, while the excretion of methanol by the kidneys is facilitated by diuretics and/or by dialysis. In addition, the acidosis is treated by replenishing the body's supply of HCO_3^-.

Another commonly used alcohol that may give acidosis is ethylene glycol, perhaps most familiar as antifreeze in the coolant of cars. Ethylene glycol is approximately as toxic to man as methanol, and a large number of poisonings occur yearly. As it is often present in private garages it is accessible by children, who mistake antifreeze for syrup because of its color (often red) and sweet taste (glycol comes from glukus, which is Greek for sweet). It is oxidized in several steps to the end-product oxalic acid (see Figure 7.15), which, besides being a strong organic acid, (pKa 1.27) is toxic to the kidney (see below). Oxalic acid is known from the kingdom of plants; it is, for example, present in rhubarb, which consequently should not be consumed in too large quantities. Lately, propylene glycol has been launched as a replacement for ethylene glycol, a highly recommended substitution as propylene glycol is just as efficient as an antifreeze and considerably less toxic.

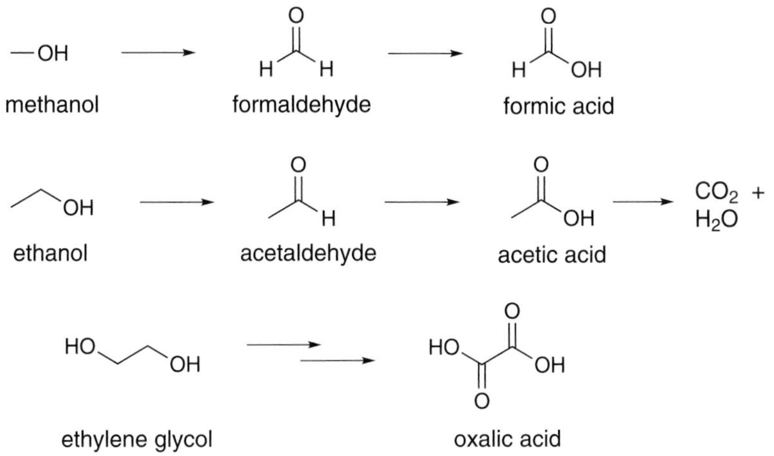

Figure 7.15 Oxidation of primary alcohols to carboxylic acids.

7.2.4
Oxygen Deficiency Due to Effects on the Blood

A human will not survive more than a few minutes without oxygen. The energy necessary to make most biochemical reactions proceed smoothly at body temperature is produced in the mitochondria by oxidizing nutrient to carbon dioxide and water, while the energy originally contained in the chemical bonds is transformed into chemical energy of ATP (adenosine triphosphate) and GTP (guanosine triphosphate). No oxygen means that energy production will stop, and when the stores of ATP are emptied the life processes will stop. Nutrients of various kinds – sugars, fats, proteins, etc. – are converted to acetic acid units in the cytoplasm of the cells, and these are brought into the mitochondria as conjugates with a coenzyme called coenzyme A (CoA). In the mitochondria, acetyl CoA enters the citric acid cycle, a cyclic process that combines the acetyl group with oxaloacetic acid to form citric acid, which is converted via isocitric, α-ketoglutaric, succinic, fumaric, and malic acids back to oxaloacetic acid (see Figure 7.16).

Two carbons, accounting for the two added with the acetyl group, are split off as carbon dioxide during the conversions. The energy of the carbon–carbon and carbon–hydrogen bonds of the acetyl group is transferred to one molecule of GTP and several molecules of the coenzymes NADH and $FADH_2$ in their reduced forms. In the reduced coenzymes the energy is stored as a high potential for reduction, or, in other words, in electrons with high energy. This will be converted to a more useful form of energy during the electron transport in a process called oxidative phosphorylation. In this, NADH and $FADH_2$ are oxidized while molecular oxygen is reduced, and this is coupled with the phosphorylation of ADP (adenosine diphosphate) to produce ATP (adenosine triphosphate). In the electron transport part of the oxidative phosphorylation, also called the respiratory chain, the energy stored in the electrons of the reduced coenzymes is gradually being extracted by, in all, four enzyme complexes via a series of coenzymes and cytochromes (see Figure 7.17) and is used for the phosphorylation of ADP. In the end the electrons are combined with molecular oxygen, which eventually is reduced to water, although this is not understood in detail (it takes four electrons to reduce molecular oxygen to two molecules of water, but the cytochromes only deliver one electron at a time). In ATP the energy is stored as high phosphorylation power, which can be used to activate other compounds, thereby facilitating chemical reactions by lowering the activation energy.

7.2.4.1 Carbon Monoxide

The ability of the blood to transport oxygen is remarkably refined, as the hemoglobin of the red blood cells binds oxygen strongly in an environment where the oxygen concentration is high (in the lungs) and lets go of it when the concentration is low (in other tissues). The red blood cells are formed in the bone marrow together with other blood cells, and exogenous compounds that affect the bone marrow may result in too few or malfunctioning red blood cells. Examples of chemicals that are toxic to the bone marrow are the solvent benzene, the insecticide lindane,

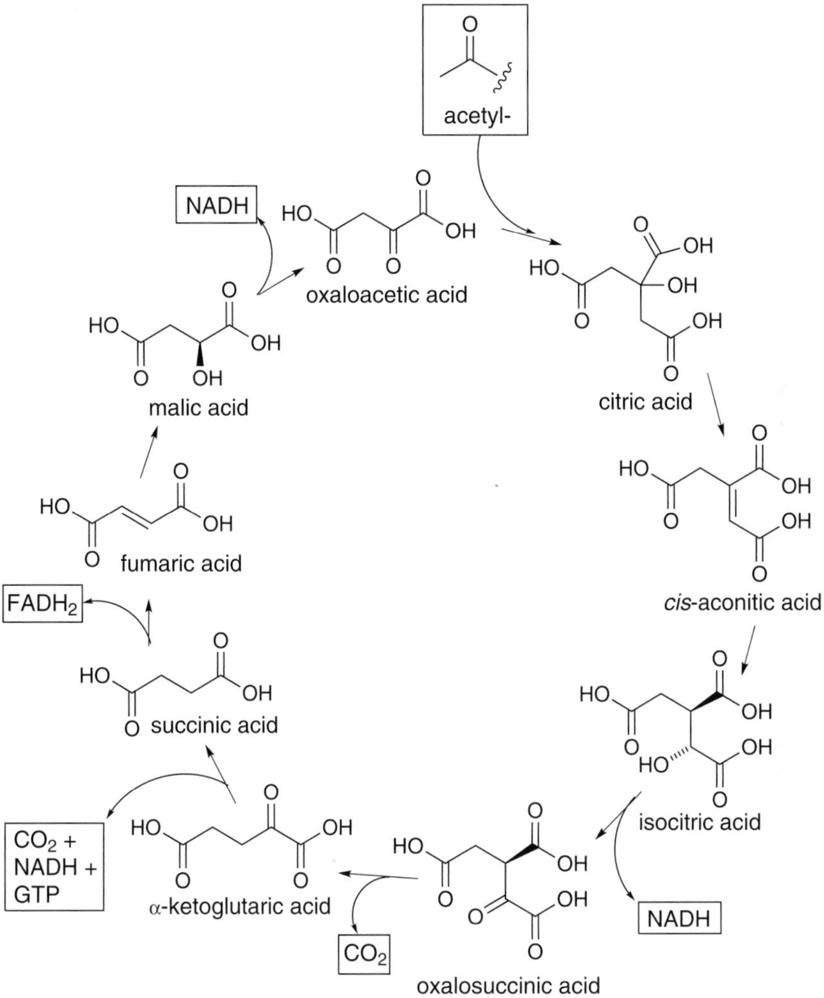

Figure 7.16 The citric acid cycle.

arsenic, and trinitrotoluene. A classic and very simple chemical that blocks the ability of hemoglobin to transport oxygen is carbon monoxide, CO, formed, for example, when organic material is incompletely burned or combusted with insufficient amounts of oxygen to oxidize the carbon to carbon dioxide. Carbon monoxide is chemically very similar to molecular oxygen and will also be bound to hemoglobin, although much (approximately 200 times) more strongly, and hemoglobin having carbon monoxide bound to it will not be able to transport molecular oxygen (see Figure 7.18). If 50% of the blood is blocked by carbon monoxide, the life of a human is in danger. This corresponds to approximately 0.1% carbon monoxide in the respiration air, and higher concentrations are lethal. The toxicity

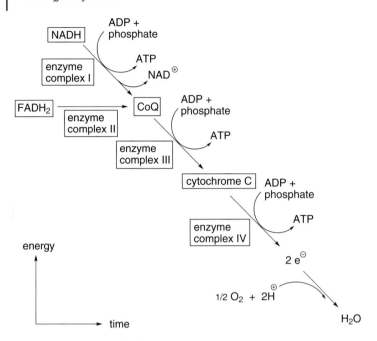

Figure 7.17 Oxidative phosphorylation.

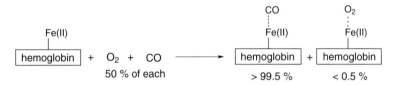

Figure 7.18 Carbon monoxide competes with molecular oxygen for hemoglobin.

of carbon monoxide is caused by the lack of oxygen that it will cause if a sufficiently high proportion of the blood's oxygen transporting capacity is blocked, and it will be excreted by the lungs if fresh air or oxygen gas is provided (an example of receptor antagonism as the two compounds compete for the same binding site). A common source of carbon monoxide is combustion engines. The exhausts of older cars contain several percent carbon monoxide, and to be trapped inside a closed room with an idling car was formerly a not unusual lethal accident. However, the exhausts of modern cars equipped with a catalyst normally contain 0.0–0.1% carbon monoxide. Malfunctioning heating systems based on the burning of fuels are often responsible for fatal accidents, and it regularly happens that campers bring the hot barbecue into the tent on a chilly evening and are found dead from carbon monoxide poisoning in the morning. We have also seen that carbon monoxide may be formed from the metabolism of exogenous chemicals, for example, methylene chloride, and the poorer physical condition that smokers usually have is to some extent explained by their constant exposure to carbon monoxide. Actually, approximately 1% of our blood is constantly blocked by carbon monoxide, as small amounts of the compound are formed in the normal metabolic processes. The molecular target for carbon monoxide is the Fe(II) in the heme group in the hemoglobin protein (other proteins containing a heme group with iron in oxidation state 2, for example, cytochromes involved in the generation of ATP, are also affected by carbon monoxide, see above).

7.2.4.2 Methemoglobinemia

Besides binding carbon monoxide strongly, the Fe(II) of the heme group of hemoglobin is also, as Fe(II) always is, easily oxidized to Fe(III). This is not surprising: everyone is aware of that anything made of iron will sooner or later corrode, because Fe(III) is the most stable form of iron in the earth's atmosphere. If the iron of hemoglobin is oxidized it is transformed to methemoglobin, which is unable to transport molecular oxygen, and the condition is called methemoglobinemia. In fact, the oxidation is so facile that it to some extent takes place spontaneously in our blood cells, and we all have a small portion of our hemoglobin as useless methemoglobin. Normally, this autoxidation is balanced and kept in control by an enzyme, NADH-methemoglobin reductase, present in the blood cells that will reduce methemoglobin back to functional hemoglobin again. Certain exogenous chemicals will be able to accelerate the oxidation of hemoglobin, and exert a toxic effect by doing so. An example is the nitrite ion (NO_2^-), used as a preservative in food, for example. Nitrite is also formed when the nitrate ion (NO_3^-) is reduced by bacteria present in the mouth and the intestines, and drinking water containing high concentrations of nitrate (not unusual in drilled wells) may therefore be toxic, especially to infants. (Infants are more sensitive because the bacterial flora in their intestines is different.) Nitrate is also present in, for example, spinach, and a soup or a stew prepared with spinach should not be left too long at room temperature because this will give bacteria a good chance to reduce the nitrate to nitrite. However, it should be noted that lethal doses of nitrite (as for example $NaNO_2$ or KNO_2) for a healthy adult are in the order of several grams. The molecular mechanism by which nitrite oxidizes hemoglobin is not understood, but aromatic amines and aromatic nitro compounds will catalyze the reaction according to Figure 7.19.

The corresponding hydroxylamine, formed by oxidation of the amine and reduction of the nitro compound, is responsible for the reduction of molecular oxygen bound to hemoglobin to hydrogen peroxide. Hydrogen peroxide formed in the vicinity of something as easily oxidized as Fe(II) will react immediately, oxidizing Fe(II) to Fe(III). As we have already noted, the oxidation of Fe(II) by hydroperoxides will generate the hydroxyl radical, which will immediately start radical reactions in the membranes of the red blood cells. This leads to characteristic damage to the blood cells, forming so-called 'Heinz' bodies' that can be observed

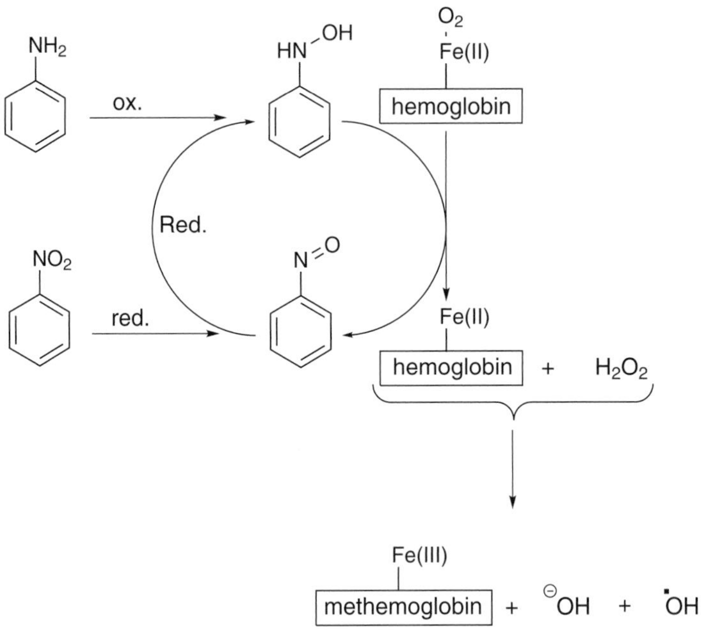

Figure 7.19 The oxidation of hemoglobin to methemoglobin by aromatic amines.

under a microscope. During the process in which a hydroxylamine transforms a hemoglobin to methemoglobin, it will itself be oxidized to the corresponding nitroso derivative. If this is reduced back to the hydroxylamine by the metabolic processes discussed in Chapter 5, it can function as a catalyst for the oxidation of hemoglobin.

It should be noted that some individuals are much more sensitive to conditions that transform hemoglobin to methemoglobin because they have a hereditary deficiency in the NADH-methemoglobin reductase that will normally reduce methemoglobin back to hemoglobin.

7.2.5
ATP Deficiency Due to Inhibition of the Primary Metabolism

The biochemical processes continuously need energy in the form of ATP, which is an efficient phosphorylating agent, and phosphorylation is used to activate chemical functionalities so that they participate in chemical reactions more readily. Shortage of ATP will hamper all cellular processes and eventually injure all tissues, but some cell types (i.e., nerve cells) are extremely dependent on ATP and will only survive for minutes without it. Other cells, for example, muscle cells, can operate under anaerobic conditions for some time, and are considerably less sensitive.

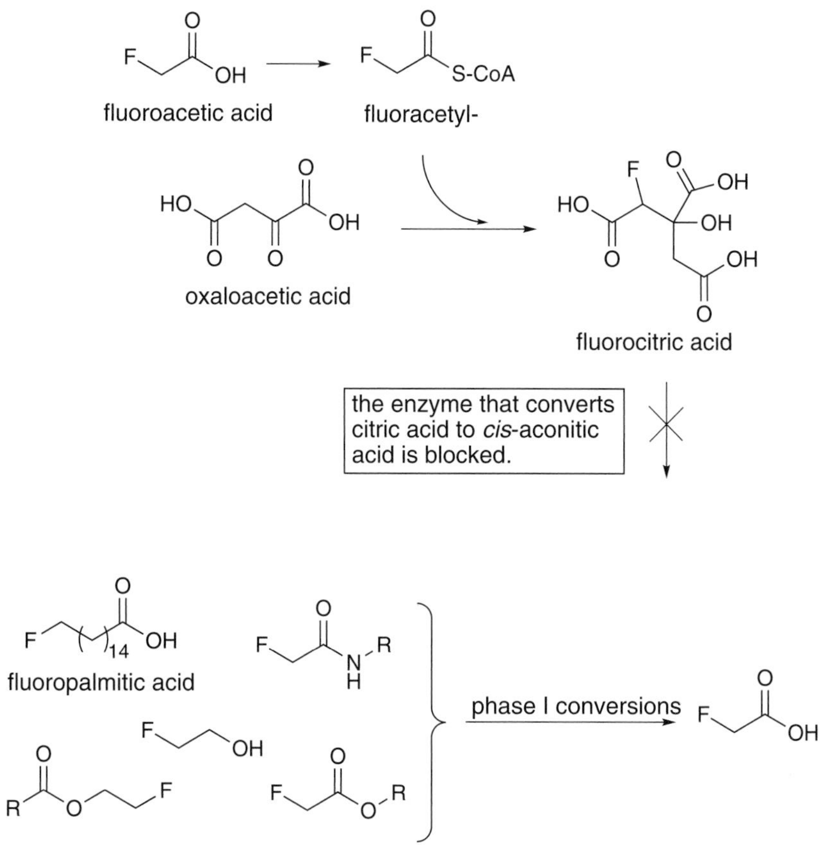

Figure 7.20 Fluoracetic acid will block the citric acid cycle.

7.2.5.1 Inhibition of the Citric Acid Cycle

An example of a chemical that efficiently inhibits the citric acid cycle is fluoroacetic acid, which resembles acetic acid so much that it can be converted to its CoA-thioester and as such accepted by the enzyme that combines acetic acid with oxaloacetate. The fluorocitric acid formed is, however, not a suitable substrate for the next enzyme in the cycle. Instead this is blocked, and the whole citric acid cycle comes to a halt. The LD_{50} value for fluoroacetic acid is 5 mg/kg (rat, or), which is comparable to that of the classic poison potassium cyanide (see below). It was originally isolated from some African plants (genus *Dichapetalum*) because it was noted that animals that grazed in areas where those plants grew often died, and it has since been used as a rat poison. Interestingly, some plants have the ability to absorb fluoride ions from the ground even in fluoride-poor soils and incorporate them into its metabolites, especially giving fluoroacetic acid, but also into fatty acids (e.g., 16-fluoropalmitic acid, which after β-oxidation will eventually yield fluoroacetic acid and is equally toxic), and inorganic waste containing fluorides should therefore be taken care of carefully. Chloroacetic acid has the same effect, although it is less potent, with an LD_{50} value of 75 mg/kg (rat, or). The reason for the difference between the two compounds is that the chlorine atom is bigger than a fluorine and the chlorocitric acid formed is a poorer enzyme inhibitor. Observe that chemicals that will be converted to fluoroacetic acid and chloroacetic acid (e.g., 2-fluoro- and 2-chloroethanol, esters, and amides, see Figure 7.20 for a few examples) by the metabolism will also affect energy production.

7.2.5.2 Inhibition of the Respiratory Chain

Classic poisons like hydrogen cyanide (HCN) and hydrogen sulfide (H_2S) bind to the Fe(III) in the heme groups of cytochromes in the respiratory chain (see above), stop the flow of electrons, and thereby stop the generation of ATP. Cyanides are used for many applications, for example, in the metal industry, and, as the sodium or potassium salts (NaCN or KCN), cyanides are fairly easy to handle. However, should a solution of cyanides be acidified or come into contact with an acidic medium, hydrogen cyanide will immediately be formed (see Figure 7.21). Hydrogen cyanide has a boiling point of 25 °C, can be considered to be gaseous at room temperature, and will be rapidly absorbed in the lungs. This is the compound used to execute people in the gas chambers and, because of its rapid absorption and distribution to the CNS, death caused by depression of the respiration regulation occurs within minutes. If a solution (or the crystalline form) of cyanide salts is swallowed, the transformation of the water-soluble cyanides to hydrogen cyanide will take place in the acidic environment in the stomach, and hydrogen cyanide will thereafter be absorbed in the intestines. However, this will take more time and require larger amounts. Drinking a glass of champagne containing potassium cyanide does not result in the immediate and dramatic death shown on TV in the mystery stories. Instead, it is actually possible to treat such intoxications successfully.

As indicated in Figure 7.21, the cyanide salts are not completely stable and may react slowly, for example, with water. If stored improperly, the crystals will absorb water and carbon dioxide from the surrounding air and be degraded to, for example, formic acid and ammonia. This is actually a common reason for failed suicide attempts, which when conducted with aged sodium or potassium cyanide may produce, instead of death, local effects in the mouth and throat due to the corrosive ingredients. The famous Russian religious mystic Rasputin, who strongly influenced the Tsar's family in the beginning of the twentieth century, was blamed

$$NaCN + H_3O^{\oplus} \longrightarrow Na^{\oplus} + H_2O + HCN$$
sodium cyanide   hydrogen cyanide

$$KCN + H_2O \longrightarrow HCN + KOH + HCO_2H + NH_3 + K_2CO_3$$
potassium cyanide

Figure 7.21 Reactions of cyanide in acid and in water.

for military set-backs during the first world war and murdered by a group of Tsar court officials. Initially he was offered tea with biscuits containing large amounts of potassium cyanide, but surprisingly he was not affected at all. Eventually he was first shot and then drowned in the river Neva, proving that he was not immortal, but it was still considered to be a mystery that he could withstand such large amounts of potassium cyanide. One suggestion has been that he was unable to produce sufficient amounts of hydrochloric acid in the stomach to produce lethal amounts of hydrogen cyanide from the potassium cyanide he ate. However, the pKa of hydrogen cyanide is 9.1, which means that cyanide will be protonated even at physiological pH. Instead it is likely that the potassium cyanide used to assassinate Rasputin was too old.

A completely different source of hydrogen cyanide is the glycoside amygdalin, present in bitter almonds and the stones of peaches, apricots, plums, and cherries. Amygdalin is most toxic if ingested, because the bacteria of the intestines possess the enzymes (glucosidases) that can hydrolyze amygdalin to its corresponding α-hydroxynitrile (see Figure 7.22). As already mentioned, α-hydroxynitriles will spontaneously liberate hydrogen cyanide, and the amounts of amygdalin present in a couple of bitter almonds will generate enough hydrogen cyanide to kill a child. Another similar glucoside is linamarin, which is present in several edible beans and roots consumed in large amounts (e.g., manioc or cassava). The consumers have to prepare these products in the correct way, including soaking to permit the enzymatic hydrolysis of the glucosidic bond, followed by boiling

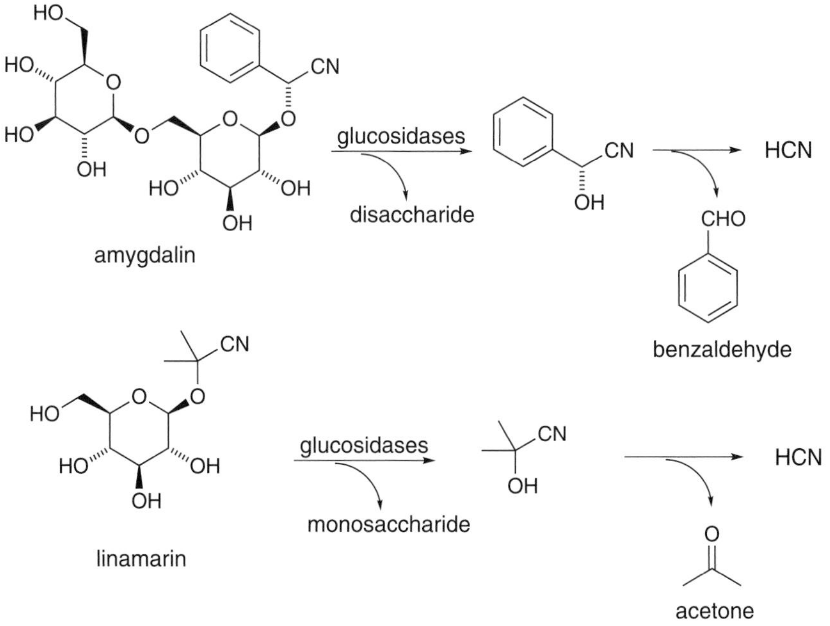

Figure 7.22 Enzymatic hydrolysis of cyanogenic glucosides.

to make the hydrogen cyanide evaporate, in order to avoid lethal intoxication. In Chapter 5 we saw that nitriles can be hydroxylated by cytochrome P450 to the α-hydroxynitrile, and hydrogen cyanide is also formed when urethane plastics are burned.

One way of treating hydrogen cyanide poisoning is to administer an appropriate amount of sodium nitrite ($NaNO_2$, approximately 300 mg for an adult), which will cause a moderate methemoglobinemia (see Section 7.2.4.2) that by itself is not dangerous. However, even a moderate methemoglobinemia will produce large amounts of oxidized heme groups [with Fe(III)] in the blood. This will extract hydrogen cyanide from the mitochondria of the cells to the methemoglobin in the blood and thereby (if the amounts of hydrogen cyanide are not too large) give the body a chance to cope with the acute situation. Our bodies have a fairly good capacity for degrading HCN with the enzyme rhodanese, which converts it to a thiocyanate ion (SCN^-) which can be excreted by the kidneys, and this metabolic detoxification is facilitated by the administration of sodium thiosulfate ($Na_2S_2O_3$). Interestingly, although HCN is one of the most well-known poisons, it is also an endogenous compound produced in small amounts by the phagocytes of the immune system and present in low concentrations in our exhaled air. However, its physiological function, if any, is not yet understood.

Hydrogen sulfide (H_2S) produces the same toxic effect as hydrogen cyanide. It is actually more toxic, but the fact that it smells so strongly (of rotten eggs) will normally warn anyone who is exposed to it. However, the sensitivity to the odor of hydrogen sulfide decreases rapidly during exposure. Even if the concentrations increase the odor fades away, and one should therefore not rely on this warning signal. Hydrogen cyanide is said to have a characteristic smell (of bitter almonds), but approximately 30% of the population are unable to detect this and therefore will not be alarmed by it. Hydrogen sulfide may be encountered in high concentrations in natural and volcanic gases. Low doses will be irritating to the eyes and the respiratory system, and may cause edema in the lung. Again, sodium nitrite may be used as an antidote, because the hydrosulfide anion (SH^-) forms a complex with methemoglobin called sulfmethemoglobin. However, hydrogen sulfide is oxidized to sulfite (SO_3^{2-}) and sulfate (SO_4^{2-}) in the body and does not accumulate.

7.2.5.3 Uncoupling of the Oxidative Phosphorylation

Chemicals that have this effect are called uncouplers, because their effect is not to bring the respiratory chain to a halt but instead to let the energy produce heat instead of being used for the production of ATP. This will result in a lack of ATP as well as in a rise in temperature, and as the lack of ATP results in stimulation of the energy production in the cells more and more heat will be produced. (One can compare this with the clutch of a motor engine as it uncouples the engine and replaces the flow of useful energy to the wheels with the production of heat.) Lethal poisoning by uncouplers will cause death by fever. The exact molecular mechanism for the inhibition of phosphorylation is not understood, although it is known to be a specific effect on the oxidative phosphorylation as other phosphorylations are not affected. A number of phenolics having electron-withdrawing groups in the benzene ring are efficient uncouplers, for example, dinitrophenol and pentachlorophenol, and have been used as both slimming agents and pesticides. They work as slimming agents because the metabolism increases, and more nutrients (fat) than necessary are then oxidized to carbon dioxide and water. The side effect at therapeutic doses is a slight fever; however, frequent overdosing has resulted in a large number of accidental deaths.

7.3
Specific Toxic Effects to Organs (or Equivalents)

7.3.1
Effects on the Blood

The blood is one of the most important targets for toxic chemicals, but as the major acute effect is inhibition of the ability of the blood to transport oxygen we shall confine the discussion to the effects already described in Section 7.2.4.

7.3.2
Effects on the Nervous System

The nervous system, especially the CNS, is frequently the target organ for toxic, exogenous compounds. We have already noted that nonspecific effects of lipophilic and amphipathic compounds in high concentrations will affect membranes in general and nerve cell membranes in particular, causing, for example, dizziness or unconsciousness. Although it is difficult to prove, it is nevertheless reasonable to assume that such compounds will be partitioned to the membranes and that they will change the chemical properties of the membranes slightly. As the three-dimensional form of a membrane protein, for example, a receptor for a neuro-transmitter in the nerve cell membranes (see below) is very dependent on the constitution of the membrane, such a change could affect their function. This is believed to be the way ethanol and anesthetics like chloroform, diethyl ether, and halothane affect us. In addition, many compounds that are toxic to the nervous system interact specifically with the receptors and affect communication between nerve cells. Toxic chemicals that in one way or another inhibit the production of ATP (see above), in principle a general toxic effect, can also be said to have a specific effect on CNS, because this is the organ first damaged.

The role of the nervous system is to relay signals from one end of a nerve cell to the other (i.e., within a cell) as well as between cells (mainly between different nerve cells but in general between cells with excitable membranes). This is done by two mechanisms, an electrical and a chemical one (see Figures 7.23 and 7.24). A resting nerve cell will create a chemical concentration gradient between the inside and the outside of the cell, by pumping out sodium ions in exchange for potassium ions (the sodium/potassium pump). As the potassium ions to some

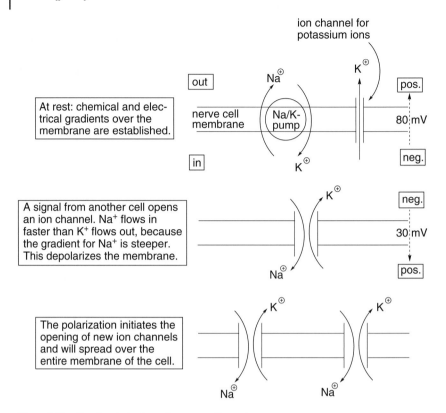

At rest: chemical and electrical gradients over the membrane are established.

A signal from another cell opens an ion channel. Na⁺ flows in faster than K⁺ flows out, because the gradient for Na⁺ is steeper. This depolarizes the membrane.

The polarization initiates the opening of new ion channels and will spread over the entire membrane of the cell.

Figure 7.23 The depolarization of a nerve cell membrane.

extent may diffuse out of the cell through special ion channels, a small electrical potential (approximately 80 mV) is created over the cell membrane. The initiation of a signal at the membrane of a nerve cell is accompanied by the opening of ion channels for both sodium ions and potassium ions, which quickly diffuse through the membrane in order to equalize the concentrations on the two sides. As the gradient for sodium ions is slightly steeper, sodium diffuses more rapidly, and this causes a local electrical depolarization of the membrane. This in turn results in an opening of further ion channels in the vicinity and the depolarization propagates over the entire cell membrane, as depicted in Figure 7.23.

After the depolarization, the ion channels close, and the sodium/potassium pump re-establishes the original membrane potential, the whole process taking only a few milliseconds. The depolarization is normally initiated by a signal from another cell, and normally results in the discharge of a signal to other cells. The chemical signal between cells is conveyed by a special compound called a neurotransmitter, which is released by the signaling cell and binds to specific receptors in the cell membrane of the receiving cell (see Figure 7.24). The neurotransmitter/receptor complex may then initiate a response in the receiving cell (for example,

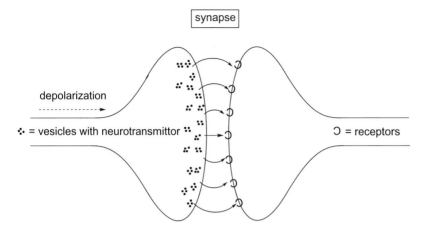

Figure 7.24   The transmission of a chemical signal between two nerve cells.

the opening of ion channels leading to the depolarization of its membrane). However, it should be noted that neurotransmitters are not always activating (stimulating a depolarization of the membrane of the receiving cell), but also can be inactivating (making the receiving cell less prone to depolarize its membrane). The junction between two nerve cells where the chemical signal is transmitted is called a synapse, and the distance between the two cells is short in order to keep diffusion times for the neurotransmitter at a minimum. The number of synapses that a nerve cell has is considerable, normally more than 1000. A nerve cell receives a number of both activating and inactivating signals, and whether it fires or not depends on the balance between the incoming signals. This is, of course, very complex and not understood in detail.

In addition, the neurotransmitter must be removed from the synapse after the signal has been transmitted, otherwise it will continue to stimulate the receiving cell as soon as it has been restored. This is done either by resorption and reuse of the neurotransmitter by the signaling cell, or by converting it to an inactive form. One of the most important neurotransmitters, acetylcholine, responsible for the signals in the cholinergic system, is inactivated by the hydrolysis of the ester bond by the esterase acetylcholine esterase (see Figure 7.24). The conversion is very rapid and the choline formed is completely inactive.

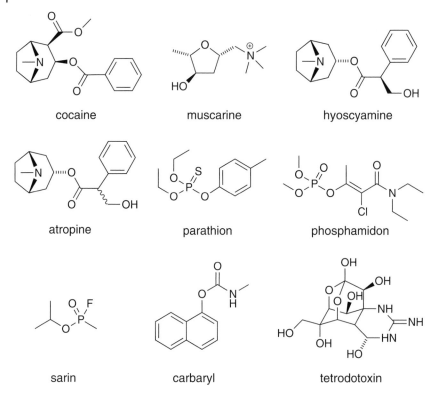

Figure 7.25 Some compounds that are toxic to the nervous system.

Exogenous chemicals may specifically affect the nervous system in a number of ways:

1) The release or reabsorption of the neurotransmitter by the signaling cell may be affected. The botulin toxins have this effect, and these will be discussed in detail in Chapter 9. Cocaine (see Figure 7.25 for structure) is active on the CNS because it blocks the resorption of noradrenalin, another neurotransmitter. The effect of cocaine is stimulating, as it makes noradrenalin stay longer than needed in the synaptic cleft.

2) Chemicals may have the same effect on the neuroreceptor as the natural neurotransmitter, and be able to stimulate cells to fire (if it is an activating neurotransmitter). Such compounds are called agonists, and a typical toxic effect that agonists may give is convulsions. Muscarine, one of the toxic principles of the mushroom fly agaric (*Amanita muscaria*), binds to the receptors that normally bind acetylcholine (which is reasonable when the two structures are compared). Not only does it bind, it also triggers the receptors to respond (i.e., it is an agonist) and results in an additional stimulation of the cholinergic part of the nervous system (see also Chapter 9).

3) The receptors may also be blocked by compounds that resemble the neurotransmitter enough to bind to the receptor but not enough to provoke the normal neurotransmitter/receptor complex response. Such compounds are called antagonists and are toxic because they prevent the normal signals going through, and this typically causes paralysis. Examples of compounds that are antagonists include hyoscyamine, isolated from the plant *Atropa belladonna*, and similar alkaloids, which bind to the acetylcholine receptor. Hyoscyamine and atropine (atropine is a mixture of hyoscyamine and its benzylic epimer) are still used to relax the muscle controlling the iris in the eye, causing the iris open maximally and facilitating inspection of the retina. The plant got its name 'belladonna' because women in Roman times used to place drops of the juice of *Atropa belladonna* in their eyes to make the pupils large, which was considered to make them beautiful. They would not be able to see much, but beauty has its price!

It is interesting to note that although the activity of hyoscyamine and atropine is quite different from that of cocaine, their chemical structures show several similarities. This is another example of how nature exploits the same molecular constructions over and over again but maintains selectivity by a molecular fine tuning that is admired and respected by pure chemists.

4) If the neurotransmitter is converted to an inactive form in the synapse, any compound that inhibits the enzyme responsible for the conversion may of course have dramatic effects. If the neurotransmitter is activating, it will not disappear from the synaptic cleft but continue to stimulate the receiving cell to fire, and a normal toxic effect of such compounds is again convulsions. Very efficient inhibitors of the enzyme acetylcholine esterase have been discovered, among the most active being the organophosphates (e.g., derivatives of phosphoric acid and phosphonic acid esters and corresponding phosphorothioic acids) of which many have been and are still used as insecticides (e.g., parathion and phosphamidon). Sarin is a chemical warfare agent that was first produced during the second world war but fortunately has not been used in any large conflict because it is so extremely toxic (lethal dose for man may be as low as 0.01 mg/kg). The structure of sarin is shown in Figure 7.25, while the mechanism by which organophosphates in general inhibit acetylcholine esterase is presented in Chapter 9. Another class of compounds that have the same effect is the carbamate esters, and these will also be discussed in Chapter 9.

5) The ion channels through which the sodium and potassium ions flow in the depolarized membrane can be blocked by compounds that fit with respect to both size and charge (like a cork in a bottle) but also affect the ion channels by interacting with the membrane. Several local anesthetics and barbiturates are active in this way, and another well-known example is tetrodotoxin, the poison of the puffer fishes (also present in other fishes and octopuses). The puffer fish (Fugu in Japanese) is considered to be a delicacy in Japan, but because of the extreme toxicity of tetrodotoxin it can only be prepared

by specially trained cooks who know exactly which organs containing the poison should be discarded. Still, several fatal accidents occur every year. Tetrodotoxin is not produced by the fish itself, but by a bacterium that is part of the same food chain as the puffer fish. Tetrodotoxin efficiently blocks the flow of sodium ions into the cells, which inhibits the depolarization of the membrane. The insecticide DDT acts on the same ion channel, but affects its closing and thereby the time it takes for a repolarization of the membrane.

7.3.3
Effects on the Liver

The liver plays a central role in both the excretion of chemicals from the body and the metabolism of both endo- and exogenous compounds. The metabolic capacity is also responsible for two typical toxic effects on the liver, fatty liver and liver cancer, because chemicals that are metabolized to reactive compounds are normally formed in the largest amounts in liver cells. Chemicals that are reactive as they are, not requiring metabolic activation, have no preference for the liver as a target organ. Synergistic effects are commonly observed with liver toxicity, as the mechanism for synergism often depends on the induction of cytochrome P450.

Certain compounds excreted by the liver via the bile damage the bile ducts as they pass on their way to the gall-bladder. Some compounds, for example, allylic alcohol, *para*-dimethylaminoazobenzene (butter yellow), coumarin, ethionine, and thioacetamide (see Figure 7.26 for structures) show toxicity to both liver and bile ducts, while other compounds, for example, α-naphthylisothiocyanate and sporidesmin A, are only toxic to the bile ducts. In a few cases (e.g., sporidesmin A), the effect on the bile ducts is considered to be caused by the parent compound, while in most cases metabolites formed in the liver are the true toxic principle. The bile flow may also be reduced by gallstones (normally composed of cholesterol and calcium salts), by inflammatory processes in the bile ducts, and by interference with the production of bile in the liver cells. A number of chemicals, for example, anabolic steroids such as testosterone, the immunosuppressive drug cyclosporine A (a cyclic undecapeptide, structure not shown), and manganese (Mn), are known to give this toxic effect, but little is known about the mechanisms.

The accumulation of fat in the liver (steatosis or fatty liver) is produced in an individual whose liver is producing more fat (triglycerides) than it can export. To produce fat from surplus nutrients is a normal task for the liver, the fat being transported to the adipose tissue for storage, to be reused when, for example, the need for energy exceeds production. The fat is released from the liver cells into the systemic circulation combined with a lipoprotein, and a toxic effect that lowers the concentration of this lipoprotein may result in a fatty liver. As the proteins are synthesized in the endoplasmic, where exogenous compounds are

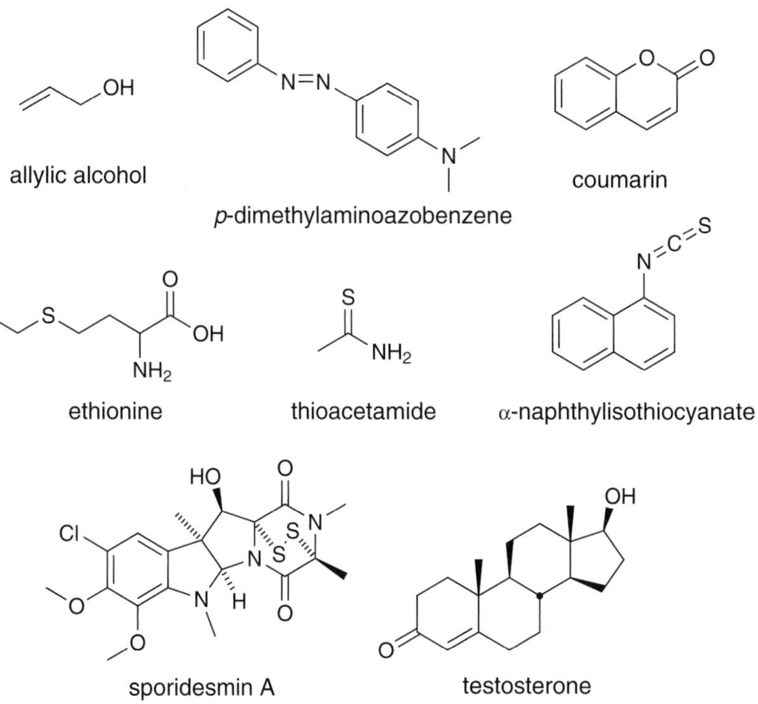

Figure 7.26 Compounds toxic to the liver.

also metabolized, reactive metabolites may of course affect the protein synthesis. Especially efficient are compounds generating radicals, as these give rise to lipid peroxidation that may destroy the membrane and the proteins associated with it. This is the mechanism by which, for example, carbon tetrachloride, which is reduced by the metabolism to the trichloromethyl radical, gives fatty liver. A fatty liver is generally considered to be a reversible condition, and not serious, but chronic exposure to a compound such as carbon tetrachloride will eventually result in liver cirrhosis (an irreversible liver damage) and liver necrosis (death of liver tissue). The reactive metabolites formed enzymatically or chemically as a result of the metabolism of exogenous chemicals in the liver may also give rise to liver cancers. Besides cancer, reactive compounds may of course also kill liver cells and induce serious liver damage (e.g., paracetamol in large amounts, see Chapter 9).

7.3.4
Effects on the Kidneys

The kidneys are the most important excretion organ, and their function has been briefly discussed in Chapter 4. Because of the intense activity of the kidneys, they require delivery of substantial amounts of oxygen and nutrients and are sensitive to chemicals that produce anoxia. A decrease in the blood pressure or blood volume (as a result of hemorrhage, for example) will also be more harmful to the kidneys than to other organs. The fact that the kidneys resorb approximately 99% of the primary urine permits the body to control the volume and composition of extracellular fluids, but small effects on this resorption by toxicants will have large consequences (1% depression results in the additional daily loss of approximately 2 litres of water from an adult). During the concentration of the urine there is also a possibility that chemicals may eventually become present in toxic concentrations that may injure the tubular cells, or no longer be soluble and form crystals. Ethylene glycol, frequently used as antifreeze in cars, may, in addition to producing acidosis, also be toxic to the kidneys. Oxalic acid, the end product of ethylene glycol metabolism, reacts with calcium ions to form calcium oxalate, which is excreted by the kidneys as small crystals together with low concentrations of oxalic acid. As the urine is concentrated, the crystals may adhere and form stones that can obstruct the tubuli. Several metals, in particular Hg and Cd, give severe toxic effects in the kidneys, and will be further discussed in Chapter 10.

7.3.5
Effects on the Respiratory System

The lungs are responsible for the exchange of oxygen and carbon dioxide between the blood and the air, and some important toxic effects associated with chemicals in the respiration air have already been discussed in Chapter 6. Especially important is lung cancer, which is one of the major causes of mortality among cancer deaths. This is largely due to smoking, although in approximately 20% of all lung cancers other factors are believed to be responsible. In addition, other factors (e.g., exposure to asbestos fibers in the workplace) act synergistically with tobacco smoke. There are many carcinogenic principles in tobacco smoke, most of which require metabolic activation, and the lung cells are consequently able to metabolize and activate exogenous compounds. An interesting example of a specific toxic effect on the lung cells is the nonselective contact herbicide paraquat. Regardless of the route by which paraquat is administered it always damages the lungs, and this specificity is caused by two effects: First, the alveolar cells are able to accumulate paraquat by a special transport system (the diamine/polyamine transport system, a system that was not discussed in Chapter 5), and second, the concentration of oxygen is relatively high in the lung cells. Paraquat is readily reduced by the coenzyme NADPH to form a radical, which rapidly is oxidized back to paraquat by molecular oxygen, which in turn is reduced to the superoxide anion (see Figure 7.27). The toxic effect depends on a combination of the depletion of NADPH in the affected cells and the generation of radicals with subsequent lipid peroxidation.

The pyrrolizidine alkaloids (e.g., monocrotaline in Figure 7.27) are plant metabolites that injure the liver and the lungs, and a large number of intoxications have

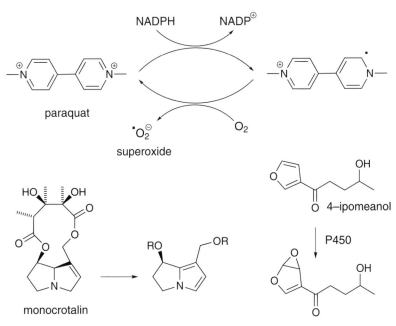

Figure 7.27 Compounds that are toxic to the lung.

been reported. They require metabolic activation via the pyrrole derivative shown in Figure 7.27 and further by cytochrome P450 oxidation of the ring to a bifunctional electrophile of yet unknown structure. This activation is mainly carried out by the liver cells, and not in the lung cells. The reason that the lungs are the target organs for some of the pyrrolizidine alkaloids is instead that the electrophilic metabolite is fairly unreactive and survives transport from the liver to the lungs – the next organ in the circulatory system, (although high doses will also damage the liver). Another example of a compound that is toxic to the lungs is 4-ipomeanol, a furan that is responsible for the toxic effect of moldy sweet potatoes given to cattle. 4-Ipomeanol is oxidized (probably to the epoxide shown in Figure 7.27) by cytochrome P450 isoenzymes present in high concentrations in lung cells.

7.3.6
Effects on the Skin

The skin is our barrier to the rest of the world: it protects us from external threats such as microorganisms, chemicals, and radiation, and it helps to preserve the body fluids (quantitatively as well as qualitatively). It is an organ system that frequently comes into contact with exogenous chemicals, which in some cases can penetrate the skin and be absorbed by the body, and it is most frequently the target for chemicals that give local effects. Strongly reactive chemicals, electrophiles, acids, bases, and so forth, will corrode the skin (chemical burns), which in severe cases can be life-threatening. In some cases chemicals react with the moisture of the skin to produce reactive products (e.g., acetyl chloride, acetic anhydride, and $TiCl_4$ and $SnCl_4$, both of which liberate HCl in contact with water), or generate heat that will burn the skin on contact with water (e.g., quicklime, CaO). In such cases one has to take care if rinsing with water is used to get rid of the chemical from the skin.

An important toxic effect on the skin is allergy in the form of contact dermatitis, and this topic is covered in Section 7.4. In addition, chemicals that are absorbed through the skin or by any other route may be activated by exposure of the skin to sunlight, leading to excited states that may affect the skin. This process is called photosensitization. When the result is a general toxic effect on the skin (e.g., blistering), it is specifically called phototoxicity, while the term photoallergy indicates that a delayed-type hypersensitivity reaction (see below) has taken place. Phototoxic chemicals in their excited forms can either react directly with macromolecules and produce a toxic effect or react with molecular oxygen, which in turn generates oxygen radicals. Examples of phototoxic chemicals are the psoralenes (plant metabolites, e.g., bergapten, present in bergamot oil), acridine (used for the manufacture of dyes), benoxaprofen (an analgesic), and nalidixic acid (an antibacterial agent), while halogenated salicylanilides (e.g., buclosamide, an antifungal agent), promethazine (an antihistamine agent), and coumarin derivatives (e.g., 6-methylcoumarin) are photoallergenic (see Figure 7.28 for structures).

7.4
Chemical Allergens

Chemical allergens are toxic to the immune system. The central functions of the immune system are to protect the organism from infectious agents (microorganisms and viruses) and to survey the cells of the organism and neutralize cells that have changed significantly (e.g., tumor cells). As well as infectious agents the immune system will also sense the presence of and attack certain structures including large molecules that do not belong in our bodies. When the immune system reacts too strongly, so that the response not only takes care of whatever should be taken care of but in addition is experienced as troublesome, this is described as an allergy. The mechanisms that are involved are too complex to be described in any detail here, but consist in principle of a nonspecific and a specific part. The nonspecific part reacts immediately to chemical signals, for example, emitted by the invading microorganism or by other cells of the immune system, and attacks and destroys the agent by phagocytosis. This causes what we recognize as an inflammation. The specific part of the immune system has aquired the ability to determine whether chemical components are 'self' or 'foreign', and will

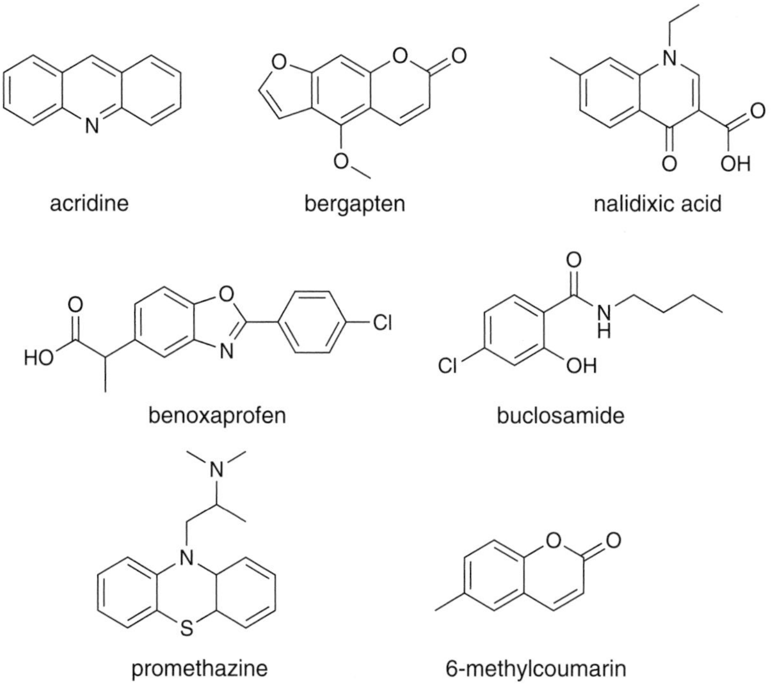

Figure 7.28 Phototoxic and photoallergic compounds.

attack anything considered foreign (containing an antigen which is the chemical component that is recognized as foreign). There are many mechanisms for the specific part of the immune system to react and deal with antigens, but they all depend on either the production (by so-called B cells) of specific antibodies, or immunoglobins, which are proteins that bind to antigens, or the development of specifically sensitized lymphocytes (T cells, T for 'thymus dependent') that contain receptors in their cell membrane that recognize antigens. Antibodies will bind to the antigens of, for example, a bacterium and attract effector cells that attack it, and this response comes quickly. Examples of allergic responses that depend on the production of antibodies against an antigen are the immediate hypersensitivity, or anaphylaxis, resulting, for example, in asthma or rhinitis (in the respiratory system), food allergies (in the gastrointestinal tract) and anaphylactic shock (in the vasculatory system). The allergic response of the T lymphocytes is called delayed-type hypersensitivity and results mainly in contact dermatitis (a rash or irritation, possibly including blistering of the skin that comes into contact with the allergen). This is the most common form of allergy to chemicals. (The immune system is described schematically in Figure 7.29.)

For obvious reasons, the proteins that identify chemical structures as 'self' or 'foreign' are designed to recognize large structures such as proteins and polysaccharides, and cannot identify small molecules. A certain size and a certain

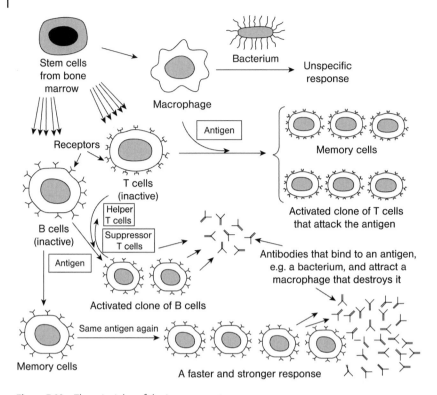

Figure 7.29 The principles of the immune system.

chemical complexity is required, and as a general rule it is usually considered that only molecules with a molecular weight of more than 5000 will be able to become antigens. Nevertheless, a number of small molecules are known to be allergenic, and they must then combine with bigger, endogenous molecules (e.g., proteins) before they are able to stimulate the immune system. Small molecules that do this are called haptens, and by reacting with and/or binding to, for example, a protein this will be transformed from a 'self'-protein to a 'foreign'-protein. Therefore, the general characteristic of an allergenic chemical is that it, either directly or after spontaneous chemical transformation or enzymatic conversion, is able to change the structure of a macromolecule. Electrophilic or pre-electrophilic compounds are potentially allergenic, as are radical-generating compounds and metal ions that are able to form stable complexes with, for example, proteins.

The mechanisms by which the immune system learns which chemical structures are 'self' are not understood, but the process probably takes place during the functional maturation of the immune system in the fetus. In general, an antigen (e.g., macromolecules from a microorganism, or an endogenous protein modified by a reactive chemical) will be picked up by special cells of the immune system that process and present them to lymphocytes in, for example, the lymph nodes or the thymus. If the antigen is recognized as foreign, the lymphocytes will develop into effector cells (either B or T cells) that are specific to that very antigen. In addition, long-lived memory cells are formed that in principle enable the body to remember that this antigen is foreign for the lifetime of the organism. This process is called sensitization and may take anything from days to years, but any contact with the antigen after sensitization will result in a quick response because the immune system has a 'memory' (the memory cells) of this antigen. The immediate hypersensitivity (e.g., hay fever) responds very fast (minutes), while the delayed-type hypersensitivity takes between 12 and 48 hours to respond. An individual who is sensitized to a chemical is often extremely sensitive to its presence even in trace amounts, and for chemicals like formaldehyde and nickel that are difficult to avoid, the allergy may become a life-long problem. In addition to the B and T lymphocytes responsible for antibody production and recognition of antigens, a number of other subpopulations of lymphocytes that play important roles in the immune response exist. Examples are the killer cells (K cells) and natural killer cells (NK cells) that are important for defense against cancer cells and helper (T_H) as well as supressor (T_S) cells that modulate the response of the immune system so that it is appropriate and balanced. In addition, there are a number of chemical factors (e.g., interleukins and interferons) that activate, amplify, and regulate various functions of the immune system.

7.4.1
Immediate Hypersensitivity

Immediate hypersensitivity is a quick response of the immune system to an agent that previously has been recognized as 'foreign', and may in some instances be life-threatening (e.g., the anaphylactic shock resulting from a bee-sting). Some allergens encountered in work-places are sufficiently big to function as antigens as they are (for example dust of flour that may give bakers rhinitis and asthma), but most act as haptens and conjugate in some way to an endogenous protein. Trimellitic anhydride (see Figure 7.30) is used for the manufacture of certain plastics and paints, and is allergenic. The most common response is antibody-mediated asthma, but it also gives rise to delayed-type hypersensitivity. Phthalic anhydride will give a similar response. The isocyanates are notorious reactive chemicals, used extensively for the preparation of polymers and paints. Especially the diisocyanates, containing two isocyanate groups (e.g., diphenylmethane-, hexamethylene- and toluene-diisocyanate) are potent allergens causing both asthma and contact dermatitis. Interestingly, individuals allergenic to toluenediisocyanate may develop cross-reactivity to other diisocyanates to which they never have been exposed. Ethylene oxide and formaldehyde, both highly reactive, also induce antibody-mediated immediate hypersensitivity. Treatment of infections with penicillin and other β-lactam antibiotics (see Figure 7.31 for the structure of benzylpenicillin) has shown that these may be allergenic in rare cases. Especially the immediate type of allergy resulting in anaphylaxis is serious and may be life-threatening, and hypersensitivity to penicillin is actually the most common cause

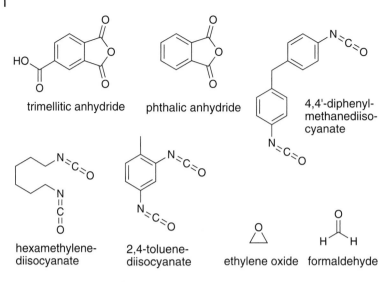

trimellitic anhydride   phthalic anhydride   4,4'-diphenyl-methanediiso-cyanate

hexamethylene-diisocyanate   2,4-toluene-diisocyanate   ethylene oxide   formaldehyde

Figure 7.30 Examples of chemicals that give immediate hypersensitivity.

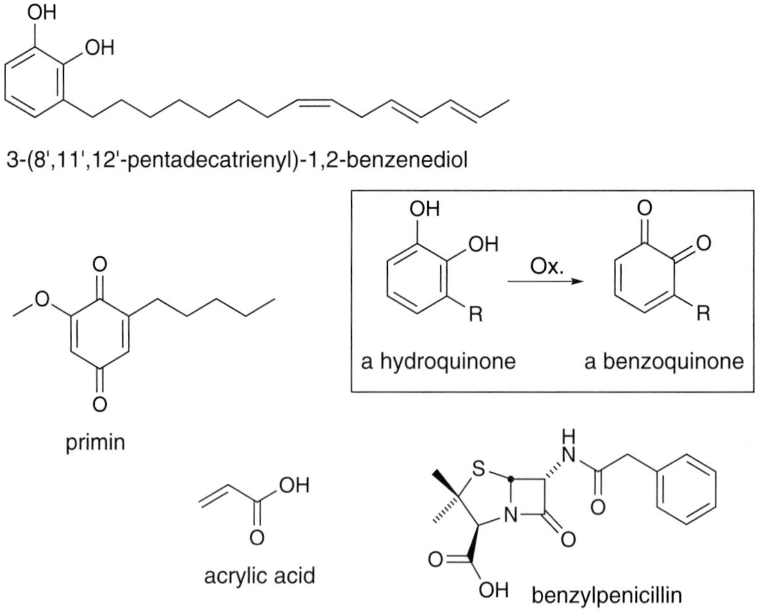

3-(8',11',12'-pentadecatrienyl)-1,2-benzenediol

a hydroquinone $\xrightarrow{\text{Ox.}}$ a benzoquinone

primin

acrylic acid   benzylpenicillin

Figure 7.31 Examples of chemicals that give delayed-type hypersensitivity.

of anaphylaxis in man. It is probably metabolites of the β-lactams that are responsible for their allergic effect, although this has not been clarified in detail. Several metals are also important allergens that give both types of responses, and this will been discussed further in Chapter 9.

7.4.2
Delayed-Type Hypersensitivity

Whether a compound is allergenic by this mechanism and not by immediate hypersensitivity depends on a multitude of factors, and not only on the chemical itself. As has been noted above, many allergenic chemicals will induce both types of hypersensitivity, which is reasonable. Several of the most allergenic compounds that give contact dermatitis are present in preparations that are applied on the skin, for example, plasters, ointments, perfumes, and so forth, and one may, guess that other reactive (or proreactive) chemicals would also give the same effect if used so that they come in close and prolonged contact with the skin. Important chemicals that may induce a delayed-type hypersensitivity are metals like Ni, Cr, Co, and organomercury compounds (see Chapter 10), catechols (for example 3-(8′,11′,12′-pentadecatrienyl)-1,2-benzenediol from the plant poison ivy), primine from the plant primrose, aromatic amines, epoxides, derivatives of acrylic acid, and formaldehyde, but also penicillin and other β-lactam antibiotics (see above). As indicated in Figure 7.31, hydroquinones are relatively easily oxidized to reactive benzoquinones (also discussed in Chapter 5), and hydroquinones (as well as compounds that yield hydroquinones after metabolic activation) are frequently allergenic.

In addition to allergy, chemicals can also be toxic to the immune system in other ways. Immunosuppression, which eventually may lead to immunodeficiency, may be caused by exposure to solvents like benzene, by environmental hazards like PCBs and PBBs, dibenzodioxins, other halogenated aromatic hydrocarbons, ozone, PAHs, several pesticides, and so forth. It is possible that the immunosuppressive activity of such agents is important for their toxicity, although it is not known to what extent it modulates other toxic effects. Another toxic effect to the immune system is autoimmunity, resulting in the attack of the effector cells of the immune system on endogenous components. Little is known about the mechanisms for autoimmunity, but in several cases it has been linked to the exposure to chemicals (e.g., vinyl chloride, perchloroethylene, and epoxy resins). In such cases it is believed that the chemical modifies an endogenous protein that is recognized as foreign, and that the response to this is incompletely specific and also attacks nonmodified protein.

7.5
Chemical Teratogenicity

Teratogenic agents are toxic to the embryo (the term used to describe the human offspring during the first 8 weeks after implantation in the uterus) or the fetus (the term used for rest of the time up to the birth) and cause structural or functional abnormalities in offspring after exposure of either parent to these agents before the conception or to the female during the pregnancy. The abnormalities, or birth defects, can be anything from extensive malformations to minor structural defects and metabolic disorders as well as growth or mental retardations, and can result in anything from the death of the embryo to insignificant defects. Obviously, it is difficult to define exactly what is a birth defect, as well as to quantify small and apparently insignificant structural or functional abnormalities. It is also hard to decide whether an abnormality was the result of an event taking place before or after birth. An example is the cancers that affect children, which are probably initiated already during the gestation although they are normally impossible to observe at the time of birth. It is known that a large proportion of all fertilized ova will not develop normally, and this in most cases results in the reabsorption or a spontaneous abortion so early that the pregnancy is not recognized. About 3% of live-born infants have birth defects that are recognizable during the first year of life. Obviously, as is the case for mutations, spontaneous factors are responsible for a significant portion of these effects, but the ratio between spontaneous and external factors is not known. Also potentially responsible for the interruption of gestation are the inheritable genetic defects that many of us, perhaps all, have. They may be the result of mutations that took place several generations ago. They are normally not expressed and therefore not noticed but are incompatible with certain genetic combinations that may arise in conception.

The gestation takes approximately 38 weeks (humans) and can be divided into the following major processes; histogenesis (formation of body tissues, during the first weeks), organogenesis (formation of organs, during the first months), functional maturation and growth of organs and tissues (last two thirds of the gestation). It starts with the fertilization of the ovum by a sperm cell, and after a number of cell divisions the blastocyst (the term used before differentiation has started, and before the term embryo) formed is implanted in the uterus approximately one week later. The organogenesis, a period when the basic structure of the organs and tissues are established, takes place from week 3 to week 9, and it is during this period that teratogenic agents give rise to malformations. The whole gestation is of course characterized by intense cellular activities in the embryo/fetus, besides the rapid cell division necessary, and these are directed by signals (e.g., hormonal) both from the mother and generated internally. The histogenesis and the organogenesis require that the correct parts of the genome are turned on or off at a precise moment (day), the maturation of the various functions at the right time require that things happen in a predetermined order, and there are really no margins for the disturbance or delay of any process.

The nature of a teratogenic agent can be physical and biological, as well as chemical, just as is the case with mutagens and carcinogens (see Chapter 8). The physical and biological agents are ionizing radiation (e.g., X rays) and viruses, an example of the latter being the rubella virus, which is a fairly common source of birth defects. Of the women infected with the rubella virus at the beginning of their pregnancy, 15–20% will have a miscarriage or a stillbirth, and the risk of severe malformations in the surviving fetus is considerable. Also, infection as late as during the third or fourth month frequently results in, for example, hearing defects. A large number of chemicals that are teratogenic in animals are known (>1000), but only a few (approximately 30) are proven to be human chemical teratogens. However, it is important to remember that chemicals do not have only negative effects on the development of an embryo; some are absolutely necessary for the biochemical processes to take place. Consequently, insufficient amounts of essential chemicals due to maternal nutritional deficiencies (e.g., fasting) may also be teratogenic. For example, lack of iodine may result in cretinism, a birth defect that is unusual today because iodine is added to table salt.

Besides interfering with the specific processes that take place in the embryo/ fetus, chemical teratogens may also produce effects due to normal toxicity, and these will also affect adults. The embryo/fetus is often more sensitive to toxic effects as its energy consumption and cell division rates are high. Mutagenic and cytotoxic chemicals are therefore generally teratogenic, as are nucleic acid analogs and chemicals affecting ATP production and oxygen transportation. Several commonly used solvents, for example, toluene and ethers of glycol, have been shown to give teratogenic effects in animals, as has even moderate alcohol consumption and the use of cocaine. Several phenoxy acids (e.g., [2,4,5-trichlorophenoxy]acetic acid) used as herbicides are teratogenic to several animals, and TCDD is a potent teratogen in animal experiments. So far, probably because of the limited exposure that has taken place, there is no evidence that TCDD is a human teratogen also, but it is strongly suspected to be so. Inorganic forms of Hg, Pb, Cd and As are also teratogenic in animal experiments, and the exposure of pregnant women in Minamata (see Section 10.7) to methyl mercury resulted in severe birth defects (CNS malformations). Many of the human chemical teratogens known are drugs, the most well known being thalidomide, marketed under a number of different trade names as a sedative and hypnotic and prescribed to calm the nausea associated with pregnancy. It appeared to be nontoxic to man and was introduced in 1956. However, it was withdrawn at the beginning of the 60s as it caused abnormalities foetuses, more than 10 000 being affected. It will be discussed in detail in Chapter 9.

The effect of teratogenic chemicals depends not only on how the teratogenic agent interferes with the biochemistry of the embryo/fetus, but also on the period of exposure, and this complicates the study of teratogenic agents. For example, the limb malformations in humans caused by thalidomide are caused by exposure only during a fortnight, between day 23 and day 38 after conception. While malformations are obviously caused during the period of organogenesis, cancer cells are predominantly formed later during the growth period, and the different effects could very well be caused by the same chemical. Early childhood tumors are generally considered to be formed already in the fetus, which actually is quite sensitive to carcinogenic agents. This is because the cell division rate is much higher in fetal cells than in normal somatic cells (a factor that facilitates the transformation of a normal cell into a cancer cell) and because the immune system

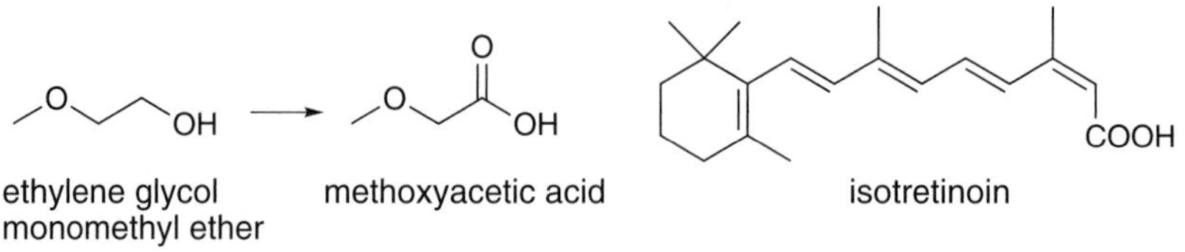

ethylene glycol   methoxyacetic acid   isotretinoin
monomethyl ether

Figure 7.32 Examples of teratogenic compounds.

that can detect and kill abnormal cells (e.g., cancer cells) is poorly developed in the fetus.

The maternal and fetal blood are not mixed, but are separated by the placenta, through which nutrients and waste products are transported by the mechanisms discussed in Chapter 4. Lipophilic exogenous chemicals diffuse rapidly through the placenta, but so also do water-soluble molecules with molecular weights less than 800 (although much less efficiently). The latter transport takes place by diffusion through water-filled channels with fixed diameters. Lipophilic compounds that are transformed/converted into more polar forms may be concentrated in the embryo/fetus, and this has been discussed as a possible mechanism for the action of thalidomide (which is relatively easily hydrolyzed to more polar compounds in contact with water). It has been shown that the pH of the blood of rodent embryos is higher than that of the maternal blood, and weak acids would consequently be trapped and concentrated in the embryos. Several chemical teratogens are in fact weak acids, for example, methoxyacetic acid (the teratogenic metabolite of the solvent ethylene glycol monomethyl ether, see Figure 7.32), isotretinoin (a synthetic retinoid used to treat certain forms of acne, see Figure 7.32), thalidomide (the hydrolysis products being weak acids), and valproic acid (a drug against epilepsy, see Subsection 5.3.1.4 for its structure and metabolic activation), and this pH gradient may play an important role for these teratogens. The placenta has a limited capacity to metabolize exogenous compounds and is normally considered to be of minor importance in this respect, although it has been shown that smoking, for example, will induce the metabolic capacity (e.g., cytochrome P450) of the placenta considerably. The fetus also has limited metabolic capacity, although this increases constantly as it grows. Normal concentrations of, for example, cytochrome P450 will be attained approximately one year after birth. Elimination from the fetus takes place via the placenta or kidneys to the amniotic fluid, the latter only being a temporary solution as this fluid is constantly swallowed.

Although few human teratogens are known, the same principle that applies with toxicity in general also applies to teratogenic effects: Any chemical is able to harm the development of the embryo/fetus if the dose is sufficiently high. The thalidomide tragedy has made us aware of the potential potency of chemical teratogens, and has had an enormous impact on the science of teratology. The toxicity of thalidomide was missed by the warning systems of the 50s, but although the test protocol for new drugs has been changed today and constitutes a considerably more fine-meshed net, there are no guarantees that the story will not repeat itself. It should be remembered that the reason for the rapid identification of thalidomide as a teratogen was due to the fact that it causes unusual malformations (limb deformations). If it had caused more common malformations it would have been much more difficult and have taken a much longer time to establish the connection, and even more children would have been afflicted.

7.6
Endocrine Disruptors

The endocrine system consists of the endocrine organs, which are glands that produce hormones that are released inside the body, where they regulate a variety of bodily functions. Examples are blood pressure and body temperature. They are also crucial for our reproduction, as well as for building muscle mass and bone. Some of the most important human endocrine glands are the hypothalamus, the pituitary gland, the thyroid and parathyroid glands, the pineal gland, the pancreas, the adrenal glands, the ovaries, and the testes (see Figure 7.33).

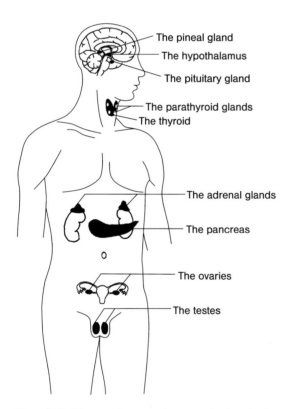

The pineal gland
The hypothalamus
The pituitary gland
The parathyroid glands
The thyroid
The adrenal glands
The pancreas
The ovaries
The testes

Figure 7.33   The most important human endocrine glands.

The hypothalamus	This is located between the two halves of the brain and produces hormones which regulate the secretion of the pituitary hormones, thus controlling the endocrine system. In addition, the hypothalamus gland is also an important part of the CNS and is involved in the regulation of metabolism, reproduction, the biological clock, and our emotional life.
The pituitary gland	This is a central endocrine organ. It releases hormones that regulate the hormone production of the thyroid, the adrenal glands, the ovaries, and the testes. In addition, the pituitary gland produces growth hormone that stimulates body growth, prolactin, which stimulates milk secretion from the mammary gland, oxytocin, which stimulates uterine contractions, and vasopressin, which reduces the water excretion in the kidneys.
The pineal gland	Also known as the epiphysis, this is a small gland in the brain of all mammals that produces melatonin, a hormone that controls the circadian rhythm. It is produced when it is dark, but when it is light the production decreases. In addition, melatonin influences the levels of a number of other hormones in the body.
The thyroid gland	This is controlled by the pituitary gland, and can produce the iodinated hormones triiodothyronine and thyroxine (see Section 9.4.4 for chemical structures), which increase cell metabolism by increasing oxygen uptake. The iodine is absorbed by the thyroid from the blood, and it is stored in the gland as a gel. If you are exposed to radioactive iodine, for example in a nuclear war or after a nuclear reactor meltdown, the thyroid is particularly exposed.
The parathyroid glands	These are four pea-sized glands at the back of the thyroid which produce parathyroid hormone that raises the levels of calcium in the blood by increasing the absorption of calcium in the intestines and by dissolving a small portion of the bone. The calcium concentration of the blood in turn controls the amounts of the hormone released, so this is a self-regulatory system.
The adrenal glands	These are positioned on the top of the kidneys (adrenal = at the kidneys) and produce corticosteroids and catecholamines. For example, cortisol, which regulates the turnover of sugar, fat, and protein, and aldosterone regulate the excretion of water and salts by the kidneys.

Also epinephrine (epinephrine) and norepinephrine, which are stress hormones that increase your heart rate and raise blood pressure and blood sugar levels, are produced here. Epinephrine and norepinephrine are also neurotransmitters in the nervous system.

The pancreas   Primarily, this produces an enzyme cocktail that is secreted in the small intestine and takes part in the conversion of fats, starches, sugars, and proteins. A small proportion (about 3%) of the pancreas, called the islets of Langerhans, produces the hormones insulin and glucagon that together and in different ways regulate the concentration of sugar in the blood.

The ovaries   These are the organs in which egg cells in female mammals mature. They produce the primary female steroid sex hormones, the estrogens, which stimulate the development of egg cells, and progesterone, which regulates the menstrual cycle and the changes that take place during pregnancy.

The testes   These are the male counterpart of the ovaries. As well as producing sperm they are also part of the endocrine system and produce male steroid sex hormones. The most important of these is testosterone, which controls the development of male sexual characteristics and stimulates the development of the musculoskeletal system. It is consequently an anabolic steroid.

Hormones are therefore chemical substances that regulate the development and function of various tissues and body functions on the molecular level. They are defined as either endocrine, produced in an organ and released to the blood which transports them to the part of the body where they have their effect, or exocrine, which is produced in the tissue where they act. From a chemical point of view they are classified as proteins and peptides (made from amino acids, such as the growth hormone), steroids (such as the sex hormones based on terpenes), simple amines or amino acid derivatives (such as epinephrine and norepinephrine), or derivatives of fatty acids (such as prostaglandins and leukotrienes, typical exocrine hormones). The hormones provoke their effects by binding to receptors in the cell membranes of target cells, which in various ways leads to a response, which often depends on a cascade of molecular events initiated by the interaction of the hormone and the receptor. The receptor for a hormone thus functions as a sort of relay, transmitting the signal through the cell membrane. However, certain hormones, such as the steroids, may themselves diffuse across the cell membrane. They instead act by binding to specific proteins inside the cell, and the steroid/protein complex has the ability to promote or inhibit the use of very specific parts of the DNA of that cell. In this way, minute amounts of a hormone can cause dramatic effect in a tissue, and the endocrine system is consequently extremely sensitive to the impact of other chemicals. Compounds having an effect on the endocrine system without being hormones are called endocrine disruptors.

Just like the nervous system, the endocrine system can be affected by compounds that act as agonists, binding to a hormone receptor/protein and provoking the same effect as the hormone itself. Antagonists, on the other hand, will bind to and block the receptor/protein so that the actual hormone has no effect. In addition, endocrine disruptors can be toxic by affecting the levels of hormones in the blood, for example, by disturbing their release, by inhibiting or promoting their degradation, or by affecting the receptor density on target cells. Endocrine disruptors that interfere with the sex hormones have received special attention, as they can cause impaired sexual development (for example, feminization of young men and malformation of the genital organs), reduce fertility, and promote tumors of the genital tract. Because the embryos are extremely sensitive to imbalances in sex hormones, the defects frequently appear in the children of exposed pregnant women, and in such cases the links between the exposure and effect are very difficult to pinpoint. Even if it is the exposed person that is affected, the effects are so vague and so difficult to measure (fatigue, etc.) that it is almost impossible to prove that the endocrine system of humans is affected. Consequently, our knowledge about endocrine disruptors is very limited, although we strongly suspect that a number of chemicals used on a large scale do have at least some effect on the endocrine system.

Examples of chemicals to which humans are exposed simply by consuming normal food and drinking water and that are considered to affect the endocrine system are:

- PCBs, which have been shown to possess estrogenic as well as anti-estrogenic activity in experimental animals and to modulate the effects of the thyroid hormones. The chemistry and hazards of the PCBs are discussed in Section 10.6.2.

- DDT, which has an estrogenic and an anti-androgenic effect in experimental animals. DDT is further discussed in Section 9.2.2.

- Brominated flame retardants, for example polybrominated biphenyls and brominated diphenyl ethers, which have roughly the same effect as PCBs (further discussed in Section 9.4.4).

- Phthalates, used as plasticizers in plastics, and alkylphenols (starting material for the manufacture of surfactants) both have estrogenic effects in animal experiments (see Section 9.4.3).

- Heavy metals like lead, which disrupt the hormones that control ovulation, and cadmium, which disrupts hormones that regulate calcium uptake into cells. The effects of metals are discussed in Section 10.7).

In addition, we are involuntarily exposed to hormones and hormone analogs that are given to animals that are bred for slaughter to make them grow faster. Some people, however, take anabolic steroids voluntarily in order to become more muscular, and the use of hormones for this kind of 'doping' is a huge problem. The synthetic compounds, such as the anabolic steroid methandrostenolone, are easily detected by analysis of a blood sample, but the naturally occurring steroids are more difficult to monitor. In some cases it has been necessary to define what natural levels of hormones in the blood are acceptable and to penalize athletes that have excessive amounts.

8
The Molecular Basis for Genotoxicity and Carcinogenicity

Genotoxic chemicals will affect the genetic material in particular. The effects may be that a gene is injured so badly that it can no longer fulfill its tasks, or that the information encoded in the sequence of the nucleic acids in DNA is altered so that the protein product is modified. The first effect, if the gene affected is essential for the cell, will lead to severe malfunction or death of the cell, while the second is called a mutation and generates permanently changed cells with slightly different properties. Note that there is a significant difference between 'damage to DNA' and 'mutation'. Mutations are the basis of a number of very serious effects of chemicals, for example, hereditary defects, tumors, and teratogenic effects, and the fact that they affect not only individuals but also the whole of mankind gives genotoxic chemicals their special importance. Although mutations have played a key role in the evolution of life on earth, and may continue to do so even though other factors today are believed to be more important, we should consider mutations to be harmful in general and particularly to individuals. Mutations happen spontaneously, although at a very low rate, because of built-in 'shortcomings' in the chemistry of nucleic acids and in the enzymatic systems synthesizing and handling DNA. Such mutations are inevitable, but the additional ones, caused by external agents, should be avoided as far as possible.

8.1
Chromosomes, Genes, and Mutations

The chemistry of DNA has been discussed in Chapter 2, and Figure 8.1 shows how the DNA is organized in larger structures called chromosomes. (How the genetic information is extracted and used for the synthesis of proteins will not be discussed here. Anyone who is not familiar with this is recommended to consult a textbook on biochemistry or molecular biology.) The chromosomes are visible in a light microscope when they are condensed, that is, when they are compact structures. They must be compact when cell division is taking place, when it is absolutely necessary that they should be transportable structures that can be physically moved from the mother cell to the daughter cells. In preparing for a cell division, the genetic material of a cell is duplicated, and in the 'chromosome

Chemistry, Health, and Environment. Olov Sterner
© 2010 WILEY-VCH Verlag GmbH & Co. KGaA, Weinheim
ISBN: 978-3-527-32582-5

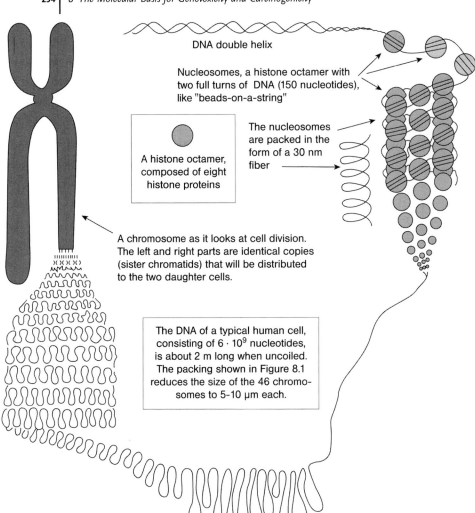

DNA double helix

Nucleosomes, a histone octamer with two full turns of DNA (150 nucleotides), like "beads-on-a-string"

A histone octamer, composed of eight histone proteins

The nucleosomes are packed in the form of a 30 nm fiber

A chromosome as it looks at cell division. The left and right parts are identical copies (sister chromatids) that will be distributed to the two daughter cells.

The DNA of a typical human cell, consisting of $6 \cdot 10^9$ nucleotides, is about 2 m long when uncoiled. The packing shown in Figure 8.1 reduces the size of the 46 chromosomes to 5-10 μm each.

Figure 8.1 The construction of a human chromosome, down to the DNA double helix.

man' (see Figure 8.1) with two arms and legs that we can see in a microscope, one arm and one leg is a copy. Our cells contain 23 pairs of chromosomes, one chromosome in each pair originating from our mother and the other from our father.

A defect, caused by a genotoxic event, affecting a specific function in a cell is naturally considered to be associated with the gene(s) that encode(s) for the protein(s) involved directly in the function in question. However, it should be remembered that, for example, the loss of the ability to carry out a certain enzy-

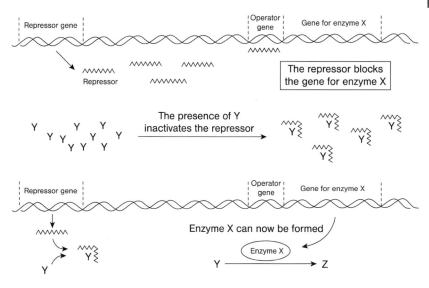

Figure 8.2 How the encoding of a gene for an enzyme can be regulated.

matic conversion because the cell has been exposed to a genotoxic chemical does not necessarily mean that the gene for that enzyme has been destroyed or mutated. Another possibility is that genes that are involved in the regulation of when and how much of a protein should be formed have been affected. Such genes may be located in completely different chromosomes, and we shall see examples of this later. In Figure 8.2, an outline of a simple regulation of the encoding of the gene for an enzyme is presented.

Say that the enzyme is called X and that its task is to convert compound Y to compound Z. As long as the cell contains no Y, there is no need for the enzyme and it is not formed. In this situation the encoding of the gene is blocked by a repressor, which binds to the stretch of DNA just in front of the gene (this stretch is called the operator of the gene) and prevents the encoding enzymes from doing their job. However, if the concentration of Y increases, the repressor leaves (for example by forming a complex with Y that is chemically different and no longer can bind to the operator), and the information in the gene is translated to produce enzyme X (see Figure 8.2). If the conversion of Y to Z in a cell does not work because of a mutation, it could of course be the gene for enzyme X that has been mutated. But it could also be the operator that has been changed by the mutation, and if the new operator does not bind the repressor, enzyme X will always be formed, which can be a big problem for a cell. Alternatively, the repressor may bind too well, not letting go as the concentrations of Y increase, resulting in a constant lack of enzyme X. The repressor, a protein that is encoded by a gene that is situated somewhere else in the genome, may also be changed if it is the repressor gene that has mutated. The mutated repressor may bind too loosely or too strongly, creating similar problems, as already mentioned. The fact that we in principle have two copies of all genes (maternal and paternal copy) is valuable; if one copy is damaged, the other can often cope alone.

Another technique that the cells use to mark out which genes should be encoded and which should not, is to methylate certain positions (e.g., the carbon opposite to the carbonyl group in cytosine, something that will not affect the base-pairing of cytosine), making use of SAM (Section 5.4.5) and specialized enzymes. Other enzymes can demethylate DNA, so it is a reversible process. The methyl groups function as 'flags' that direct the encoding machinery, which of course may be confused if an exogenous methylating agent inserts methyl groups at random.

8.2
DNA as a Molecular Target

DNA is an extraordinary molecule, cleverly designed for its function as the carrier and mediator of information in cells and organisms, and is therefore also the master. One can imagine that evolution has tested many different molecular constructions for the job before settling on DNA! DNA can be damaged not only by chemicals, but also by various forms of electromagnetic radiation (e.g., UV light) or by ionizing radiation emitted from radioactive materials. At the molecular level, DNA is characterized by a rich content of heteroatoms, oxygen in the backbone, and nitrogen in the DNA bases. The bases are reasonably efficient nucleophiles and react with electrophiles forming chemically modified bases. In addition, DNA is an extremely large molecule, and a portion of DNA can form a complex with another molecule or with metal ions. This may be significant for the mechanisms by which some mutagenic and carcinogenic materials act (e.g., metals like chromium, nickel, and cadmium).

While our exposure to genotoxic chemicals certainly has increased during recent decades, radiation has always been around and has constantly posed a serious threat to life. As a response, our cells have developed several enzymatic systems that detect and repair damage to DNA. As the double helix is such a condensed structure, an additional group attached to a base will give rise to a distortion of the helix, and the repair systems are very skillful in detecting this. DNA repair is an important aspect of genotoxicity, and Figure 8.3 gives a schematic representation of two such repair systems. The so called excision repair will probe along the double helix and sense structural irregularities caused by a chemically modified base, and the affected nucleotide (possibly together with a few neighbors) will be enzymatically removed. Because it leaves the opposite unmodified strand as it is, other enzymes can use this as a template, add the missing nucleotides, and restore the DNA completely. This repair is almost faultless, and will take care of the majority of all alterations of DNA. However, some types of damage are not so easily repaired, for example, those that take place immediately before the DNA of a cell is duplicated in preparation for cell division or that are caused by bifunctional electrophiles crosslinking the two DNA strands. There are repair systems capable

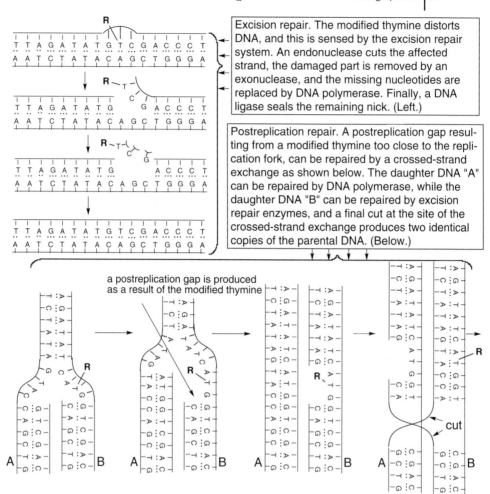

Excision repair. The modified thymine distorts DNA, and this is sensed by the excision repair system. An endonuclease cuts the affected strand, the damaged part is removed by an exonuclease, and the missing nucleotides are replaced by DNA polymerase. Finally, a DNA ligase seals the remaining nick. (Left.)

Postreplication repair. A postreplication gap resulting from a modified thymine too close to the replication fork, can be repaired by a crossed-strand exchange as shown below. The daughter DNA "A" can be repaired by DNA polymerase, while the daughter DNA "B" can be repaired by excision repair enzymes, and a final cut at the site of the crossed-strand exchange produces two identical copies of the parental DNA. (Below.)

Figure 8.3 Excision repair of DNA.

of handling such events also (see Figure 8.3), but they are more error-prone, and the risk of an unsuccessful repair leading to a mutation increases.

8.3
Chemical Effects on DNA and DNA-Handling Systems

8.3.1
Modification of the DNA Bases

DNA bases that have been modified chemically have problems fitting into the double helix in the correct way and forming the hydrogen bonds to the

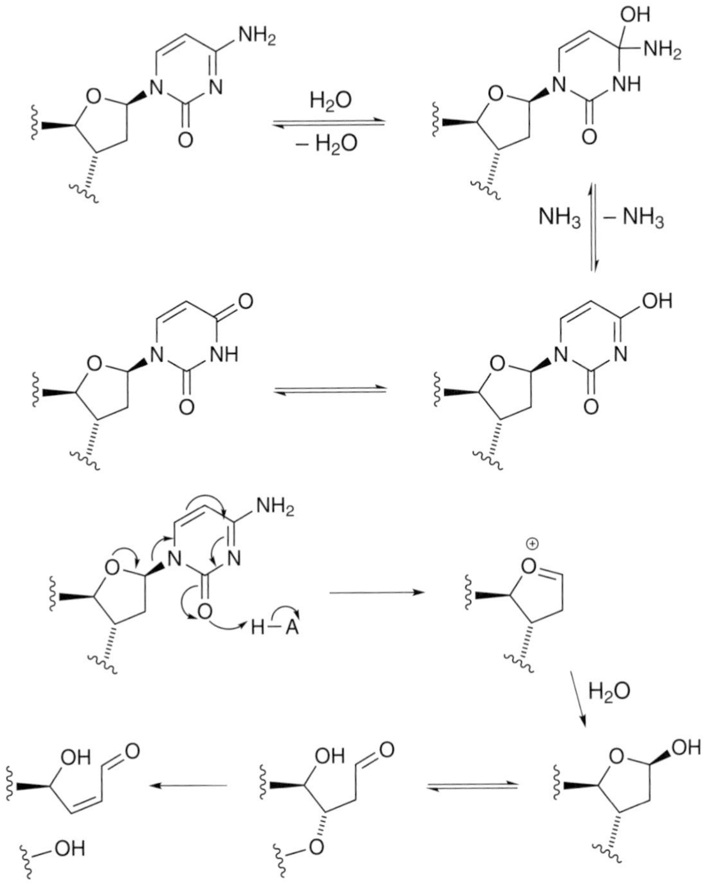

Figure 8.4 Spontaneous chemical reactions that may take place in DNA.

complementary base in the opposite strand. There is a certain probability that such chemical changes take place spontaneously, for instance by hydrolysis (deamination) of cytosine via the addition of water to the double bond followed by elimination of ammonia to form an enol, which is rapidly converted to a ketone (demethylthymine). Another possibility is the spontaneous hydrolysis of a nucleoside followed by the β-elimination of the open form of the deoxyribose (the two forms are in equilibrium with each other). This will result in a break in one strand as shown in Figure 8.4 (see Figure 8.8 for a discussion about the mechanism).

These and other spontaneous chemical conversions of DNA are not rapid reactions, but, in view of the high number of DNA bases present in our cells, a few per day will nevertheless be affected. Such spontaneous changes may lead to mutations according to the mechanisms discussed in the following section, and this is part of the evolution of life on earth. More important are exogenous chemicals

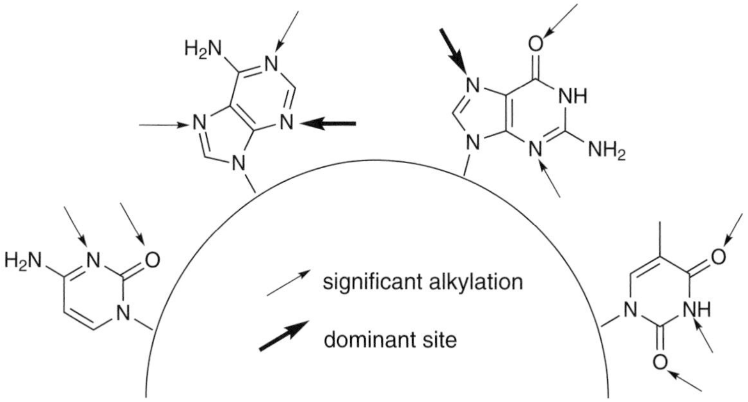

Figure 8.5 The most frequent sites for alkylation of the bases in double-stranded DNA.

that are able to react with and modify DNA, and, as the DNA bases are nucleophilic, any electrophile will in principle be genotoxic. A wide range of electrophilic chemicals and compounds that are converted to electrophiles by the metabolism are known to react with the DNA bases, both *in vivo* and *in vitro*, and the properties of electrophiles and their formation during the metabolic conversions have been discussed in Chapters 3 and 5. The positions in the DNA bases in double-stranded DNA that most frequently react with nucleophiles are indicated in Figure 8.5.

Because of the closed structure of double-stranded DNA, the bases will only react when the helix is dissociated, for example, during decoding. The monomeric nucleosides, which are building blocks for the enzymatic systems synthesizing and repairing DNA, are more prone to reaction with electrophiles, and although the enzymes normally reject modified nucleosides they could in principle be used, resulting in a chemically modified DNA. However, the likelihood of this is so small that the reaction between a free nucleoside and an electrophile can instead be regarded as a protection for more sensitive biomolecules. The fate of unnatural or modified bases in DNA is in most cases that they are detected by the repair systems (see above) and exchanged. Some modified bases will be chemically unstable and eliminated from DNA (see below) and will be replaced by the repair systems. Chemical lesions to DNA that for some reason are difficult to detect, difficult to repair, or formed just before the cell replicates its DNA in preparation for a cell division, may cause a mutation by the mechanisms discussed in the next section. Examples of DNA lesions that are difficult to handle are those caused by bifunctional electrophiles, as they as well as modified bases can form intra- and inter-strand cross links (see Figure 8.6), which in general are more difficult to handle for the repair systems.

In addition, although little is in fact known about this, it is reasonable to believe that chemicals that are reactive, as, for example, radicals (formed for example

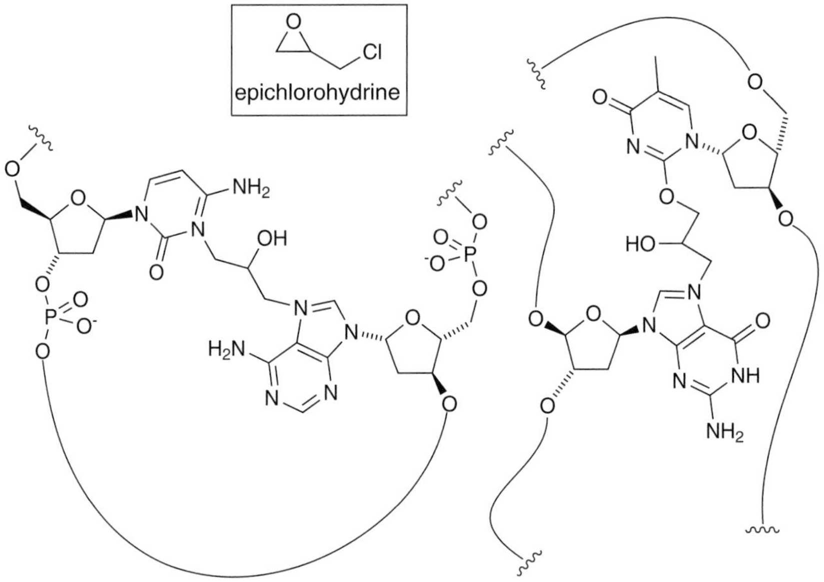

Figure 8.6 Intra- and interstrand cross links by the bifunctional epichlorohydrin, which to the left has reacted with adenine and cytosine in the same strand and to the right with guanine and thymine in opposite strands.

during lipid peroxidation) or oxidizing agents (e.g., metal ions) may also modify DNA in various ways.

8.3.2
Intercalation

Intercalators are flat molecules, often consisting of several aromatic rings, that can fit between two base pairs in the double helix, as the meat in a hamburger is positioned between two slices of bread. Intercalation as a phenomenon has been known for quite some time, as the length of DNA increases in the presence of intercalators and can easily be measured. As many polyaromatic hydrocarbons, as, for example, aromatic amines, are mutagenic and carcinogenic, it was initially believed that their genotoxicity is caused by their ability to intercalate with DNA. However, the expanding knowledge about the metabolic activation of such compounds to reactive electrophiles has changed our view of interrcalation. We know that it takes place although it is no longer considered to be important *per se*. It can, of course, direct a compound containing an additional electrophilic functionality to a suitable position and facilitate its reaction with DNA, something that is believed to happen in the case of the carcinogenic mycotoxin aflatoxin B$_1$ (see Figure 9.30 for structure). An example of intercalation is shown in Figure 8.7.

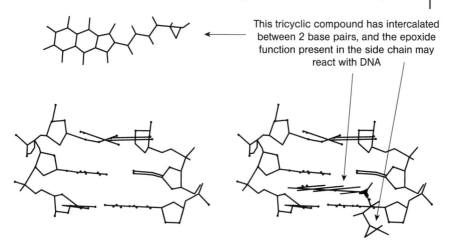

This tricyclic compound has intercalated between 2 base pairs, and the epoxide function present in the side chain may react with DNA

Figure 8.7   Intercalation.

8.3.3
Loss of Modified Bases

In DNA, the bases are linked to deoxyribose by an N–C bond, and the carbon atom also has an oxygen bound to it. Molecules containing a carbon atom that has two heteroatoms bound to it by single bonds will be hydrolyzed in contact with water, generating the free carbonyl group. Consequently, DNA in water should be hydrolyzed, and in Figure 8.4 a mechanism for this is shown. However, the rate of hydrolysis is very slow. At pH 7 and 37 °C the half-life for adenine and guanine in DNA is approximately 15 000 000 h. (This appears reassuring until one calculates how many adenines and guanines our DNA contains ...) Chemical modification of a DNA base may decrease the stability of a nucleoside dramatically, and the corresponding half-life of adenine and guanine (in DNA) that has been alkylated in position 7 (the major position for alkylations, see Figure 8.5) is less than 200 h. As free nucleosides, 7-alkylated adenine and guanine are even less stable, with half-lives of a few hours. The reason for this is evident from the deficit of electrons that such an alkylation will cause, and the mechanism for the hydrolysis is shown in Figure 8.8. Note that the hemiacetal produced by this hydrolysis is in equilibrium with its open form, the γ-hydroxyaldehyde, which may undergo β-elimination according to the mechanism shown in Figure 8.8 (via the corresponding enol). This will result in a strand break, serious DNA damage for any cell to handle.

8.3.4
Modification of other Macromolecules

A large number of enzymes and other proteins are involved in the handling of DNA, the synthesis and repair of DNA, and the transport of the chromosomes to the two daughter cells during cell division. If these enzymes are not fully

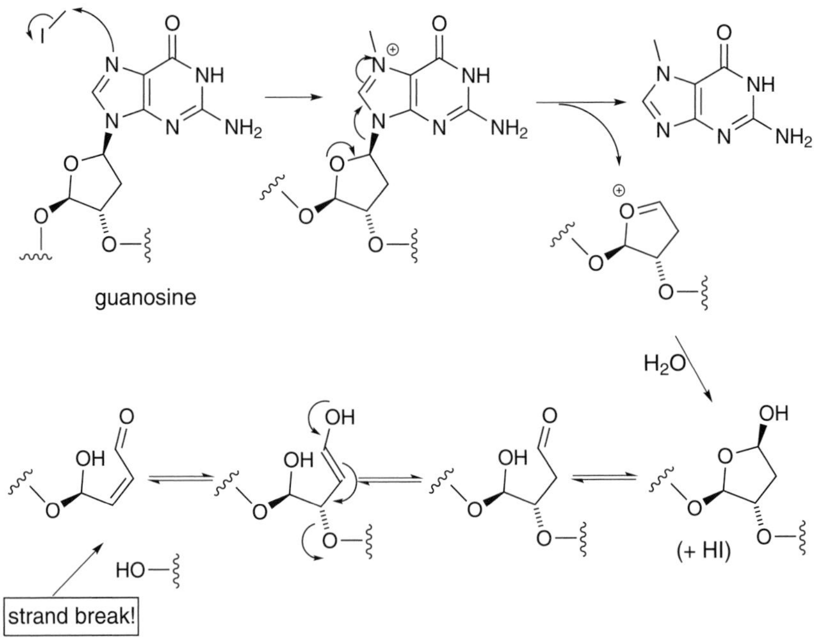

Figure 8.8 The methylation of guanosine with iodomethane promotes hydrolysis.

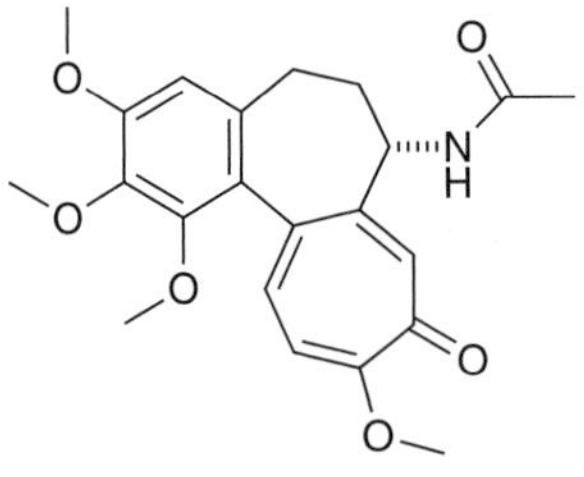

Figure 8.9 Colchicine.

functional, for example, because they have been affected by chemicals, they may make mistakes leading to mutations. Again, little is known about this, but it is reasonable to assume that the fidelity of enzymes will decrease if they are slightly modified by chemicals. Electrophilic compounds will of course be able to affect proteins by reacting with the nucleophilic functionalities present, and so will radicals. In addition, these enzymes are considered to be important targets for mutagenic and carcinogenic metals (e.g., Cd, Cr, and Ni). Metal ions may bind specifically to proteins by complexing, leading to a change in the three-dimensional protein structure and consequently to a modification of its function.

The transport systems for the chromosomes are still poorly understood: they are complex and can be disturbed by many chemicals. One of the best studied is the toxic plant alkaloid colchicine (see Figure 8.9). Colchicine disturbs the microtubules, which are protein structures that function like threads during cell division and pull the chromosomes to the two cell nuclei being formed. Any exogenous chemical affecting these mechanisms may, as well as being very toxic, also be genotoxic.

8.4
Different Types of Mutations

8.4.1
Point Mutations

Point mutations are small, affecting only one or a few base pairs. They can be divided into two types; base-pair substitutions and frame shift mutations.

8.4.1.1 Base-Pair Substitution Mutations

As the name implies, base-pair substitutions arise if one base-pair in the base sequence, for example, G–C, is exchanged for another, C–G, A–T or T–A. A base-pair substitution will only change the sequence in one triplet of bases, and if the triplet is part of a gene encoding for a protein, one of the amino acids in this protein (consisting of perhaps several hundreds of amino acids) will at worst be exchanged in the mutated cell. In the case that this amino acid plays a crucial role for the function of the protein and the protein plays a crucial role for the function of the cell, the mutation will be serious and possibly lethal for the cell. However, many individual amino acids in a protein may be exchanged without affecting the function of the protein significantly. Base-pair substitutions are therefore, as indicated in Figure 8.10, in most cases not very serious for a cell.

8.4.1.2 Frame Shift Mutations

A shift in the frame of a double helix means that one or several base-pairs are added or removed, resulting in a mutated cell with one or more bases added or removed. A frame shift mutation in a gene will in most cases be much more serious for the cell than a base-pair substitution. If the number of base-pairs added or removed is not divisible by three, the reading frame for the encoding enzymes will be distorted from the position of the mutation, and in principle all triplets and thereby all amino acids will be changed. This will result in a useless protein, and the effect on the cell is more obvious, as indicated in Figure 8.11.

If the number of base-pairs added or removed is a multiple of three, the immediate effect of a frame shift mutation will be more comparable with that of a base-pair substitution. Besides affecting genes for proteins, mutations can also change base sequences used as, for example, a 'stop' signal or to which enzymes or other proteins bind in order to initiate or block the encoding of a certain gene (see above). However, there are no general rules about the effects of various point mutations in such cases.

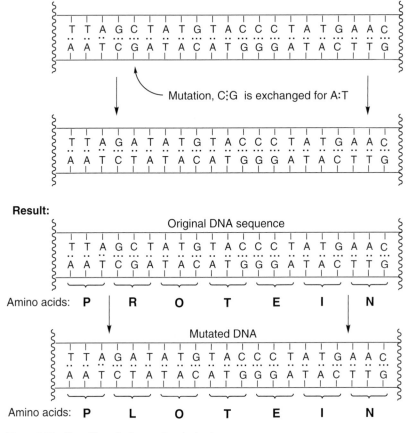

Figure 8.10 The effect of a base-pair substitution.

8.4.2
Chromosome Aberrations

Genetic alterations that are so extensive that they can be seen when the condensed chromosomes are observed under a microscope are called chromosome aberrations. They include the loss of a chromosome or a part of a chromosome from the genome of a cell, the transfer of a part from one chromosome to another, or the duplication of a part or a whole chromosome (see Figure 8.12).

Chromosome aberrations normally result in the death of the afflicted cell, and this is certainly to be expected if genetic information has been lost. However, although no information has been lost after a transfer of a piece of a chromosome from one position to another, this is nevertheless a serious mutation because the cell's ability to control how the information in the transferred part should be used has changed. Even if one whole chromosome has been duplicated and exists as a free-standing perfect copy, this will give rise to a slight imbalance in the genetic

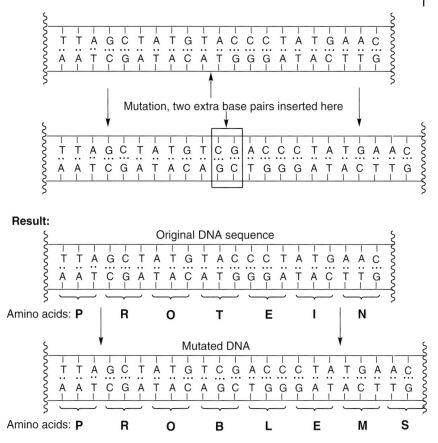

Figure 8.11 The effect of a frame shift mutation.

information that does not leave the cell untouched. However, chromosome aberrations are not necessarily lethal; for example, in the case of C-21 trisomy (Down's syndrome) in which all cells of a human have an additional copy of the 21st chromosome, the differences from a normal individual are not that conspicuous.

In addition to point mutations and chromosome aberrations, it is reasonable to assume that intermediate mutations affecting one or a few hundred base-pair(s) also take place, but we have at the moment very little knowledge about such mutations.

8.5
Chemical Mutagenesis

As discussed above, the most likely event that will result from a chemical modification of DNA is that it is repaired, and very few chemical lesions actually

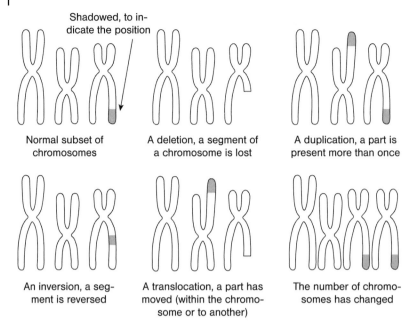

Shadowed, to indicate the position

| Normal subset of chromosomes | A deletion, a segment of a chromosome is lost | A duplication, a part is present more than once |

| An inversion, a segment is reversed | A translocation, a part has moved (within the chromosome or to another) | The number of chromosomes has changed |

Figure 8.12 Examples of chromosome aberrations.

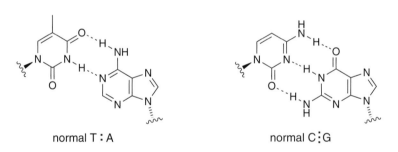

normal T : A   normal C : G

Figure 8.13 The hydrogen bonds between the normal DNA base pairs.

give rise to mutations. This happens when the repair systems fail because the lesion is too complicated, when there are too many injuries so that the repair systems are saturated, or when the cell is dividing rapidly. The process in which chemically modified DNA is transformed to mutated DNA may proceed via different mechanisms, and this is another area where our knowledge is not complete. Base-pair substitutions can obviously be the result of a mismatch, when a chemically modified base pairs up with an incorrect base. Modifications that affect the hydrogen bonding ability of the normal base pairs (see Figure 8.13) could in principle lead to a mismatch that eventually results in a base-pair substitution.

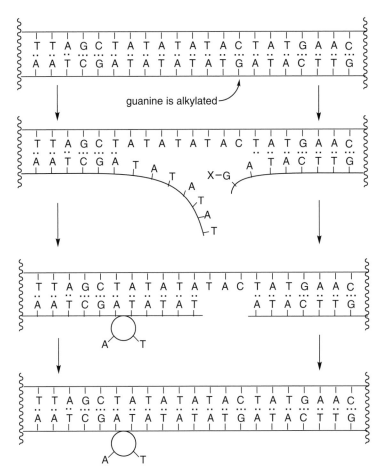

Figure 8.14 Frame shift mutations may be the result of failed repairs.

Chromosome aberrations could be the result of damage that cuts both strands in DNA, and an example of a compound that is able to do this is the antitumor antibiotic calicheamicin γ_1^{Br} (Section 7.1.4.2).

Any type of mutation can be caused by unsuccessful attempts to repair a chemical injury to DNA, and this is how frame shift mutations are probably induced. Frame shift mutations are most common in repetitive stretches in the DNA containing the same or the same combination of bases over and over again, as exemplified in Figure 8.14. If one base in such a stretch is modified and subjected to the efforts of, for example, the excision repair system (see above), the deoxyribose-phosphate polymer is hydrolyzed close to the modified base. This induces some motility of the ends close to the cut, and if one end dissociates for a second and then binds back again, a loop containing one or several bases can be formed. Meanwhile, if the modified base has been hydrolyzed, the scene is set for adding the missing bases (with the complementary strand as template) and sealing the deoxyribose-phosphate chain. As shown in Figure 8.14, the result is that one strand in the double helix now contains additional bases, and, if the cell containing this DNA enters the cell division cycle before this has been recognized, one of the daughter cells will have a frame shift mutation.

8.6
Effects of Mutations

We have already noted that chromosome aberrations are more serious for a cell than point mutations and that frame shift mutations are more serious than base-pair substitutions. However, we are composed of many cells of many different functional types (see Chapter 2), and the risks for multicellular organisms are obviously not the same as the risks for individual cells. For a discussion about the effects of mutations, it is convenient to distinguish between the three different cell types: germ cells (in mammals, sperm and ova and their precursors), embryo cells (of which an embryo consists), and somatic cells (the cells that make up a complete individual) (see Figure 8.15).

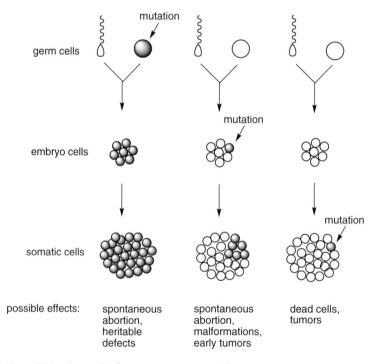

Figure 8.15 The result of mutations in various cell types.

8.6.1
Effects of Mutations in Germ Cells

Germ cells (sperm and ova in mammals) participate in conception and may result in new individuals. In a conceived ovum, which will develop into an embryo and further into a fetus (discussed in Chapter 7), mutations in either the ovum or the sperm that participated in the conception are very likely to create difficulties. The major reason for this is that in principle all genetic information present in these cells is important and will be used immediately. Mutations in germ cells therefore have a relatively high probability of creating a problem, and problems in an embryo frequently lead to a miscarriage within a short time after conception. In some cases, if the mutation is insignificant for the development of the embryo and the fetus, it may survive and is then, of course, present in every cell of the whole individual, including its germ cells (see Figure 8.15). If the mutation does not prevent the individual reaching sexual maturity and does not interfere with reproduction, the mutation will be transferable to future generations and is in principle eternal.

8.6.2
Effects of Mutations in Embryo Cells

The cells in a growing embryo need more of the genetic information available than somatic cells, and are therefore more sensitive to mutations. In addition, the fact that they divide more frequently enhances the risk that a chemical lesion will be transformed into a mutation. The earlier during a pregnancy a mutation takes place, the more likely it is to result in a miscarriage. Various malformations can also be the result of mutations in embryo cells, and this was discussed in Chapter 8. In addition, it is believed that many cancers afflicting children are caused by mutations disturbing the control of the cell division that occurs in embryo cells. Cancers, as will be discussed below, normally appear in older people because it takes time for most cancer cells to develop, but if the transformation starts in the womb the process may be finished at an early age.

8.6.3
Effects of Mutations in Somatic Cells

The cells of a complete individual are generally the least sensitive, simply because they need little of the available genetic information to fulfill their tasks and because most somatic cells are relatively unimportant for the organism. If a mutation occurs in a somatic cell, the most likely outcome is that because it takes place in a part of the genome that is not used by this cell it will pass unnoticed. If the mutation takes place in an active part of the genome and affects information that is vital for the cell, the cell will not work as well as it should and may die. In this case it will most likely be replaced by a neighboring cell that divides. Some cells, for example, nerve cells, are not replaced, and lethal mutations in such cells will slightly deteriorate the function of the individual. The worst thing that can happen

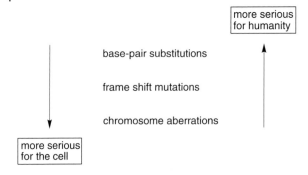

Figure 8.16 The effect of different kinds of mutation.

is that mutations in somatic cells change them in a way that they evade the control systems necessary for the cooperation of various cells in a multicellular organism. For example, the rate of cell division has to be kept under strict control, this control being based on genetic information, and mutations may transform a cell into a tumor cell (see below). However, although the diseases that result from mutations in somatic cells may be lethal, it is important to remember that the effect is restricted to that one individual.

8.6.4
Which Mutation is the Most Serious?

For any cell, chromosome aberrations are more serious than point mutations. However, if one instead considers organisms, such as human beings, the question is not so easily answered. Is a chromosome aberration in an epithelial cell in the lung that results in the transformation of this cell to a tumor cell worse than a frame shift mutation in a germ cell that will induce a miscarriage? If the perspective is further expanded to the human species or to life on earth in general, the question is simplified again (see Figure 8.16). For humanity, the base-pair substitutions in germ cells are the most serious, because they have the greatest chance to pass from one generation to another and become part of the human genome. These are the mutations that really matter in the long run and that we should be most concerned about!

8.7
Chemical Carcinogenesis

While mutations have been absolutely necessary for the development of life on earth and a prerequisite for the existence of humans, there are no positive effects in the short run. In organisms that reproduce sexually there are also other mechanisms for creating new sets of the genome. Man would therefore continue to evolve also in the absence of mutations, and it is reasonable to assume that mutations are harmful. One of the most dreaded effects for man is that an individual cell loses its ability to control its rate of division and is transformed into a cancer cell. The loss of control over the rate of division is typical of a tumor cell, although it should be remembered that tumor cells can be benign or malignant. While benign tumors, made up of benign tumor cells that divide frequently, are seldom dangerous because they only grow to a certain size and do not spread, the malignant tumor cells grow faster and infiltrate other tissues (metastasis) and will generally kill its victim if not treated. Cancers are malignant tumors. Anything that is carcinogenic is then by definition able to speed up the process of transformation of a normal cell to a cancer cell. A cancer cell is completely different from a normal cell, even compared with the cell that it originates from. Apparently, several major changes must have taken place, either as a result of multiple mutations or from the activation/deactivation of silent/active parts of the DNA. Remember that somatic cells still contain the entire genome although they only make use of a small part of it, and the parts that are not used are 'shut off' in various ways. However, if reactivated, for example, by a chromosome aberration, proteins that the cell should not produce are suddenly formed and this may have dramatic consequences. The fact that it normally takes a long time (decades) between the exposure to carcinogenic chemicals to the appearance of a diagnosable tumor may reflect the necessity for several discrete changes in the developing cancer cell, although the time scale is also influenced by the ability of the immune system to detect and kill abnormal cells, thereby probably eliminating most tumor cells formed or at least constantly reducing the number of cells in a growing tumor. A central feature of a cancer cell is that it is dividing continuously. Normal cells can also divide, but do so only when there is a need, for example, during the healing of a wound after an injury. The information regulating cell division is stored in the genome, and, although it is not understood in detail how cell division is regulated, external signals must, through some kind of signal transduction, be translated into an effect on the cell's DNA that makes it respond by initiating the process of cell division. This can be imagined as either the activation of genes that produce proteins that are important for the process or the deactivation of genes producing proteins that repress it, or both, and will be further discussed in Section 8.7.2 (repressors are discussed above). As the changes that transform a cell into a cancer cell are hereditary, they must have a genetic base and in fact be mutations. Chemicals may cause such changes either by being reactive (directly or after metabolic conversion/chemical transformation), interacting directly with DNA to produce mutations (as described earlier in this chapter), or may affect DNA in indirect ways (discussed below). However, it should be remembered that many chemicals are carcinogenic by more than one of the mechanisms discussed in this section.

8.7.1
Reactive or Pre-reactive Chemical Carcinogens that Act on DNA

There is obviously a strong link between chemical mutagenesis and chemical carcinogenesis, and compounds that are mutagenic according to the mechanisms discussed earlier in this chapter should at least be strongly suspected to be carcinogenic as well. Differences may occur when mutagenicity has been assayed in other organisms, for example bacteria or yeasts, which are more convenient to use as test organisms than animals or mammalian cells, because of differences in the secondary metabolism, cell membrane, and so forth. To summarize, reactive or pre-reactive chemicals that act on DNA may be divided into the following categories:

8.7.1.1 Electrophiles
The term electrophiles means electrophilic compounds either as they are or after metabolic activation. Such compounds will react with DNA to transform it to chemically modified DNA, which may result either in base-pair substitutions, frame shift mutations, or chromosome aberrations. Most chemical carcinogens known today belong to this class of compounds.

8.7.1.2 Radicals or Reactive Products Formed after Radical Reactions
In general, radicals are too reactive to be responsible for systemic effects if they are not formed inside the cells, for example, by the metabolism of xenobiotics or by lipid peroxidation. The radical reactions in the cells will produce not only new radicals that by themselves may react with, for example, enzymes and DNA, but also products of the oxidative degradation of the unsaturated fatty acids, which may be electrophilic and mutagenic.

8.7.1.3 Metal Ions
Several metal ions are known to possess mutagenic and carcinogenic activities, for example, Ni^{2+}, Cd^{2+}, and Cr^{6+}. The chemical mechanism for their activity is, however, not yet understood. One can imagine several possibilities, for example, that the metal ions make a strong three-dimensional complex with the heteroatoms of DNA and that this complex is recognized as chemically damaged DNA by the repair systems, which initiate a repair process that results in a mutation. Other possibilities are (a) that the metal ions, while bound to DNA by electrostatic forces, will catalyze redox reactions that result in a chemically modified DNA, possibly via the formation of reactive oxygen species, and (b) that the metal ions are complexed by the enzymes that synthesize/repair DNA (see below).

8.7.2
Chemical Carcinogens Affecting DNA by Indirect Mechanisms

While chemical mutagens are likely to be carcinogens as well, the reverse is not necessarily true. Chemicals may also be carcinogenic, for example, by generating mutations by indirect mechanisms or by influencing the way the genetic information is apprehended and interpreted. There are several ways for our cells to regulate which parts of the genome should be active or inactive, for example, by the selective methylation of certain bases or by the influence of external factors, which may be hormones. Anything that interferes with such processes, for example, by demethylating methylated DNA-bases, may activate 'fetal' parts of the genome (used during the early developments), which in turn may accelerate cell division. As well as the obvious effect that such events may lead to the loss of control of how the cell should divide, cell division *per se* may generate mutations and further develop the emerging cancer cell.

8.7.2.1 Chemicals Acting on the Components of DNA Metabolism

The biochemical components involved in the synthesis, transport, protection, and repair of DNA and the nucleic acids may also be the targets. Reactive compounds in general may, as well as affecting DNA, also react with other biomolecules, and if they react with the proteins and enzymes that take part in the synthesis, transport, or repair of DNA this may result in the loss of their fidelity and consequently an increase in the risk of mutations. In addition, one can imagine that the free DNA-nucleotides that are used for synthesis or repair of DNA are modified by reactive compounds, and if these pass the proof-reading systems they may induce mutations as well. Chemicals that reduce the level of glutathione, for example, by simply reacting with it, will also be harmful and will indirectly enhance the possibility of another chemical carcinogen to interact with DNA.

8.7.2.2 Alteration of Gene Expression

Mg^{2+} and Zn^{2+} play important roles in the transcription of genetic information, and several metals are able to induce and/or inhibit the formation of proteins by interacting with the regulation of genes. Any chemical that has such an effect may be carcinogenic, for example, by inducing oncogenes/repressing anti-oncogenes (see Section 8.8) or inducing enzymes that activate procarcinogens to carcinogens.

8.7.2.3 Mitogenic Chemicals

Several carcinogenic chemicals have been shown to possess virtually no chemical reactivity, but instead stimulate the cells to divide. This may be caused by a specific mechanism by which compounds bind to receptors on the outside of the cell and convey a signal to the nucleus that 'it is time for a division'. Such signals are absolutely critical when cells should divide, but only then. Examples of compounds that may fit into such receptors are phorbol esters and TCDD (Fig. 8.17), both of which are carcinogenic without binding significantly to DNA.

The phorbol esters are present in the oil of croton seeds (*Croton tiglium*, Fam. Euphorbiaceae) and the latex of other plants of the same family. They are extremely irritating and will cause blisters and burns if they contact the skin. Ingested in very small amounts, croton oil is strongly cathartic and has been used for this purpose; in larger doses (approximately 1 mL) it will cause severe and potentially lethal effects on the esophagus.

Although in principle all chemical mutagens should be mitogenic, chemical mitogens do not have to be mutagenic, and mitogenesis will stimulate cell division without necessarily resulting in critical mutations. In this case, if the exposure to the chemical mitogen stops, things will simply go back to normal

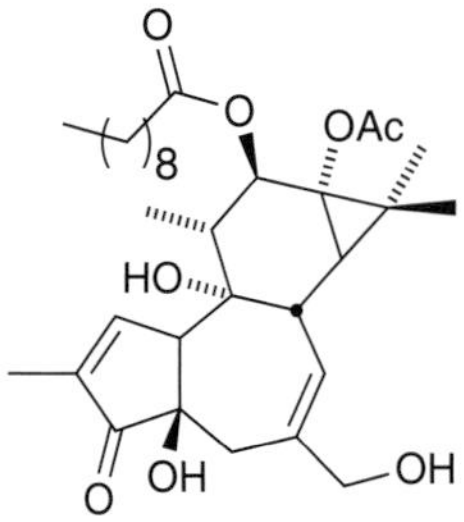

one of several naturally
occurring esters of phorbol

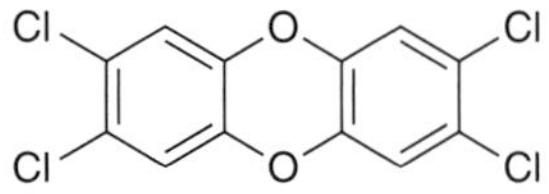

TCDD

Figure 8.17 Specific mitogenic compounds.

(i.e. as they were before the exposure to the mitogen). However, the increased cell division rate will increase the risk of spontaneous mutations. It has been calculated that approximately 10 000 chemical modifications of DNA take place in every cell every day (which is not many if the size of DNA is considered). These are normally efficiently repaired, but if the cell constantly divides there will not be time to carry out all repairs with the required fidelity. One can therefore say that cell division is genotoxic! Besides being specific, chemical mitogens can also be nonspecific and irritate cells so that they divide. This is nothing but a general toxic effect, and such compounds are mitogenic while the concentration is cytotoxic. Many compounds, if not most, will have this effect in high doses, and the question arises whether it is reasonable to assay carcinogenic effects at cytotoxic doses.

If a mammal is exposed to mitogenic compounds together with or after exposure to a genotoxic compound, the effect of the genotoxic compound will be enhanced because the cell's ability to repair DNA damage is restricted by the mitogen. In such cases, the mitogen is called a cocarcinogen or a promoter (see below).

8.7.2.4 Cocarcinogens and Promoters

A cocarcinogen is something that enhances the effect of a genotoxic compound if the two are given together, while a promoter enhances the effect of a genotoxic compound if it is given later. There are several mechanisms for cocarcinogens and promoters. Compounds that induce cytochrome P450 or other enzymes that may be involved in the metabolic activation of a genotoxic compound are cocarcinogens, and ethanol together with other simple alcohols is an example of this. In addition, compounds that react chemically with another compound to form a genotoxic adduct are also cocarcinogenic, and we have previously seen that sodium nitrite can react with secondary amines to form carcinogenic *N*-nitrosoamines. All genotoxic compounds will be cocarcinogens and promoters, because they are toxic, and for example liver infections will promote liver cancer while asbestos and tobacco smoke will promote the genotoxic activity of certain components in tobacco smoke leading to lung cancer. The classic picture of chemical carcinogenesis, established by extensive animal experiments decades ago, is shown in Figure 8.18. The process starts with the initiation, which is an irreversible mutation brought

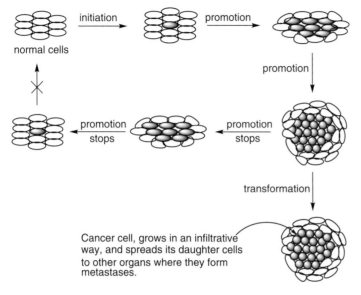

Figure 8.18 The classic view of chemical carcinogenesis.

about by a genotoxic compound. The mutation strikes at a critical disposition that is central in the transformation of a normal cell to a cancer cell, but this mutation alone is not enough. Other changes are also required, and these can be caused by the action of genotoxic compounds as well, or by 'spontaneous' mutations happening if the mutated cell is stimulated to excessive cell division. This is where promotion comes in, and although it was never understood exactly what cellular changes were required and in which order they should take place, promotion was considered relatively harmless if it did not go beyond a critical point. The discontinuation of promotion will then only lead to the loss of the redundant cells and in principle the reversal of the process to the situation after the initiation (mutations will of course not be corrected). However, if promotion (or exposure to genotoxic agents) continues, perhaps for years or decades at a slow rate (e.g., smoking), it may eventually induce all the changes that are necessary to convert the cell to a cancer cell (a process called transformation). After transformation it is all down-hill from the point of view of the cancer cell, and it will rapidly divide and spread via the blood vessels and the lymphatics to other tissues and organs where metastases (daughter tumors) are formed. The metastasis is essentially the reason why cancers are so difficult to treat.

Today the events that are important for the transformation are better understood, and a brief outline of this is given in Section 8.9.

8.7.2.5 Immunosuppressive Agents
Immunosuppressive agents are compounds that decrease the ability of the immune system to respond to stimulation, and are used to prolong allograft survival after organ transplantations and to treat autoimmune disorders. Although

it has limited importance, it has nevertheless been observed that, for example, renal transplant patients treated with immunosuppressive drugs are more likely to develop tumors. As one of the duties of the immune system is to control and limit the growth of new tumor cells, it is not surprising that the unbalancing of the surveillance functions of the immune system increases the possibilities of cancer cells to proliferate. Several of the compounds discussed above that are believed to be carcinogenic by indirect mechanisms, for example, TCDD (see Figure 8.17) also possess immunosuppressive activity, which in fact may increase its carcinogenicity.

8.7.2.6 **Hormones**
Several estrogens have been shown to promote the liver carcinogenicity of mutagenic compounds, either by increasing the mitotic activity of the liver cells or by inducing certain cytochrome P450 enzymes. However, estradiol, which is the most potent of the mammalian estrogenic hormones produced by the ovary, and estrone, which is its oxidation product, may also be metabolized themselves to reactive products (see Figure 8.19).

8.8
Oncogenes, Proto-oncogenes and Anti-oncogenes

Oncogenes are genes that will make a cell lose control over its division rate (i.e., turn a normal cell into a tumor cell) when activated (i.e., being transcribed from DNA to RNA and translated from RNA to protein) in the cell. They were first discovered when carcinogenic viruses were investigated. Because a virus only contains a few genes it was possible to separate the viral genes and determine which one (i.e., the oncogene) was associated with the loss of cell division control when the virus infected a mammalian cell. Interestingly, oncogenes were also found to be present in the genome of all human cells, as identical but inactive copies of the oncogenes were found in viruses or as almost identical copies in which one or a few base pairs are different, and such genes are called proto-oncogenes. The conversion of a normal cell to a tumor cell may then simply be a question of turning on proto-oncogenes, by activating their transcription/translation or, in the case the proto-oncogene is not an identical copy of an oncogene, by a mutation in the proto-oncogene in the correct position so that it becomes an oncogene. In the case that the expression of a proto-oncogene is suppressed by a repressor, a mutation in the DNA sequence that the repressor binds to could affect its binding and initiate the transcription. This would be a dominant mutation, and the effects of dominant mutations are observable directly (see Figure 8.20). Another possibility is that the regulation of the proto-oncogene is affected by a chromosome mutation, which moves it from an inactive region in a chromosome to an active region, perhaps in another chromosome, and this has been observed in a human cancer called Burkitt's lymphoma.

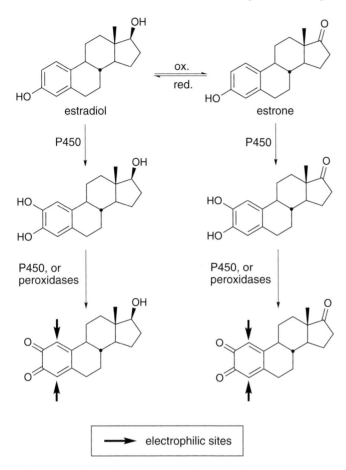

Figure 8.19 The metabolic activation of estrogenic hormones to quinone derivatives.

The proteins encoded by the oncogenes are in most cases enzymes with protein kinase activity, which phosphorylate certain amino acids (tyrosine and serine/threonine) in other proteins. Phosphorylation of proteins is an important tool used by the cells to initiate and control cellular processes, as a protein (enzyme) may behave very differently according to whether or not it is phosphorylated. One phosphorylating enzyme may phosphorylate many other proteins and thereby bring about several changes in a cell, including expression of genes. Protein kinases are certainly involved in the signals that make a cell divide, and the activation of oncogenes coding for protein kinases is believed to be a key step in chemical carcinogenesis. However, for multicellular organisms the regulation of cell division is crucial, and the evolution has provided mammalian cells with additional control systems.

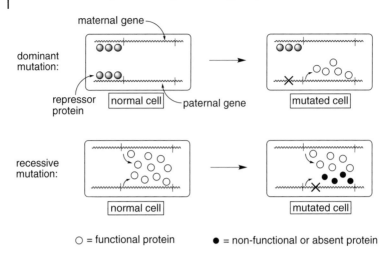

Figure 8.20 The difference between dominant and recessive mutations.

Consequently, cell division can also be suppressed by proteins produced by the cell, and the genes coding for such proteins are called anti-oncogenes or tumor suppressor genes. They were initially discovered when a hereditary tumor of the eye (retinoblastoma), which almost exclusively afflicts infants, was investigated. The cells of this cancer are characterized by the lack of a protein that normal cells produce and that acts as a suppresser of cell division. This protein is consequently a repressor, which blocks the transcription of a gene important for cell division. This tumor suppressor gene is continuously expressed in cells that do not divide, of both individuals with this hereditary defect and normal individuals. However, what runs in the families where the risk of retinoblastoma is high is a defect gene for the repressor protein, but this defect is not noticed as long as the complementary gene, originating from the other parent, is functional. The mutation that has caused this defect is called recessive; it can be transferred from generation to generation and will not be noticed until the complementary gene is damaged (see Figure 8.20). In individuals that have a hereditary tendency for retinoblastoma, one of the two anti-oncogenes discussed above is defective, and the risk that a mutation strikes at the other and thereby stops the production of the repressor protein in a cell is very much higher than the risk that two independent mutations will inactivate both anti-oncogenes in a cell of a normal individual (see Figure 8.21).

Instead of the concepts of initiation, promotion and transformation, the process when a normal cell develops into a cancer cell can now be regarded as a series of genetic changes that involves the activation of proto-oncogenes and inactivation of anti-oncogenes. It is still not understood what other changes, if any, are needed, or how many and which proto-oncogenes/anti-oncogenes have to be changed, or in what order, but at least some of the gene targets of chemical carcinogens have been identified.

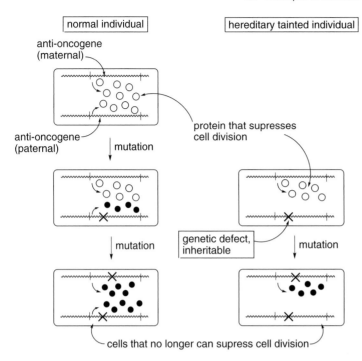

Figure 8.21 Mutations that will stop the production of a protein that suppresses cell division, in normal individuals and in individuals with a hereditary tendency for retinoblastoma.

8.9
Techniques to Establish Carcinogenicity

Obviously, a key question is how we investigate whether a certain chemical has any effect on our genetic material and whether it is genotoxic or carcinogenic or both. The historical errors made by humanity when it comes to using chemicals prior to the investigation of their toxicity are especially serious for genotoxic compounds, as such compounds are not only a threat to exposed individual but in principle to the whole species. However, nobody had any idea about the molecular basis of the genetic material in those days, and nobody imagined how an illness could be connected with something that happened 20–30 years ago. With the new theories that were introduced in the 1950s about the chemical structure of DNA and conversion of the information in the DNA into the function of proteins, our view of life changed in a fundamental way. Gradually a hypothesis that actually was able to explain grand developments, such as the evolution of life on earth, emerged, and could be evaluated by the novel scientific tools that constantly were being launched. Based on this rapid development, there followed a general understanding for how genotoxic effects depend on the chemical properties of individual compounds, and how, in principle, we should be able to predict the genotoxic/carcinogenic activity of a specific compound from its chemical structure. However, in spite of this encouraging development we are not there yet. Life on the molecular level is complicated and still holds a lot of secrets. For the time being, our judgment of the genotoxic/carcinogenic activity of a compound has to be based on experiments, and the options are basically:

- Epidemiological investigations
- Animal experiments
- Assays with mammalian cells
- Assays with bacteria

8.9.1
Epidemiological Investigations

Epidemiology as a science concerns the study of factors affecting the health and illness of populations, and it has long been one of the most important research lines of the medical research. Basically, it looks for connections between various illnesses and diseases with habits, conditions, and exposure, and it is the only way to establish that a chemical is carcinogenic to humans. Unfortunately, epidemiological investigations suffer from very low sensitivity, the major reasons being that only normal exposure levels can be studied, and an exposed group differs from an nonexposed one in a number of other ways as well. A classic example is the lung-cancer causing effect of cigarette smoking which was unheard about in the fist half of the 20th centuary, but in the 1950s when it was possible to link smoking to lung cancer (the hazards of cigarette smoke is discussed in Chapter 9). By then, millions had already died. Other examples are the fetal damaging effect of the tranquilizer thalidomide, and the carcinogenic effect of vinyl chloride (also discussed in Chapter 9). In order to establish, in an epidemiological study, that a certain compound causes a certain effect in humans, it takes at least one of three things.

- **That the group of exposed is big**
  For example, smokers constituted 30–40% of the population in some countries, and it was possible to compare million of smokers with million of non-smokers.

- **That the potency of the compound is high**
  Some of the chemicals used industrially are very efficient carcinogens and could relatively easily be pinpointed, even if very few were exposed. Examples are some aromatic amines.

- **That the effect caused by the compound is unusual**
  Vinyl chloride is not a potent carcinogen, and very few individuals have ever been exposed to the compound, but the liver cancer it gives rise to is quite rare. A sudden increase in the number of cases, of which all are employed by the PVC industry, will indicate the connection.

It is also important to recognize that 'to suspect' is not the same as 'to establish'. Because the use of chemicals often involves a lot of money, there are strong interests that will oppose any unfounded suggestions about the possible hazards connected to the use of an approved chemical, and not hesitate to go to court. It is therefore necessary to prove the relationship by an accepted statistical analysis of the primary data and this can only be done by trained and experienced epidemiologists. Altogether, some 40 compounds or defined mixtures of compounds have in this way been shown to be carcinogenic to humans. It is primarily chemicals that have been used in industry, organic compounds such as aromatic amines, benzene, ethylene oxide and vinyl chloride, and inorganic compounds like arsenic, asbestos, cadmium, chromium, nickel and radon (all these compounds are discussed in this book).

However, even non-smokers and people not working with chemicals are afflicted by tumors, and one may ask what is causing the tumors of apparently unexposed people. Besides chemicals, various forms of radiation (e.g., ultraviolet, X-ray and ionizing radiation) and some viruses may also be carcinogenic, and for some types of tumors there is a strong hereditary disposition, so the significance of chemical exposure for the total incidence of tumors is not evident. One way to study the causes of tumors is to compare how common various types of tumors are in different parts of the world. Surprisingly, some tumors are more than 100 times more common in one place (the high-risk area) compared to another (low-risk area). In general, every country has its own tumor profile, and while, for example, England is a low-risk area for liver cancer it is a high-risk area for lung cancer. Such differences may be explained by the different environment, habits and background exposure in different countries, the English have, for example, always been relatively heavy smokers, and the high incidence of skin cancer in Australia may be caused by the large proportion of fair-complexioned individuals not adapted to the strong sunlight of that continent. However, another explanation is that different people have slightly different genomes, and that, for example, the English are predisposed for lung cancer. The importance of genetic factors for the incidence of tumors has been estimated by comparing how tumor trends change for populations that emigrate from one part of the world to another. Several investigations have shown that such groups in a few generations adopt a similar tumor pattern to that of the native inhabitants of the new home country, even if they during this time keep a substantial part of their original genome intact. All this information suggests that the majority of all tumors in industrialized countries are caused by external carcinogenic agents, while approximately 20% are linked to hereditary factors. However, few tumors are solely due to hereditary factors; instead, these will modify the impact of an external carcinogen by, for example, not being able to detoxify a carcinogenic chemical efficiently or to repair damage to the DNA. Radiation in various forms is responsible for a few percent, accounting for most cases of skin cancers. The impact of viruses is not known, but a reasonable guess is a few percent. In addition, the deficiency of essential chemicals (e.g., vitamins and selenium) due to unbalanced diet is associated with an increased risk of tumors, although it is difficult to estimate its importance. Also, tumors may, of course, arise from spontaneous mutations that take place in all cells. The rest of the tumors of mankind are in principle caused by chemicals, and the two major sources of carcinogenic chemicals are tobacco smoke and food. While most people are aware of the hazards of smoking, fewer are aware about the impact of food for the incidence of tumors. These issues will therefore be further discussed in Chapter 9.

8.9.2
Animal Experiments

We normally assume that chemicals that are carcinogenic to animals are also carcinogenic to man, although as discussed in Chapter 7 this is not necessarily absolutely true. Anyway, data from reliable animal experiments increase the number of established chemical carcinogens to a few hundreds. A typical animal experiment in which the carcinogenic activity of a compound is to be investigated is carried out by comparing the number of different tumors in animals that were not exposed with those that received the tested compounds. As discussed in Chapter 7, there are many parameters that can be varied during animal experiments, but there are standard protocols that are followed if possible. The advantages with *in vivo* experiments in this respect are in principle:

- An animal has similar systems for the uptake, distribution, excretion, and metabolism of chemicals to those of humans, and it is reasonable to expect that the fate of a chemical is the same in animals as humans.

- Animals also get tumors, and the mechanism for this is in principle the same as that in humans.

- The groups of exposed and nonexposed are made up of identical individuals who are treated in exactly the same way during the experiment.

- The amounts of a tested compound can be increased in order to get a statistically more significant result.

The disadvantages of animal experiments compared to the simpler assays discussed below is that they are quite expensive and that highly skilled personnel is required to perform them.

8.9.3
Assays with Mammalian Cells

Going from whole animals to cells of animals (the cells can also be human) that are grown *in vitro* in a laboratory offers several advantages. The cost of assaying one compound is, of course, only a fraction, the capacity is more or less unlimited, and results are obtained in days or weeks. A major drawback is that the metabolism of the cell type chosen can never match that of an entire animal, not even if it is a liver cell. In addition, the end point in an assay with mammalian cells is not the transformation of a normal cell to a cancer cell; instead, it is some measurable event that takes place with the cell's genome. Assays with mammalian cell are therefore not able to determine the carcinogenic activity of a compound, only the genotoxic activity, but, as argued throughout this chapter, it is reasonable to assume that genotoxic compounds are also carcinogenic.

Three principally different assays with mammalian cell can be mentioned:

The detection of genotoxicity in mammal cells can be carried out by cultivating cells in the presence of the assayed compound as well as DNA bases labeled with a radioactive isotope. If the tested compound has a genotoxic effect on the DNA, this will induce DNA repair, which will cause the labeled bases to be built into the cells' DNA. By rinsing the cells from any remaining labeled DNA bases that are not covalently attached, the radioactivity of the exposed cells can be compared with that of unexposed cells. If the exposed cells are more radioactive, and if this is correlated to the dose of the tested compound, the conclusion is that the compound is genotoxic.

Gene mutations in mammal cells can be detected in cells that are prepared in a special way. In the HPRT assay, special hamster cells have been mutated by a point mutation so that they cannot form the enzyme HPRT, which is important for DNA synthesis. These cells are sensitive to the toxicity of the nucleoside analog 6-thioguanine, and will not be able to grow in its presence. However, a compound that is able to give point mutations may revert the damaged gene for the HPRT enzyme so that it becomes functional again (a forward mutation), and then the cells have no problems with 6-thioguanine. The test cells are cultivated with and without the compound assayed, and with 6-thioguanine. If the compound gives rise to gene mutations the exposed cells will grow, while the unexposed ones will not.

Another type of assay with mammalian cell focuses instead on the detection of chromosome mutations, and is known as the micronucleus test. Chromosome mutations may give rise to loose parts of the chromosome, and, when an affected cell divides, such parts will form their own nuclei in addition to the cell nucleus. Such micronuclei are easily detected with a microscope. There are, in fact, both *in vitro* and *in vivo* variants of the micronucleus test. In the former, cells are cultivated and exposed to the compound being tested, after which the cell is checked for the appearance of micronuclei. In the latter, a whole animal is exposed to the tested compound. In the bone marrow, stem cells develop to form, among other things, red blood cells, and in this process the red blood cells produce all the proteins they need and eject the cell nuclei with the genetic material before they enter the blood circulation. However, if chromosome mutations take place during the development of a red blood cell, forming a micronucleus, this will be left in the circulating blood cell, and a simple blood sample can be taken to look for the presence of such cells.

By combining more than one *in vitro* assay, additional information can be obtained.

8.9.4
Assays with Bacteria

As well as mammals and mammalian cells, many other organisms have been used in assays to detect genotoxicity. The most well-known of them all, the Ames test, uses a prokaryote, a bacterium. Bruce Ames is a professor of biochemistry and molecular biology at the University of California, Berkeley, and together with his coworkers he developed this assay around 1970. The test uses special strains of the bacterium *Salmonella typhimurium* which have been given point mutations in genes coding for proteins required for the bacterium to synthesize the amino acid histidine. Consequently, the test bacteria cannot make histidine themselves and therefore require external histidine to grow. If exposed to a mutagen that can give point mutations, there is a certain probability that point mutations generated in the test bacteria will correct the original mistake and give them back the ability to synthesize histidine. If they are cultivated on a histidine-free medium, only these mutated bacteria can grow, and these will in 24 h form a colony that is visible to the naked eye. Different tester strains have been constructed that have either frameshift or point mutations, and by using them it is possible to detect mutagens acting via different mechanisms. In order to simulate mammalian metabolism, it is possible to add an extract of rat liver containing cytochrome P450 and other enzymes which will perform the metabolic activation *in situ*.

However, the behavior of a *Salmonella* bacterium is very unlike that of a human being in many ways. The ability of the Ames test to predict carcinogenicity is therefore limited. Nevertheless, it is possible to identify compounds as point mutagens, and this is very valuable. As Ames test is rapid and inexpensive, it has become a standard test that is widely used. For example, before investing big money in developing drug candidates in the pharmaceutical industry, the compounds of interest are routinely assayed in an Ames test at an early stage in order to verify that no genotoxic surprises are lying in wait around the corner.

9
Hazardous Chemicals

Is a certain chemical hazardous or not? In many circumstances, that is a critical question, but it is not one that is easily answered. We have already learnt that there is in general very little knowledge about the biological effects of chemicals. In addition, we have seen that the toxicity of a chemical depends on many factors, and that different individuals perceive hazards in quite different ways. There are always risks involved with the handling of chemicals, and a solution that at least some politicians are attracted to is to ban their use. However, this is not an option. Instead, we need to use chemicals in a safe way. To do so, we need to seek out all the knowledge that is available about the chemicals we are using. A fast way that is available anywhere to anyone is simply to Google the name or the CAS number of a chemical and see what comes up. For the most common chemicals, the net will always provide some information, even if it is not always useful. One must also remember that the information on the internet has not been critically reviewed and can be misleading. A more reliable but less accessible source is the scientific literature. However, in addition to the bare facts, it is also important to try to learn from the relationships between chemical structure, chemical properties, and effects on humans and the environment, in other words, to use the general principles provided by this textbook. This chapter gives examples of various hazardous chemicals, arranged according to their uses. It is not a comprehensive selection, and the choice of chemicals is based on which, in the author's experience, are the most important and most useful for university students to learn about.

There is a difference between chemicals that are toxic to man and those that harm the environment, although they also share many properties. A chemical that is toxic to one organism is normally also toxic to others and therefore bad for the environment, although many of the classical environmental effects are caused by chemicals that are not dangerous in reasonable amounts. However, they have in many cases been spread over the environment in huge quantities, and any chemical would be an environmental hazard in such circumstances.

Toxicity in humans is normally a direct effect which frequently can be traced to chemical properties (e.g., reactivity) of a chemical. In most cases the toxic effect is observable and its mechanism can be understood. Chemicals that are hazardous to the environment may have similar effects on all living organisms, although some species are more sensitive than others, but any apparent effects on the

Chemistry, Health, and Environment. Olov Sterner
© 2010 WILEY-VCH Verlag GmbH & Co. KGaA, Weinheim
ISBN: 978-3-527-32582-5

ecosystem are more diffuse and difficult to define. A chemical discharge in a lake, for example, may be disadvantageous for some species while others are less affected and can take over, at least for a short time, until a new balance has established itself in the ecosystem. Chemicals will affect not only an organism that is exposed to them, but also organisms that are part of the food chain including the primary organism. This may be caused by direct transfer of the chemicals, but also by, for example, a change in the nutrients taken up from the soil by plants, the release of toxic heavy metals in the soil, or changes in the atmospheric composition. The latter has received particular attention because the emission of large amounts of a few chemicals into the atmosphere is a relatively immediate risk that can dramatically affect life on earth in just a few decades.

9.1
Fuels, Exhausts, and Other Air Pollutants

The emission of gases/particles from motor vehicle and power plants is a major threat to our environment, and linked to this problem are the effects caused during the extraction of petroleum and the handling of various types of fuels. However, we should not forget that the natural turnover, that is, emissions of both inorganic and organic substances from natural sources into the atmosphere is significant (discussed further in Chapter 10). For example, the hazardous chemical sulfur dioxide (SO_2) is constantly emitted from areas with volcanic activity in large amounts, and, if the additional emissions due to human activities account for only a few extra percent, this would not make any difference to the global environment (although we would of course have to consider local effects). However, the emissions of sulfur dioxide caused by human activities are bigger than the natural ones and cannot be ignored. Nitric oxide (NO) (see Section 9.1.11), present in many types of exhausts, is a severely hazardous gas, but has recently been found to be a transmitter substance in the nervous system of mammals and necessary for normal human functions. That does not make nitric oxide any less hazardous, but it certainly changes its status. An example of a natural emission of organic substances is the emission of terpenes from conifers, which is more important (in kilograms) than the total emissions of organic substances from human activities. Terpenes are hydrocarbons and thereby involved in the generation of the reactive gas ozone (see Section 9.1.10), and it has been suggested that the man-made emissions of hydrocarbons are negligible or at least less significant in this context. However, it has been shown that terpenes are not very efficient at generating ozone, but instead react directly with and transform it. It has even been suggested that conifers emit terpenes to protect themselves against ozone, which is always present in the air in low concentrations. Again, evolution has adapted organisms and ecosystems to the current, natural, chemical conditions as well as possible, but novel human activities are not covered by this protection. Any burning, whether it is natural (caused for example by lightning) or caused by man, produces exhaust gases which are harmful. Examples that will be discussed here are carbon monoxide, carbon dioxide, polycyclic aromatic hydrocarbons, ozone, and nitrogen oxides. It is a great challenge for humanity now and in the future to invent and develop engines and methods to produce energy that emit less or even no exhausts.

9.1.1
Biogas

Biogas, essentially methane, is produced by bacterial degradation of organic material in oxygen-poor environments, primarily in waste treatment plants, and is a renewable energy source. The same process takes place in many different environments, for example, in swamps and lake beds, but also in the stomachs of cattle. Methane emitted to the atmosphere from such sources makes an important contribution to the greenhouse effect and will be discussed later. New techniques to take care of the methane produced have been developed, and today we can condense it in flasks and use it as an alternative fuel for various purposes. Waste water with a high content of organic material, for example, from food industries, can be used for the production of biogas, as, of course, can any biomass that is grown and harvested. The price of biogas is, however, still considerably higher that that of natural gas, but this is simply a question of time. Methane can be used as a fuel in normal car engines without any problems. Its octane rating is high, allowing high compression in the engine and therefore high efficiency. Methane as a fuel in combustion engines is also advantageous for other reasons. As it is a single hydrocarbon, the conditions in the engine can be optimized for methane, and the exhausts will contain essentially only carbon dioxide and water.

9.1.2
Natural Gas

Natural gas has been formed by the decomposition of organic matter a very long time ago, and exists under high pressure in significant quantities trapped in pockets in porous rocks. In contrast to biogas, natural gas is a fossil fuel, consisting of a mixture of the smallest alkanes (methane, ethane, propane and butane). However, it also contains carbon dioxide, water, hydrogen sulfide (H_2S), and significant amounts (up to eight percent) of helium (He). Natural gas thus provides us with a rich source of helium, a noble gas that is safe to use and has a boiling point ($-269\,°C$) close to absolute zero ($-273\,°C$). The presence of hydrogen sulfide is a problem if natural gas is used for combustion, as the hydrogen sulfide will be oxidized to acidifying sulfur oxides in the process. The problems with sulfur oxides will be discussed below, and to avoid their formation the hydrogen sulfide must be removed from the natural gas prior to use. Apart from this problem, natural gas has essentially the same advantages as biogas as a fuel. The fact that large quantities of natural gas are present naturally on earth means that some amounts will leak out to the atmosphere. Human activities such as drilling for natural gas and extracting it from the rock, transporting and distributing it, and of course using it, have increased the leakage significantly. This is of some concern, as methane and other alkanes in the atmosphere give a strong green-house effect (see below).

9.1.3
Petrol

Petrol (gasoline) is a mixture of many hydrocarbons with boiling points in the range 30–190 °C. The most important constituents are the alkanes hexane, heptane, and octane, as well as which, alkenes and aromatic compounds, for example benzene, are present. Petrol is obtained from crude oil by distillation, the yield of petrol obtained in this way being typically 1–20% of the crude oil. However, the demand for petrol is normally higher than this, and therefore heavier distillation fractions are 'cracked', using catalysts, to transform larger hydrocarbon molecules to smaller ones. The composition of petrol may differ from time to time and from place to place, and is determined by the properties required. More volatile components will facilitate cold starts, especially in cold regions, and high octane rating will increase the efficiency. The octane rating is a measure of how much a fuel/air mixture (as a gas) can be compressed without self-igniting, something that will give rise to engine damage and must be avoided. Higher octane ratings show that the petrol can be used in engines with higher compression, which are more efficient. To increase the octane rating, tetraethyl lead used to be added to petrol (leaded petrol). The lead reacts with the radicals that initiate the self-ignition in the cylinder, but after the explosion it goes out with the exhaust and becomes an environmental problem (discussed below). As an alternative, benzene can be added, although its carcinogenic properties make it less suitable. Recently, MTBE (methyl tertiary-butyl ether) has been used as an octane rating booster. MTBE is an ether whose health hazards are somewhat unclear. The risks associated with petrol are often due to the additives, and benzene and tetraethyl lead are examples of additives that should be avoided. Leaded petrol is now being phased out, and alternatives to benzene are being introduced. Another problem with petrol, as with all hydrocarbon fuels, is of course that it is easily ignited and burns explosively.

9.1.4
Diesel

Diesel is a fraction from the distillation of crude oil that is more high-boiling than petrol. Typically, diesel is obtained between 200 and 300 °C, meaning that it is considerably less volatile. The toxic effects of diesel are more limited than those of petrol, mainly because the exposure is lower (lower volatility = slower evaporation). Diesel engines operate at higher compression, the fuel being injected directly into the cylinder, where the explosion takes place. It is actually possible to run a diesel engine with modified rapeseed oil, and this has created a market for this renewable fuel. However, although it functions satifactorily the cost of rapeseed oil is still too high for it to compete with fossil diesel.

9.1.5
Oil and Coal

In spite of the rumors that it is on its way out, crude oil continues to be a general lubricant of the global economy. It is produced in enormous quantities (several million tons per year), and is still the basis for the chemical industry and fuels. Like natural gas, crude oil is a fossil fuel and can be found in many places on earth, normally along with natural gas. The natural emissions of crude oil are relatively small, as it is not very volatile, but man's drilling, pumping, and transporting are accompanied by leaks. As the amounts handled are mind-boggling, the leaks are quite substantial. We are all aware of the effects of leaking crude oil on a sea shore, something that happens regularly all over the world because of accidents and/or greed. In addition to its more obvious effect – killing sea birds that come into contact with it by destroying their feathers – it is also toxic to the eggs and larvae of many sea organisms at concentrations as low as 50 mg oil per liter of water.

Although crude oil may appear to be something quite persistent, it can actually be broken down by microrganisms, although this takes time. Thus, crude oil will typically produces large local effects when it is discharged into the environment. Coal is normally taken from open mines that cover huge areas. Villages and towns have more than once been forced to move in order to facilitate further extraction. Both coal and oil are used in power plants to produce electricity, and the nature of the emissions from such plants depends on the presence of certain impurities in the coal (or oil). This will be discussed below. In addition, coal can contain high amounts of heavy metals, which are also expelled with the combustion gas. Modern power plants using coal and oil are therefore equipped with systems that purify the emissions.

9.1.6
Ethanol

Alcohols are an alternative and/or complement to hydrocarbons as fuels, especially in combustion engines. They do have certain disadvantages, such as being more expensive and having a lower energy content, making the engines consume more fuel, but there are advantages too. Alcohols can be produced easily and cheaply by adding water to the corresponding alkene, obtained by cracking petroleum, but if the goal is to use alcohols as fuel it is pointless to make them from hydrocarbons, which provide more energy. However, alcohols, especially ethanol, can be produced from renewable sources (plants) via chemical or enzymatic degradation of cellulose to simple sugars and fermentation of the sugars. This produces a mash that can contain up to 15% ethanol, which is distilled off. The technology of fermenting sugar-containing solutions and distilling off the ethanol is old and well known, and does not need further development. The problem is instead the production of the sugar solution. Cereals, potatoes, and sugar beet, which traditionally have been used to produce alcohol, cannot be produced cheaply enough to compete with petrol, and the cellulose in wood is not degraded easily enough. However, intensive research into developing new ways to make efficient use of cellulose for the production of ethanol is ongoing, and there are great hopes that this will be successful. The major advantage of replacing natural gas/petrol/diesel with ethanol as fuel for cars is that it reduces the proportion of new carbon added to the atmosphere as carbon dioxide, thereby, in theory, reducing the amounts of this gas. In addition, the exhaust from an ethanol-powered engine are not particularly bad (no hydrocarbons or particles), and ethanol is a relatively non-toxic and more environmentally friendly chemical than the others. The toxic effects of ethanol are discussed in Section 9.5.

In Brazil, where sugar cane is a major crop, ethanol has been used as fuel for cars for many years, either pure or in mixtures with petrol. Brazil has shown the world that the technology is there and that it works well. Today, cars running on ethanol-based fuels are sold on all the major markets around the world. The problem is, again, that ethanol costs more to produce than petrol. The production price of petrol and diesel is really astonishingly low, especially when it is considered how much of the price paid by the consumer is actually tax. Companies have invested heavily in the development of the petrol industry and everything attached to it, and there is naturally a reluctance to change to something different. This can of course be influenced by the politicians by means of taxes and regulations. Cars that consume less are promoted by lower road tax, and the tax on ethanol is considerably lower than that on petrol. Efforts to make international agreements to limit the output of greenhouse gases are being made, although their effectiveness is a matter for discussion.

9.1.7
Polycyclic Aromatic Hydrocarbons

PAHs (polycyclic aromatic hydrocarbons) are formed in small amounts when organic material is heated to temperatures up to 600 °C without burning with free access to oxygen. PAHs contain at least two aromatic rings joined side by side, and may be substituted in various positions. As always, when it comes to molecules, there is an almost infinite number of variations. As PAHs are present in small amounts and not easily distinguished from each other, they are often discussed as a group. However, some PAHs that are potent carcinogens have been studied individually, and the structures of some of these are shown in Figure 9.1. These were selected by WHO as indicator substances as they are often found in significant concentrations, have 'typical' chemical and physical-chemical properties (for PAHs), and are all hazardous. Smaller PAHs (up to four rings) are sufficiently volatile to evaporate to the atmosphere and will fairly quickly be destroyed there by reacting with hydroxyl radicals (the molecular cleaners of the atmosphere, see Section 10.2.1). Larger PAHs, such as benzo[a]pyrene, have such low volatility that they tend to be adsorbed on smoke particles, such as soot, and in this form can be transported over long distances. If such particles are inhaled they may be trapped in the alveoli of the lung, from where the PAH can diffuse into

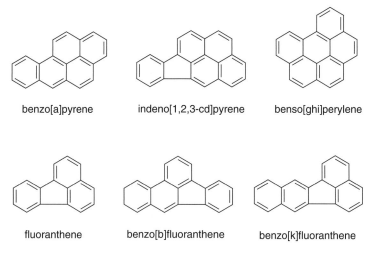

benzo[a]pyrene   indeno[1,2,3-cd]pyrene   benso[ghi]perylene

fluoranthene   benzo[b]fluoranthene   benzo[k]fluoranthene

Figure 9.1 Examples of relatively common and dangerous PAHs.

the blood. In total, it is estimated that more than 4 million tonnes of PAHs are emitted per year into the atmosphere. The major sources of PAHs are combustion engines, wood burning, grilled and smoked foods (see Section 9.3), and tobacco (see Section 9.5). Generally, diesel engine exhausts contains more PAHs than petrol engines, but the latter generate more of the larger and more carcinogenic PAHs. The worst pollution is from the exhausts of poorly maintained and tuned engines of any kind. Similarly, varying concentrations of PAHs are found in the smoke from burning wood, depending on how efficiently it burns (depending on, for example, the water content and the supply of oxygen). Another source of PAHs is rubber tires, which contain a product called carbon black that is contaminated by PAHs. When tires wear on the road, the carbon black and its PAHs are spread into the environment.

Another historically important source of PAHs has been creosote, which is a mixture of hundreds of different substances obtained from coal tar distillation (it can also be produced from wood (wood creosote), but this is a minor product). This takes place at high temperatures, typically between 200 and 350 °C, and it is no surprise that it contains PAHs. About 70 percent of the PAHs are small: phenanthrene, anthracene, pyrene, and so on, but more potent carcinogenic PAHs are found at lower levels. In addition, creosote contains neutral as well as basic heterocycles, phenols, etc. Many have a strong taste and smell, and creosote has for many years been used as an effective wood preservative using pressurized impregnation. Surprisingly, one of the first uses of creosote was as a drug to treat tuberculosis! Creosote is still used in large quantities, but its use is declining. Nevertheless, the millions of tonnes produced and used are still performing their protective function to a large extent. In any case, it is not possible to assess how hazardous it is to humans and the environment as it is such a primitive product containing so many different components.

As already noted, PAHs have low volatility and are in addition poorly soluble in water. Therefore they bioaccumulate efficiently, mainly in organic matter in the soil. The PAHs are chemically stable, and are converted/transformed primarily because they are oxidized. In the atmosphere, this can be caused by reaction with natural hydroxyl radicals, but also with ozone, nitrogen oxides (which generates highly toxic nitropolyaromatics), or sulfur dioxide. In mammals, the PAHs will be oxidized by cytochrome P450, and it is the metabolites formed that are responsible for the toxic effects (discussed in Chapter 5).

About 40 PAHs, of which eleven are considered as strongly and ten as weakly carcinogenic, have received special attention. One of the most dangerous and most common PAHs is benzo[a]pyrene, a compound that has come to be regarded as the typical PAH as it is highly toxic, carcinogenic, and teratogenic. (The 'a', and other letters and numbers inside the square brackets in for example benzo[a] pyrene indicates how the various components of a PAH are arranged in the molecule. Benzo[a]pyrene, for example, is a benzene ring attached to the side 'a' of the PAH pyrene, which itself has four benzene rings.

One of the more frightening properties of some of the more potent PAHs is their ability to damage the genetic material of our cells at a molecular level. This is initially thought to be caused by the fact that they can intercalate between the base pairs in the DNA double helix (See Section 8.3.2) rather like the piece of meat between two pieces of bread in a hamburger. They can do this because they are relatively big and completely flat, and although this effect on the DNA at the molecular level has been demonstrated it is not considered to be toxic by itself. However, if an intercalating substance also contains an electrophilic group in an appropriate position (for example an epoxide ring that happens to be positioned close to a nucleophile), the intercalating properties by may well be critical.

9.1.8
Carbon Dioxide

While carbon monoxide (CO) is an extremely dangerous gas (discussed in Section 7.2.4.1), carbon dioxide (CO_2) is in moderate concentrations a harmless gas that we are constructed to handle. Virtually all organisms produce it, and humans exhale quite a lot of carbon dioxide with every breath of air. However, when inhaled in very high concentrations it can be quite irritating, and you will feel a stinging in your nose if you inhale air containing a lot of carbon dioxide (for example above a fermentation vessel). The reason for this is that carbon dioxide dissolves in water (e.g., the mucosa of your nasal cavities) and reacts with it to produce carbonic acid, which is acidic and alerts your sensitive receptors that the acidity of the inhaled air is high. The signal is the slight irritation you sense, and you would typically react by searching for better air to breathe.

Carbon dioxide is the end product of the chemical oxidation of carbon in all organic material, and it is in this context interesting to have a quick look at the comings and goings of the element carbon on earth. The cycle of organic carbon is well known: plants absorb carbon dioxide from the atmosphere and convert it into carbohydrates, which are consumed by other organisms that use the carbohydrates to produce energy. This is done by oxidizing the carbohydrates, as well as other organic materials, to carbon dioxide, which is released into the atmosphere. Carbohydrates are also used to produce all other organic components used by organisms. When a life is finished and an organism dies, it will be degraded by other organisms. With access to air (i.e. aerobically), the organic components will in principle be oxidized to carbon dioxide, but in the absence of oxygen (for example, at the bottom of the sea) the organic matter may instead be reduced to hydrocarbons by anaerobic organisms, which after millions of years will form natural gas, oil, and coal. Any organic material in contact with air is actually oxidized slowly by atmospheric oxygen to carbon dioxide. From a chemical point of view the end products carbon dioxide and water have lower energy than organic compounds and oxygen, so this process is spontaneous. However, it goes so slowly that we normally will not notice it if we do not speed it up by adding, for example, a burning match. The fate of inorganic carbon on earth is perhaps less familiar, but very important as huge amounts of inorganic carbon is present on earth. Limestone is present more or less everywhere, and as it is weathered in the presence of water and acidic substances in the atmosphere a solution of calcium and bicarbonate ions is formed. This can be absorbed and used by organisms to build bone and shell, for example, which when the organisms die eventually will form new minerals (for example limestone). Carbon dioxide can be emitted to as well as absorbed from the atmosphere during these processes, and the net production of CO_2 on an earth in balance is in principle zero.

Life on earth constitutes a valve between the organic and inorganic carbon, and the biotic carbon cycle is powered by energy from the sun.

The concentration of carbon dioxide in the atmosphere increases by approximately 0.4% per year, mainly due to human activities such as the burning of fossil fuels. It is an important greenhouse gas, and the gas primarily associated with the greenhouse effect, which is discussed in detail in Section 10.5.

9.1.9
Methane

The second most important greenhouse gas, excluding water, is methane, which is approximately twenty times more effective in that respect compared to carbon dioxide. However, its concentration in the atmosphere is lower (less than 2 ppm). As well as efficiency and concentration, it is also necessary to take an interest in the rate of concentration change, because this is what will change the climate. The concentration of methane in the atmosphere is currently increasing strongly, at a rate of more than 1% per year. This is caused by the use of natural gas and oil, as discussed above, but even more by the anaerobic degradation of organic matter.

Two important sources are rice fields fertilized by man, and domestic animals. Earth's 1.3 billion cows produce approximately one hundred million tonnes of methane per year, and the concentration of methane in the atmosphere is increasing considerably faster than that of carbon dioxide. If this continues, methane will be the most important greenhouse gas in only a few decades. The methane in the atmosphere today absorbs only a small part of what methane theoretically can absorb, and an increase in the concentration of methane will directly generate a corresponding increase in the greenhouse effect. This is not the situation with carbon dioxide, which already absorbs most of the radiation it can absorb. Huge quantities of methane are stored in the soil as natural gas but also in other forms, and it is easy to imagine that global warming increases the leakage from such sources.

9.1.10
Ozone

Ozone is a form of oxygen, which exists either in the dimeric form (O_2) that is present in air (which we breathe) or in the trimeric form ozone (O_3). Ozone is an irritating and quite toxic pale blue gas with a characteristic odor that you can detect in environments where there are electrical discharges or strong UV lights, which convert O_2 to O_3. As will be discussed further in Section 10.4, ozone occurs naturally in the stratosphere, 15–40 km above sea level, where it protects the earth's surface from dangerous ultraviolet light. It is also present in the troposphere, where it is has other effects (see below). However, in some circumstances it is dangerous air pollutant. A major environmental problem today is that the amount of ozone in the stratosphere is declining because of the emission of halogenated chemicals such as the CFCs and the halons, while the amount in the air that people are exposed to is increasing because of exhaust emissions. In the stratosphere, an 'ozone hole', with very low amounts of ozone, is created over the north and south poles of the earth during the winter period. This will let through potentially dangerous UV light, but the effects on organisms are limited because it happens during the period when the sun is low. Since sunlight and air pollution are the conditions for the reactions that produce ozone, people living in polluted urban areas will be exposed to ozone in afternoons of clear and windless days. The effects of ozone, and other photochemical oxidants (see Section 10.2.2), is irritation of the respiratory tract and eyes at relatively low concentrations. It also damages plants and affects organic materials such as plastics, rubber, and paint. Concentrations in the air greater than 50 ppb (about 100 mg/m^3) result in damage to the respiratory tract, although plants may also be severely affected at lower concentrations. Higher concentrations can penetrate into the lung alveoli and cause a pulmonary edema (discussed in Section 7.3.5), but because of its chemical reactivity it will not be distributed to the rest of the body. In addition, ozone is a potent greenhouse gas (2000 times more effective than carbon dioxide) in the troposphere, and, as the levels of ozone increase every year, it could make a significant contribution to global warming.

9.1.11
Nitrogen and Sulfur Oxides

NO, NO_2, SO_2, and SO_3 are all highly irritating gases that in relatively low concentrations may damage the airways. While the sulfur oxides are more associated with allergies and asthma, the nitrogen oxides, which chemically are highly reactive, may give rise to pulmonary edema. The main source of sulfur oxides in the air is the combustion of fossil fuels (which contain various sulfur compounds) for heating and transport. The amount of sulfur oxides is directly proportional to the amount of sulfur in the fuel, while the amount of nitrogen oxides, which mainly come from oxidized nitrogen from the combustion air, is influenced by the combustion process (temperature, pressure, etc.). These gases, as well as affecting humans, are responsible for some of the major environmental effects that we have wrestled with during the last decades. The acidification of lakes, the eutrophication of soils and water, the degradation of stratospheric ozone, and the formation of photochemical oxidants in the troposphere are examples that are discussed further in other parts of this book.

9.1.12
Summary

The fuels, exhausts, and other air pollutants discussed here constitute only a small fraction of the total, but perhaps the most important fraction. Other chemicals could be included, but are discussed elsewhere in the book. Among the fuels, exhausts, and air pollutants, many of the major environmental threats are to be found because these chemicals are used and produced on a very large scale. In addition, they are in general gaseous or volatile and are easily spread globally in the atmosphere. Some of the problems we have discussed in this section can be relatively easily remedied, for example, by the purifying natural gas and exhausts from power plants operating on fossil fuels, but others are more difficult. The use of fossil fuels for transport is still increasing, although modern engines use less fuel and electric cars are being introduced. It is likely that we will continue to use fossil fuels until they simply become too expensive, and this will be when supplies are almost exhausted. If that is the case, the greenhouse effect is perhaps the single biggest threat for humanity in the short run.

9.2
Pesticides and Chemical Warfare Agents

Humans have always had a great need for ways to protect themselves, their belongings, and their food from other organisms that want to have a share. In general, when our protection against other organisms is by the use of a chemical, we talk about pesticides, which will kill 'pests'. This is a very broad concept, formally including, for example, the antibiotics we use as pharmaceuticals to treat infections, but we draw a line between pesticides and pharmaceuticals (the latter are discussed in Section 9.3). Actually, the killing is not always important: modern pesticides are often designed to keep pests at bay and discourage them from invading our territory. Important subgroups among the pesticides are the herbicides (against plants), insecticides (against insects), fungicides (against fungi), and rodenticides (against rodents). Humankind has in modern times become fairly skilled in all these areas except, perhaps, the insects, which are very difficult to combat. This is because insects are so many, so hungry, and so mobile, and not only will attack us, our food, and our houses, but will also spread many diseases. In the earlier times, man-made pesticides were often unsatisfactory, and there are several examples of pesticides that have done more harm than good. Often the problem is selectivity: chemicals that kill insects are easily invented, but those that kill only insects are harder to develop. The use of pesticides is nowadays somewhat politically incorrect, as there are powerful forces that argue for a society free from chemicals. However, especially in countries with a hot and humid climate, it is difficult to produce crops economically without them. On the positive side, new pesticides continue to become more selective and less damaging to humans and the environment, and as our increasing understanding of how the processes of life, including that of pests, work at the molecular level, our ability to design pesticides that are safe as well as efficient, but also are difficult to develop resistance against (this is a major problem for pesticides), increases. However, one should be aware that such modern pesticides in general are more expensive, and the use of the older, cheaper, but less selective products is continuing. As a rough estimate, it is believed that several thousand farm workers die every year from poisoning by the pesticides they are exposed to.

Chemical warfare agents are chemicals that, because of their effects on humans, animals, and/or plants, are considered potentially useful in conflicts. They are terrible weapons. Few nations have developed any at all, and most consider them to be a last resort to be used only in response to an enemy's use of such products. Only about seventy chemicals have been developed for use in war, at least this is the number of known agents. Their action may be lethal, as in the case of the nerve gases discussed below, or just incapacitating, like tear gases. Others affect the mind, making soldiers less interested in fighting. Herbicides can also be used as chemical warfare agents, and we will discuss Agent Orange that was used in the Vietnam war below. Besides chemical warfare agents there are of course also the biological ones, mainly viruses and bacteria, but these will not be discussed here.

9.2.1
Naphthalene (in Mothballs)

Naphthalene is an important industrial chemical that also has a traditional use as an insecticide against moths. The LD_{50} value of naphthalene is 490 mg/kg, and the LD_{low} value is 100 mg/kg). Its primary acute effect is methemoglobinemia, while chronic exposure will affect the lungs and eyes and may give rise to allergies. In

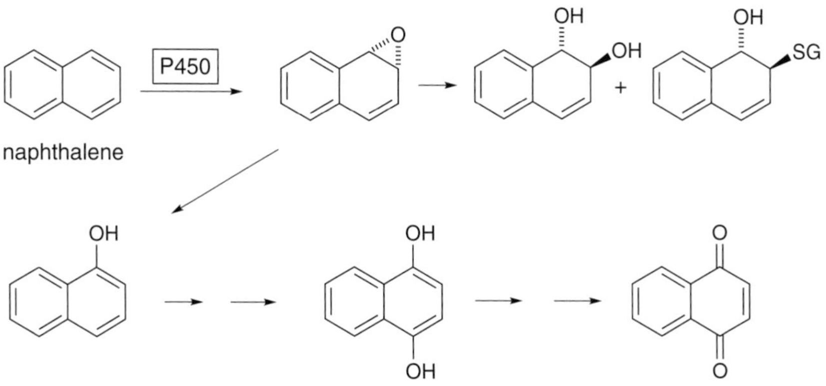

Figure 9.2 Metabolism of naphthalene.

addition, it is believed to be a carcinogen, because it is metabolized, in the same way as benzene, to reactive naphthoquinones (see Figure 9.2). It has become less popular as more modern and effective halogenated pesticides have appeared on the market (it was largely replaced by 1,4-dichlorobenzene, also discussed below), but as many of these were banned in later years naphthalene is back in business. After all, it is not very toxic and is fairly safe to use because it is solid at room temperature and has a low volatility, resulting in a low exposure. However, a drawback with naphthalene will always be that it is highly flammable. As insecticides, both naphthalene and 1,4-dichlorobenzene work similarly. They both sublimate directly from the solid state straight to a gas, and when this takes place in a closed space the concentrations can become high. The effect on the moths is probably a combination of a toxic effect from these very lipophilic compounds and the fact that they will condense on the fibers of the clothes in the closet which will make them taste bad.

9.2.2
DDT

The acronym DDT stands for dichlorodiphenyltrichloroethane (a chemically more correct name is 2,2-bis (4-chlorophenyl)-1,1,1-trichloroethane). It is one of the world's most well-known chemicals, in the beginning as a magic bullet against insects but later as the symbol for environmental damage. Although it was first discovered in the end of the nineteenth century, the fame of DDT began in 1939 when the the Swiss chemist Paul Müller, working for the company Geigy, discovered its insecticidal effect. DDT appeared to be nontoxic to humans (it could be eaten with a spoon without any side effects), but insects that came into physical contact with it died quickly. Commercial insecticides containing DDT were introduced on the market in 1942, and in a short time DDT was being used in large amounts as the need for cheap and effective insecticides was huge in the disease-ridden misery caused by the Second World War. The breakthrough came in 1944

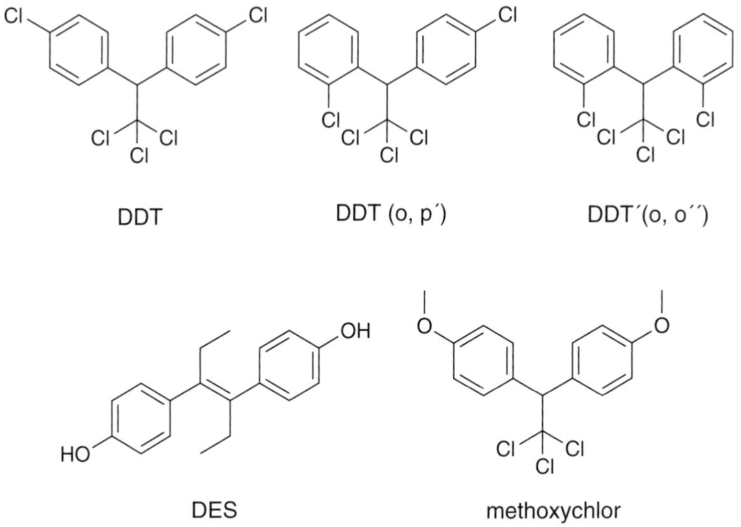

DDT   DDT (o, p′)   DDT′(o, o″)

DES   methoxychlor

Figure 9.3 Some relatives of DDT.

during a massive outbreak of typhus fever, spread by lice, in Italy. 1.3 million people were treated, and the situation was soon under control. DDT, together with penicillin, which also was introduced during this time, made the Second World War the first war in which fewer people died of war-related diseases than by direct violence. In the post-war era, the use of DDT was promoted by international efforts organized by the WHO to eradicate malaria (the malaria parasite is spread by a mosquito that is susceptible to DDT), and in 1955 almost 300 million tonnes were spread throughout the world. The program was, at the time, a fantastic success, and the number of new malaria cases dropped from millions to a few in the areas that were worst affected. Unfortunately, over time the mosquito developed resistance to DDT, which became less effective, so that even larger amounts of DDT were used. In addition, reports about DDT's adverse effects on other organisms were appearing. The first warning sign came as early as 1950, when researchers showed that roosters treated with DDT did not develop masculine traits, and they suggested that DDT acts like estrogen (female hormone) because it is similar to the synthetic estrogen DES (diethylstilbenediol, now banned because it is carcinogenic, see Figure 9.3). Other studies in the 1950s showed, for example, that birds that eat larvae fed on leaves sprayed with DDT were poisoned, indicating that DDT is bioaccumulated in food chains. However, it was not until the 1960s after the publication of the book 'Silent Spring' by Rachel Carson (in 1962) that the time was ripe for a critical re-investigation. The title of the book suggests what we can expect if the birds disappear because of the use of pesticides and summarizes the gravity of the situation. The problems with DDT suddenly got full attention, and the chemical was banned in many countries at the end of the 1960s/early 1970s. However, it should be noted that DDT is still

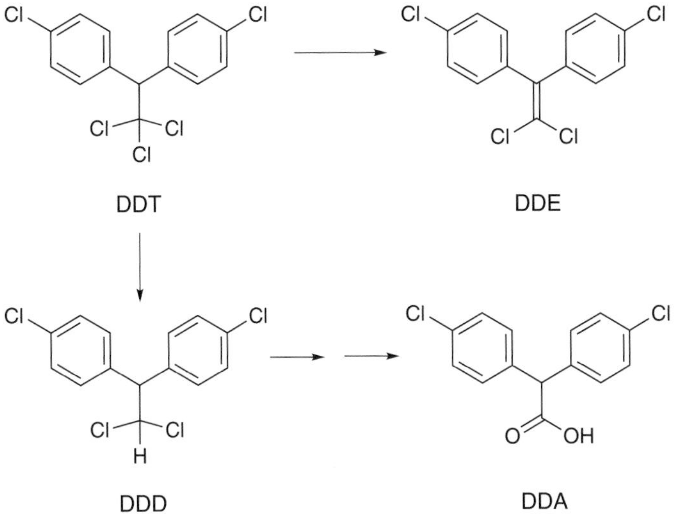

Figure 9.4 The metabolism of DDT in mammals.

a frequently used insecticide in large parts of the world because its short-term advantages (cheap and effective).

The commercial DDT used was actually is a mixture of isomers, p, p'-DDT (85%), o, p'-DDT and o, o'-DDT (see Figure 9.3). DDT is especially characterized by its extreme lipophilicity (its solubility in water being only $1 \mu g \, L^{-1}$) and its extreme chemical stability. These two properties ensure that DDT will have a long life in all possible environments and that it will be readily extracted into organisms, which are more lipophilic compared to the abiotic world. In addition, such a compound can be expected to be accumulated when one goes up in a food chain (depending on the metabolic stability of the compound in various organisms). Much of the DDT that was spread during the 1950s and 1960s is still out there, and its chemical stability has facilitated the spreading of DDT across the globe, including the poles. The exposure in the part of the world where DDT was banned early is currently low but relatively constant. The turnover in humans is mainly due to metabolism, the main metabolites being DDD and DDE (see Figure 9.4). An exception is lactating mothers, who excrete DDT and its lipophilic metabolites with the milk (containing fat). It has been estimated that a baby in the Western world, if breast-fed, will be exposed to approximately 0.5 mg DDT/DDE in the first trimester.

DDT kills insects by inhibiting the ion pumps in the nerve cell membranes. A DDT-poisoned nerve cell cannot signal as frequently as a normal nerve cell because it cannot restore the action potential after it has fired. Its potency is rather low (it cannot be compared to the nerve gases discussed below), and although human neurons are equally sensitive, 'normal' exposure to DDT does not pose a real health hazard. Exposure to dangerous quantities is hampered by the poor solubility of DDT, and if you eat it with a spoon, only very small amounts will actually be absorbed in the gut. More worrying is the effect of DDT in very low concentrations on the endocrine system. This causes the symptoms in birds discussed above. o, p′-DDT is an estrogen agonist, while p, p′-DDE is an androgen receptor antagonist, which explains why exposure to DDT during fetal development can inhibit the development of masculine features. Birds have been easiest to study in this respect, because they eat a lot and bioaccumulate DDT, and their sexual differentiation and mating behavior is determined by the androgen and estrogen levels during the embryo stage. Another result of the endocrine disruption of DDT is that the eggshells become abnormally thin and the eggs are easily destroyed during hatching. The transport of calcium and carbonate ions (shell consists of calcium carbonate) to the machinery that produces the shell is hampered.

The hazards of DDT are consequently linked to its persistence in various environments and the fact that it is accumulated in food chains, and the organisms that are most affected are consequently those at the top. Methoxychlor, with two methoxy groups instead of chlorine atoms in the rings (see Figure 9.3), is easily degradable and is used as a substitute for DDT in many countries. It is metabolized by α-hydroxylation of the methoxy carbons to form the corresponding phenols, which are conjugated and excreted. In animals, the entire dose after an acute exposure to methoxychlor is excreted within 24 h.

9.2.3
Aldrin and Dieldrin

Aldrin and dieldrin are related chlorinated insecticides banned in most countries since the 1970s, but still in use in others as they are easy and cheap to manufacture and efficient against insects. They have mainly been used for moth-proofing fabrics and carpets, and the importation of clothes and carpets treated with aldrin or dieldrin is likely to continue. Aldrin is formed by a Diels–Alder reaction between hexachlorocyclopentadiene, obtained by chlorination of cyclopentadiene, and nor-bornadiene, and dieldrin is obtained by epoxidation of aldrin with an oxidizing agent (see Figure 9.5).

Dieldrin is actually formed in humans but also in microorganisms and plants by the metabolic epoxidation of aldrin. The epoxidation is fast, making aldrin a short-lived pesticide. However, dieldrin, which from a chemical point of view looks reactive and easily degraded, is surpisingly stable and persistent, and bioaccumulates in food chains. This is surprising because we have learnt that epoxides are rapidly metabolized via phase I as well as phase II routes, but the lipophilicity of dieldrin is so high that it is not exposed to the metabolic pathways. Consequently, the excretion rate is very low. Over time, however, it is metabolized by hydroxylation. Dieldrin is also persistent in the environment, and the half-life in sandy soils, for example, is estimated to be 15 years. Both compounds are absorbed rapidly by humans, also through the skin, and the acute effects are primarily on the CNS, resulting in tremors and breathing difficulties. The chronic effects are primarily on the liver, where the metabolism takes place, and dieldrin has been demonstrated to be a liver carcinogen.

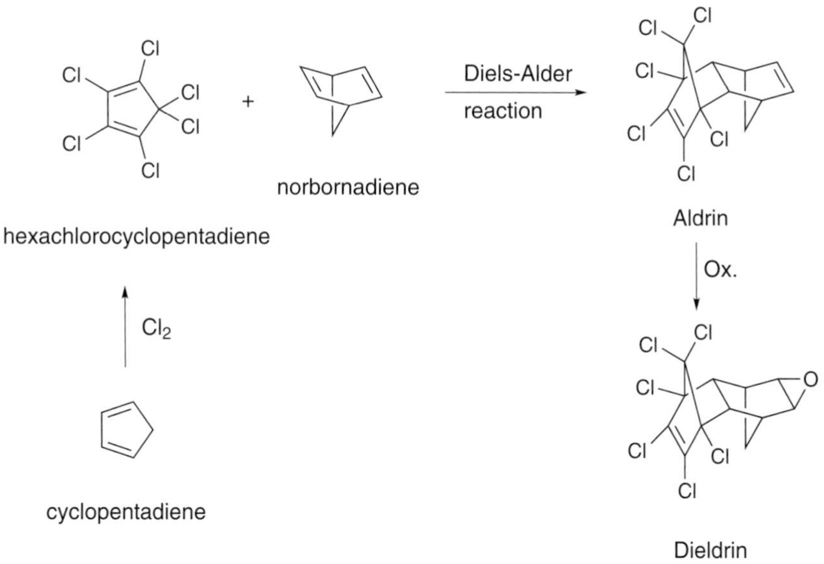

norbornadiene

hexachlorocyclopentadiene

Aldrin

cyclopentadiene

Dieldrin

Figure 9.5 The synthesis of aldrin and dieldrin.

9.2.4
Hexachlorobenzene

The chemical class of chlorobenzenes, consisting of a benzene ring substituted with between one and six chlorine atoms (see Figure 9.6), are important industrial chemicals with an estimated world production of hundreds of thousands of tonnes per year. The estimated total emission of chlorobenzenes into the environment is approximately 50% of the amount produced. Altogether there are 12 different isomers (including three each with two, three, and four chlorines), and the really cheap products, produced by unspecific chlorination of benzene, are mixture of the 12. They are used in the sanitation sector, for the production of paint and polymers, as solvents, and as insecticides. More chlorines means lower volatility: the boiling point of chlorobenzene is 132 °C while that of hexachlorobenzene is 322 °C, as well as higher lipophilicity (log P varies between 2.8 and 6.44). In addition, higher number of chlorines correlates with lower chemical reactivity and higher persistence. The acute toxicity of the chlorobenzenes is generally low, especially for mono- and hexachlorobenzene, while the trichlorobenzenes are more toxic. Chronic exposure damages the liver and kidneys and affects the CNS. Monochlorobenzene, the di- and trichlorobenzenes, and probably also the tetrachlorobenzenes, are epoxidized by cytochrome P450 in mammals (like benzene itself) to form phenols, which can be conjugated and excreted. Penta- and hexachlorobenzene are not, but need to be dechlorinated before oxidation/excretion. This can be performed by microrganisms, as discussed in Chapter 6. Hexachlorobenzene was formerly a very important fungicide, but its used has diminished as a result of

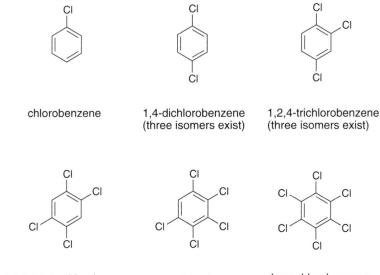

chlorobenzene   1,4-dichlorobenzene   1,2,4-trichlorobenzene
                 (three isomers exist)   (three isomers exist)

1,2,4,5-tetrachlorobenzene   pentachlorobenzene   hexachlorobenzene
(three isomers exist)

Figure 9.6 The structures of the chlorobenzenes.

regulations. It was, for example, used to inhibit fungi from attacking crop seeds, and the accidental use of such treated seeds for food production has resulted in massive poisonings. In Turkey in the 1960s more than a thousand people, mainly children, died from exposure to hexachlorobenzene. Today it is mainly used as a starting material for the manufacture of pentachlorobenzene. The hexachlorobenzene that is released into the environment today is instead produced as a by-product during, for example, the combustion of chlorinated organic compounds and in the production of chlorinated solvents. Hexachlorobenzene is still a global problem because of its extreme stability and its ability to bioaccumulate, and it may be present in alarmingly high concentrations in the body fat of organisms at the top end of food chains. The acute toxicity is limited at low concentrations, but severe effects on the CNS and the biosynthesis of hemoglobin have been reported.

9.2.5
1,2-Dibromoethane and 1,2-Dichloroethane

Both these compounds have used as lead scavengers in leaded petrol and as pesticides to kill nematodes in the soil. 1,2-Dichloroethane is also an industrial solvent and an intermediate used in the production of vinyl chloride. While their large-scale use as pesticides has stopped, 1,2-dibromoethane is still used as a fumigant for the treatment of logs to eliminate termites and beetles. Both are toxic, especially to the liver and kidneys. Both can react as primary halides with nucleophiles, they are also hydroxylated to form the highly reactive metabolites α-chloroacetaldehyde

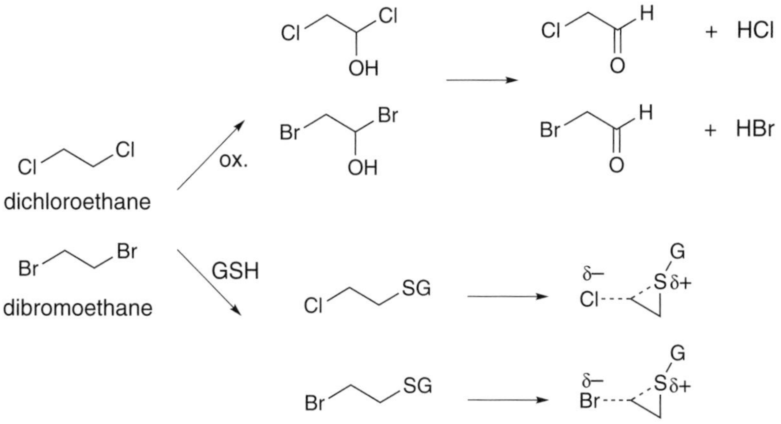

Figure 9.7 Dichloroethane and dibromoethane are electrophilic as they are (left), after oxidative metabolism (top), and after conjugation with glutathione (bottom).

and α-bromoacetaldehyde, and they can be directly conjugated with glutathione to S-(2-chloroethyl)-glutathione and S-(2-bromoethyl)-glutathione (see Figure 9.7). The latter two, either as they are or as the corresponding cysteine derivatives formed after hydrolysis of the tripeptide, can react in a similar way to mustard gas, forming an extremely reactive electrophile by an intramolecular attack of the sulfur on the carbon with the halogen atom (discussed in detail below). The toxic effects are caused by a combination of these reactive forms.

A similar halogenated alkane is 1,2-dibromo-3-chloropropane, or DBCP, used as a soil fumigant and a nematocide until 1985 (in the United States), which is nephrotoxic and genotoxic. It also causes male sterility, and has injured many workers especially in banana and pineapple plantations. As well as the reactive metabolites corresponding to those shown in Figure 9.7, 1,2-dibromo-3-chloropropane has also been shown to form reactive acrolein derivatives, epoxides (formed by ring closure of an alcohol substituted with a bromine or a chlorine on the α-carbon) and α-bromo-α′-chloroacetone (see Figure 9.8).

9.2.6
The Phenoxy Acids

The phenoxy acids, or phenoxyacetic acids, were used in enormous quantities as herbicides in the 1960s and 1970s, for example, to get rid of the vegetation on railway embankments. There are several variants, but the most famous phenoxy acids are 2,4-D (2,4-dichlorophenoxyacetic acid) and 2,4,5-T (2,4,5-trichlorophenoxyacetic acid) (see Figure 9.9). The latter compound was banned in many countries towards the end of the 1970s. In addition, a substantial number of analogs and derivatives in which, for example, a chlorine was exchanged for a methyl or the acetic acid was replaced with propionic acid (MCPA, mecoprop and

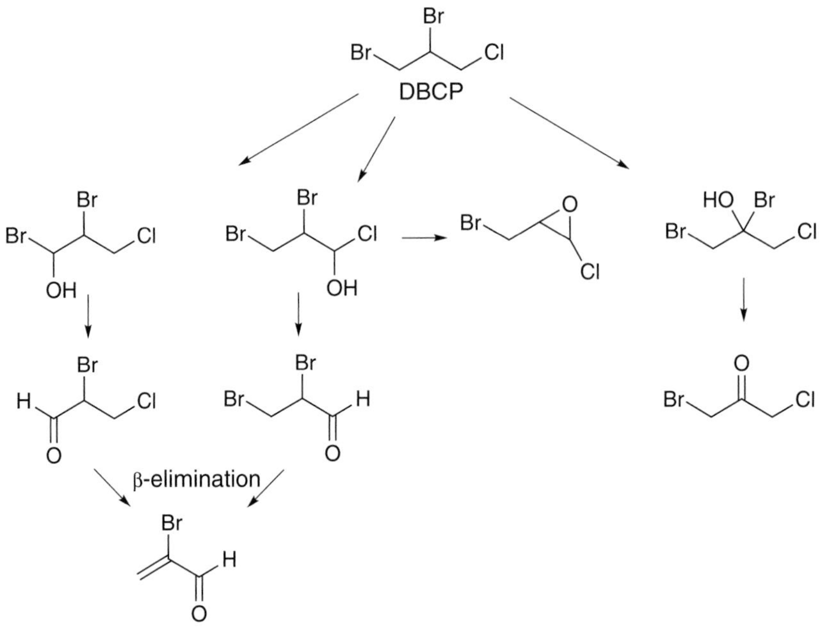

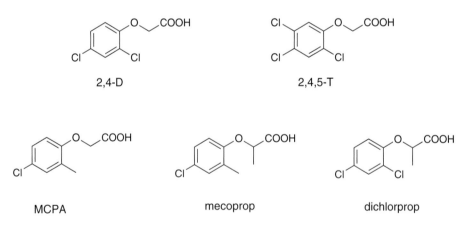

Figure 9.8 Additional reactive metabolites formed from 1,2-dibromo-3-chloropropane.

2,4-D

2,4,5-T

MCPA

mecoprop

dichlorprop

Figure 9.9 Phenoxy acids with prior and present use as herbicides.

dichlorprop, see Figure 9.9), or in which the carboxylic acid functionality was esterified, resulting in compounds with higher lipophilicity that are absorbed more easily, or converted to an amide. The infamous product Agent Orange was used in large quantities during the Vietnam War for the defoliation of forest and jungle plants in order to prevent the concealment of enemy concentrations. It was mainly a mixture of 2,4-D and 2,4,5-T, but contained side products that made it even more toxic.

It is generally believed that the phenoxy acids function as herbicides by acting as a hormone for growth, stimulating the plants literally to grow themselves to death. However, this has not been established, and the phenoxy acids show a certain degree of selectivity, being significantly more toxic to herbs than to grass. They are considered to be moderately toxic to humans, although their genotoxicity and carcinogenicity are still debated. The presence of a carboxylic acid function makes the phenoxy acids more hydrophilic, promoting their excretion from the body, and their half-life in man is short. However, the major problem with the phenoxy acids may actually be a by-product formed during their synthesis. As discussed in Section 10.6.1, the conditions employed during the transformation of chlorobenzenes to phenoxy acids produce dioxins, and these very toxic compounds pose a severe risk to anyone handling contaminated phenoxy acids. In the beginning this was not understood, and until 1965 the 2,4,5-T contained at least 30 mg of the dioxin TCDD per kilogram. In the 1970s, when the chemical process had been optimized, the amounts of TCDD in 2,4,5-T were down to levels below 0.05 mg per kg. People who were exposed to phenoxy acids during the early period have shown dioxin-related symptoms, but it is very difficult to separate the effects of the dioxins from those of the phenoxy acids themselves. The phenoxy acids have also contributed to the effect. The chlorophenols, intermediates in the synthesis of phenoxy acids and responsible for the formation of dioxins, have also been used for other purposes. They can have between one and five chloro substituents, and there are nineteen isomers. Pentachlorophenol was the most important chlorophenol and has been used, for example, as a wood preservative to protect telegraph poles and any other wood in contact with the soil. However, the compound has been banned in most countries since the 1980s. The chlorophenols are produced by direct chlorination of phenol or by alkaline hydrolysis of chlorobenzenes, and they are commonly contaminated with phenoxyphenols (also called predioxins because they are just one step from a dioxin), dioxins, chlorinated dibenzofurans, and diphenyl ethers. Pentachlorophenol is a potent decoupler of the oxidative phosphorylation (see Section 7.2.5.3) and is thus highly toxic to mammals.

9.2.7
Tributyltin Oxide

Tributyltin oxide (TBTO) is a symmetric ether between two units of tributyltin (see Figure 9.10). It is used as a preservative in, for example, cooling water, and in antifouling paint used to prevent sea organisms from attaching themselves to boat hulls. The entire amount of tributyltin oxide used ends up in the environment, where it is slowly transformed into inorganic tin. Although it has been reported

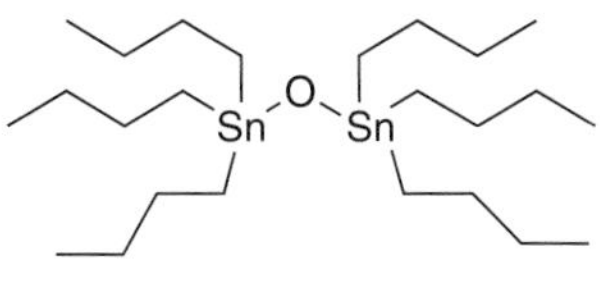

Figure 9.10 Tributyltin oxide.

to be significantly toxic to a range of marine organisms, its effects on humans and the environment is still debated. Tributyltin oxide is accumulated in plants and animals and is not readily biodegradable.

9.2.8
Carbaryl

Many insecticides have a chemical structure related to the organic phosphoric acid derivatives used in chemical warfare agents, which are discussed later in this section. They have the same basic effect, that is, they inhibit the enzyme acetyl-cholinesterase, without which, in a functional form, both humans and insects will perish. If one wants to use these compounds against insects (i.e. as an insecticide), the trick is to develop them so that insects are more sensitive to them than humans. For example, if the oxygen double-bonded to phosphorus in the warfare gases is exchanged for a double-bonded sulfur, this change introduces species selectivity. Insects have the metabolic ability to change the sulfur to an oxygen considerably faster than mammals, which means that a selective metabolic activation to the active and neurotoxic form should be present in insects. Another type of compound that has the same effect on the enzyme acetylcholinesterase and is also used as an insecticide is the carbamate group. These are also known as ure-thanes, because this is the name of their basic chemical functionality, and they can be exemplified by the substance carbaryl. The carbamates are generally less toxic than the organic phosphorus compounds, as their effect on the acetylcho-linesterase is not permanent but reversible. The insecticide carbaryl is produced by the addition of 1-naphthol to methyl isocyanate, a process that is forever associ-ated with a chemical disaster (see Figure 9.11).

This occurred in Bhopal in India, 1984, where a tank of methyl isocyanate (MIC) was pressurized and about 40 tonnes of the substance was spread in the air in a

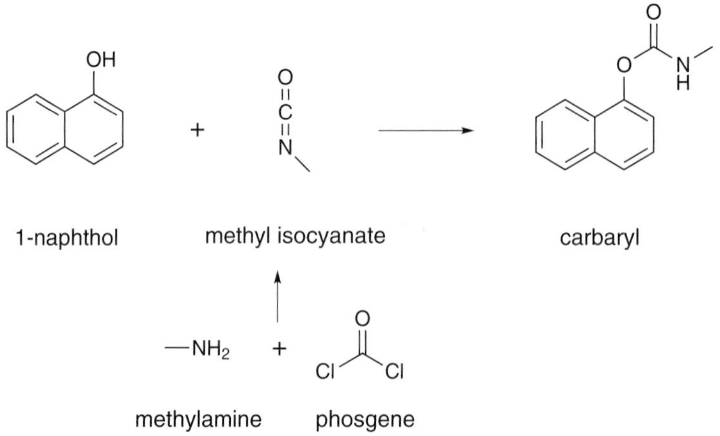

| 1-naphthol | methyl isocyanate | carbaryl |

methylamine   phosgene

Figure 9.11   The synthesis of carbaryl.

densely populated area. Around 3500 died of acute exposure to MIC, and it is estimated that the number of injured was 200 000. The majority of the survivors suffered from a number of complaints including circulatory disorders, muscle pain, and problems with the gastrointestinal tract, so as well as the acute effects there were evidently also some delayed effects. Besides its use for the synthesis of carbaryl, methyl isocyanate is also used to manufacture polyurethane. Methyl isocyanate is a highly electrophilic and very reactive substance, and it also reacts with very weak nucleophiles such as water (the half-life of MIC in water is only two minutes at room temperature). Consequently, methyl isocyanate is extremely irritating to the respiratory tract and the eyes. Higher concentrations will give local damage that can be life threatening, and this is what killed the victims in Bhopal. Those that were exposed to lower, non-lethal, concentrations later suffered from systemic effects on a variety of other organ systems, such as the circulatory, digestive, and immune systems. Methyl isocyanate can react with glutathione to produce a metabolite that is relatively stable and can be distributed in the body, but that slowly 'regenerates' methyl isocyanate.

The carbaryl factory in Bhopal itself produced the methyl isocyanate that was required for the reaction with 1-naphthol by reacting phosgene with methylamine. The phosgene was in turn made from carbon monoxide and chlorine gas. Carbon monoxide is highly toxic and explosive, as discussed previously (see Section 7.2.4.1), and both phosgene and chlorine gas were used as a warfare chemicals during the First World War (and are consequently quite dangerous). The chemicals handled in Bhopal during the manufacture of carbaryl were hazardous in many ways: it was certainly a high-risk factory, and therefore there was several security systems. If the storage tank for methyl isocyanate became pressurized its contents were to be led into an alkaline scrubber, and any gas passing through the scrubber were to be burned. Unfortunately, neither of these two systems worked. The pressure in the methyl isocyanate storage tank was probably caused by a leak of water that entered into the tank, and the water reacted with the methyl isocyanate to produce gaseous carbon dioxide, which built up the pressure.

9.2.9
Glyphosate

Glyphosate is the active ingredient in the herbicide Roundup. It is very effective, and because it is cheap it has been and is used in huge quantities. It is selectively toxic to foliage, without notably affecting other organisms (even conifers). For the plants with leaves it is completely unselective, killing everything. The effect is caused by a disturbance of the plant's synthesis of amino acids. It is a systemic effect that appears when the glyphosate has been absorbed and distributed in the plant. However, glyphosate is poorly absorbed, only about 25% in plants, and an even smaller proportion is actually distributed. It only functions on plants that are in a vegetative phase, and it typically takes a week or two to cause death. A major advantage of glyphosate is that it not only removes the parts above ground, but it also kills the root.

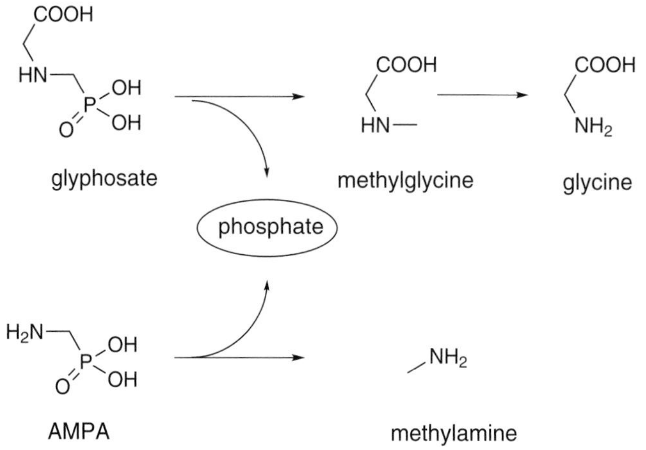

Figure 9.12 Glyphosate and its degradation products.

Glyphosate is an organic phosphonic acid derivative, with the amino acid glycine linked through the nitrogen to a phosphonomethyl group. It is ionized at neutral pH and therefore very water soluble. It is stable in a water solution but is sensitive to sunlight. In the soil, it will be adsorbed on the soil particles, but also slowly follow the rainwater on its way down. The presence of glyphosate in groundwater has been reported in low concentrations but this is still of some concern. There is a relatively efficient microbial degradation of the compound, depending on the ability of microorganisms to use glyphosate as a source of phosphorus. The metabolites are aminomethylphosphonic acid (AMPA) which is hydrolyzed further to methylamine and inorganic phosphorus, as well as *N*-methylglycine (see Figure 9.12).

Glyphosate shows relatively low toxicity in mammals, mainly due to the fact that it is poorly absorbed because it is so hydrophilic. The uptake in humans through the skin is close to zero, and only a small proportion of any glyphosate swallowed will be absorbed in the intestines (less than 30%). Again, because of its hydrophilicity, almost all of the absorbed amounts are excreted unchanged via the kidneys and liver. Small amounts of AMPA are formed, but they represent only a fraction of a percent. However, if the compound is given intraperitoneally, the toxicity to mammals is not negligible (LD_{50} approximately 200 mg/kg). The toxic effect in mammals is caused by glyphosate's ability to act as a decoupler of the oxidative phosphorylation in the mitochondrial energy production (see Section 7.2.5.3). Glyphosate has been tested for allergenic, carcinogenic, teratogenic, and genotoxic effects, but no evidence of such properties has been reported.

9.2.10
Dinoseb

Dinoseb, which belongs to the group of dinitrophenols, is a classical decoupler of the oxidative phosphorylation. It has been used extensively to control insects, but

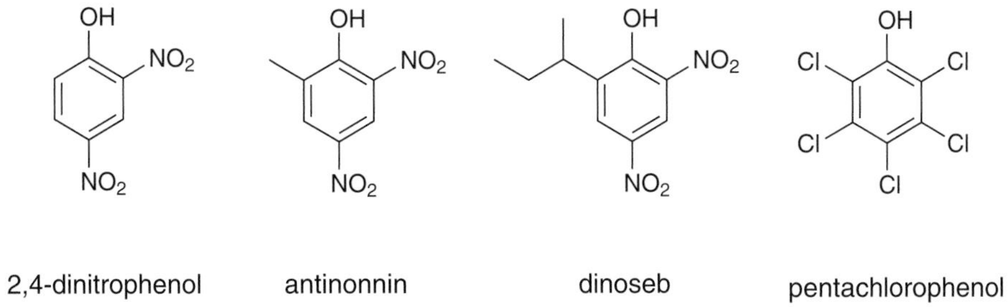

Figure 9.13 Dinoseb and other decouplers.

because of its severe toxicity to virtually all organisms it is now banned in most countries together with similar and equally toxic dinitrophenols (see Figure 9.13). However, some less toxic dinitrophenols are still used. The term 'decoupler' relates to the mechanism by which these compounds act: they literally decouple the oxidative phosphorylation from the electron transport chain with the result that the energy released in the electron transport chain is not used for phosphorylation but instead generates heat; this is discussed in Section 7.2.5.3. Phenols are general decouplers, but more efficient if they have electron-withdrawing substituents, such as nitro groups, in the ring. An interesting effect of decouplers is that they, at non-lethal concentrations, will make you lose weight. The metabolism of nutrients, for example, fat, to produce energy is accelerated, and as long as the fever is not too high the kilos will drop away fast. In the 1920s 2,4-dinitrophenol was marketed, and very popular, as a weight-controlling agent, but banned after too many fatal accidents. Apparently, the possibility to lose another kilo before the weekend tempted many to overdose. In spite of the ban 2,4-dinitrophenol is still on the market, although now on the illegal one. It is especially popular with body builders, as it efficiently burns off all remaining traces of fat in the body.

9.2.11
Chemical Warfare Agents Containing Sulfur

Sulfur dioxide (SO_2) was the first documented chemical warfare agent used in battle. It was produced *in situ* by the combustion of sulfur and was used to subdue the defenders of the besieged city of Plataea during the peleponnesian war between Athens and Sparta (400 BC). However, the breakthrough for chemical warfare came during the first world war, during which approximately 50 different chemicals were used with the purpose of killing or incapacitating the enemy. The first large-scale attempt was made with chlorine gas, or dichlorine (Cl_2), at Ypres (Belgium) in 1915, when 30 tonnes were released along a front line 6 km long by German troops. It was a 'success' because the French troops that were the subject of the attack were completely unprepared and could not escape. The French lost 15 000 men at a stroke, and in a war that appeared impossible to finish in a conventional way the dramatic effect of the chlorine gas raised expectations. It prompted the testing of other gases with potential use on the battlefield, and in addition to chlorine gas, phosgene, arsine, mustard gas, and various tear gases were used. However, in parallel with these tests, new ways to protect the soldiers were developed, and the gas mask became a standard issue for the soldiers on both sides. Soon it became evident that in the absence of the element of surprise in the attack in Ypres, the first generation of war gases was not always very efficient. Their effect was based on their chemical reactivity, and they had to be inhaled in order to produce an acute toxic effect on the respiratory system. Furthermore, the attacker was completely dependent on the direction of the wind, and if it changed the attacker would suddenly be the attacked. The new gas masks gave full protection for gases of this kind. New agents were developed, mustard gas being perhaps the most famous example of the new generation of chemical warfare agents. It is an oily liquid at room temperature, melting at 14 °C and boiling at 217 °C, which is an advantage as it can be transported in normal containers. It was dispersed as an aerosol, not as a gas, and it was used for the first time at Ypres (France) in 1917 by German troops against British soldiers. Mustard gas was also called Yperite. After an exposure to mustard gas, it took several hours before the effects became apparent. Few actually died from mustard gas; instead, the soldiers became incapacitated, which was considered an advantage. Not only were the affected soldiers useless, they required help from their fellow soldiers. Nitrogen mustard is an analog that works in the same way and that was also produced in large quantities. The mechanism of the chemical reaction of mustard gas with nukcleophiles is shown in Figure 9.14. This reaction would take place with any tissue with which it came into contact, and it destroyed the skin as well as the respiratory tract in a horrible way.

The central sulfur atom functions as an internal nucleophile, making the bond between chlorine and carbon weaker and facilitating the reaction of the external nucleophile. In addition, mustard gas is both genotoxic and sensitizing at low doses, and an important explanation for why it is so harmful is that it is a

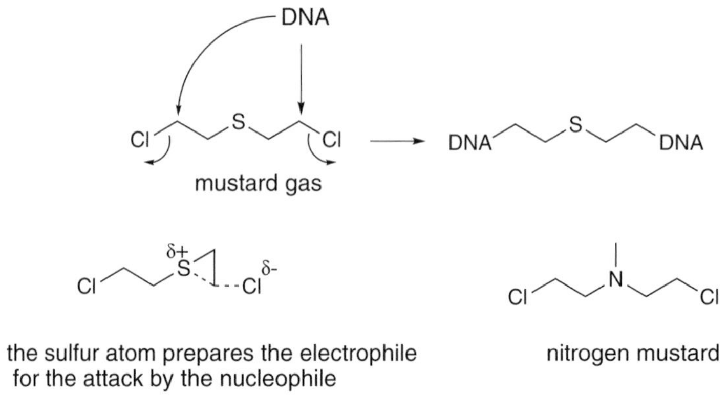

the sulfur atom prepares the electrophile for the attack by the nucleophile

nitrogen mustard

Figure 9.14 Mustard gas is a bi-functional electrophile.

bi-functional compound that can cross-link both proteins and DNA. We have previously discussed (Section 9.2.5) how the conjugation of certain compounds, for example, 1,2-dichloroethane, with glutathione will result in mustard gas-like metabolites.

9.2.12
Chemical Warfare Agents Containing Phosphorus

The main focus of chemical warfare agents in the 20th century has been on the organophosphorus compounds. These were discovered in the early 1930s by the German chemical industry IG Farben during their search for new and potent insecticides. The invention caught the attention of the German military, which during the Second World War developed about 200 potent organophosphorus compounds. Already in 1940 the mechanism for their effect, the inhibition of the enzyme acetylcholine esterase that hydrolyzes acetylcholine in the cholinergic nervous synapses (discussed in Section 7.3.2), was understood, and it was found that atropine is an antidote. Examples of organophosphorus compounds that were produced during the war are sarin, soman, and tabun (see Figure 9.15), all of which are deadly if you breathe air containing more than $0.1\,\mathrm{mg\,L^{-1}}$ for a minute. Sarin and soman are both relatively volatile (boiling points 147 and 167 °C, respectively) and will evaporate and therefore disappear from a contaminated area, while tabun is less volatile (boiling point 247 °C) and will remain for a long time. All can be absorbed via the lungs, the skin and the gastrointestinal tract. Fortunately, these compounds were not used, at least not in bigger battles, in the war.

The researchers in England and the United States followed the activities in Germany, and after the end of the war the United States decided to develop new and more potent compounds. VX (see Figure 9.15) is an example. Its low volatility (boiling point 300 °C) and high lipophilicity makes it suitable for contact exposure and absorption via the skin, and 10 mg (one-fifth of a drop) is enough to kill a man. Areas contaminated with VX will be very dangerous to visit for a long time.

As indicated in Figure 9.16, the acetylcholine esterase inhibitors react covalently with the enzyme as it attempts to hydrolyze them as well. They add to the hydroxyl group of a serine moiety, and even though the acetyl group that normally ends up in this position when acetylcholine is hydrolyzed is easily removed by the enzyme, other groups will be more long-lived. The alkylphosphates especially react almost irreversibly, and are consequently very toxic. The search for effective antidotes

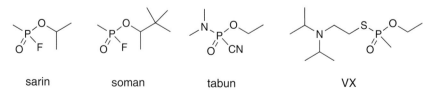

sarin   soman   tabun   VX

Figure 9.15 Some highly toxic organophosphorus compounds.

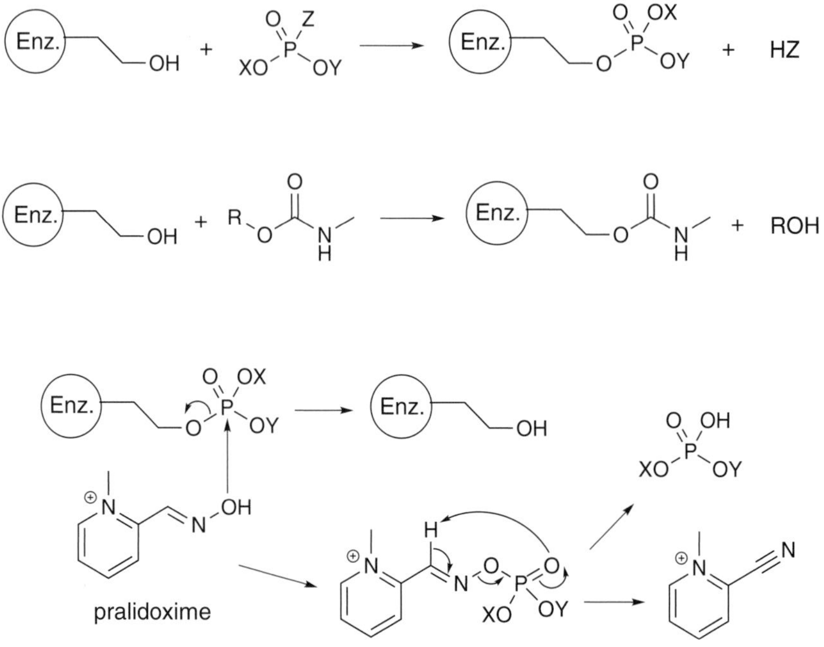

Figure 9.16 The inactivation of acetylcholine esterase by an organophosphate and by a carbamate ester, and the reactivation of the inhibited enzyme by pralidoxime.

against these nerve poisons inhibiting acetylcholine esterase was carried out in parallel, and not only led to the identification of atropine acting as antagonist of the receptor for acetylcholine and thereby inhibiting the effect of high levels of acetylcholine, but also to the development of compounds with the ability to regenerate acetylcholine esterase that had been destroyed by organophosphorus compounds. These so-called reactivators may chemically be oximes, and the mechanism of the reactivation of acetylcholinesterase by one of them, pralidoxim, is shown in Figure 9.16. It fits to the same binding site as acetylcholine and is able to react with the organophosphate residue so that the serine becomes free again. Pralidoxim is relatively effective in sarin and VX poisoning but less so when it comes to tabun and almost ineffective against the effects of soman. To treat victims of the two latter poisons there are other oximes that work better. The oximes are given together with atropine, which is an acetylcholine esterase antagonist (discussed in Section 7.3.2) that reduces the effect of the surplus acetylcholine present.

Although there are other anticholinergic substances that are considerably more potent than atropine, this substance has nevertheless proved to be the most effective. Atropine is a poison by itself, but a dose of a few milligrams is not dangerous. When protecting against the effects of organophosphorus compounds, atropine most importantly protects the respiratory center in the CNS. It does not appear to have a significant impact on the toxic effects of the peripheral nervous system, but these are less life-threatening. The oximes cross the blood-brain barrier with difficulty, and will therefore have their main effect in the peripheral nervous system.

The use of chemical warfare agents is strictly regulated by international agreements, and it is unlikely that they will play any role in a traditional conflict. However, as these chemicals are relatively cheap and easily made, there is always a possibility that they will be used by states or organizations that do not follow international agreements. This threat is quite significant and must be taken seriously.

9.2.13
Summary

In the section covering pesticides and chemical warfare agents, we find a number of substances that are well known both as poisons as well as environmental threats. This is actually not surprising, since they were designed to disrupt biochemical pathways in various organisms and to be used in ways that spread them in the environment. The use of pesticides varies with the location of the growing area and the local climate. In general, the use of pesticides is becoming more and more restricted, and the authorities constantly raise their requirements. Tomorrows pesticides must be nontoxic and selective, they must not promote the development of resistance, and they must be easily degraded to harmless end products. However, we are not there yet, and still thousands die every year after having been in contact with pesticides. Chemical weapons should be banned completely in the civilized world, but they exist and are relatively easy to produce by terrorist organizations.

9.3
Food, Pharmaceuticals, and Natural Products

The suggestion that hazardous compounds are present in our food and pharmaceuticals might sound strange, but it is nevertheless a reality. It is believed, for example, that as many as one third of all human tumors are linked to chemical substances that we ingest together with food. Toxic substances in food may be natural (so-called natural products), they may be chemicals that are added during the preparation of the food, such as preservatives, and they may be new products that are formed from the natural ingredients and/or are added during the processing of the food (for example, by cooking). Especially the latter has been shown to play an important role, and we are today aware about the formation of a number of toxic compounds from common amino acids and carbohydrates during cooking at elevated temperatures. Especially hazardous are roasting, grilling, and frying, and to some extent also smoking of food. Three substance classes are particularly interesting in this context: the polycyclic aromatic hydrocarbons, the heterocyclic amines and the N-nitroso compounds, which all are associated with carcinogenic effects. Pharmaceuticals are designed to cure and mitigate, and always to have a positive effect on humans. However, in unfortunate cases, and in high doses, they can also be hazardous. The development of a modern drug takes more than 10 years and costs enormous amounts of money, and one of the most important goals is to obtain safe drugs. However, drugs are biologically active and they can never be completely safe. Natural products are low-molecular weight compounds produced by organisms (in contrast to synthetic chemicals made by humans in an artificial way) which have some kind of use for them. They can be colored compounds that attract pollinators, compounds that smell interesting to a mating partner, or compounds whose taste discourages hungry neighbors. Normally we have no idea about the function of natural products, but even if they play an important ecological role they may also be toxic. In fact, the most toxic compounds that we know are natural products, although they are not, by definition, environmental hazards.

9.3.1
Safrole, Myristicin, Elemicin and Estragole

The related compounds safrole, myristicin, elemicin, and estragole have been and are used as a flavorings in perfumes, food, and liquor. All four are obtained from various plants, for example, anise and fennel, and are also present in a variety of spices such as parsley and dill. Safrole is an important component of many essential oils and the main component (approximately 75%) of sassafras oil (*Sassafras officinale*), with weak insecticidal and antimicrobial activity. It is used for various purposes in the perfume industry, but since it was discovered that safrole is a carcinogen it has not been allowed as an additive in food. Since these compounds can be consumed in large quantities, their toxicity is of some concern. Their metabolic activation is exemplified by that of safrole (see Figure 9.17). Safrole is metabolized by epoxidation of the double bond in the side chain, by aliphatic hydroxylation of the carbon, which is both benzylic and allylic, and by α-hydroxylation of the acetal carbon. The epoxide is reactive as it is, and the benzylic alcohol is at least partially conjugated with sulfate, whereby the hydroxyl group is converted to a relatively good leaving group, and the catechol formed spontaneously in the third route can be oxidized to a reactive *ortho*-quinone. It is not known which of the activated metabolites is mainly responsible for the carcinogenic properties of safrole, but they probably all contribute.

9.3.2
Nitrite

Nitrite (NO_2^-) is one of the oldest preservatives,and is used in the form of the sodium ($NaNO_2$) or the potassium (KNO_2) salt mainly to preserve meat and fish. Its primary task is to prevent the growth of bacteria that cause botulism (see below), and despite the fact that the use of nitrite is not entirely risk-free we have not found a better replacement. Moreover, ham without nitrite would have a gray and unappetizing appearance. Nitrite (as its acid, nitrous acid) is somewhat unstable and is reduced to nitric oxide, NO, which binds to the heme group in the

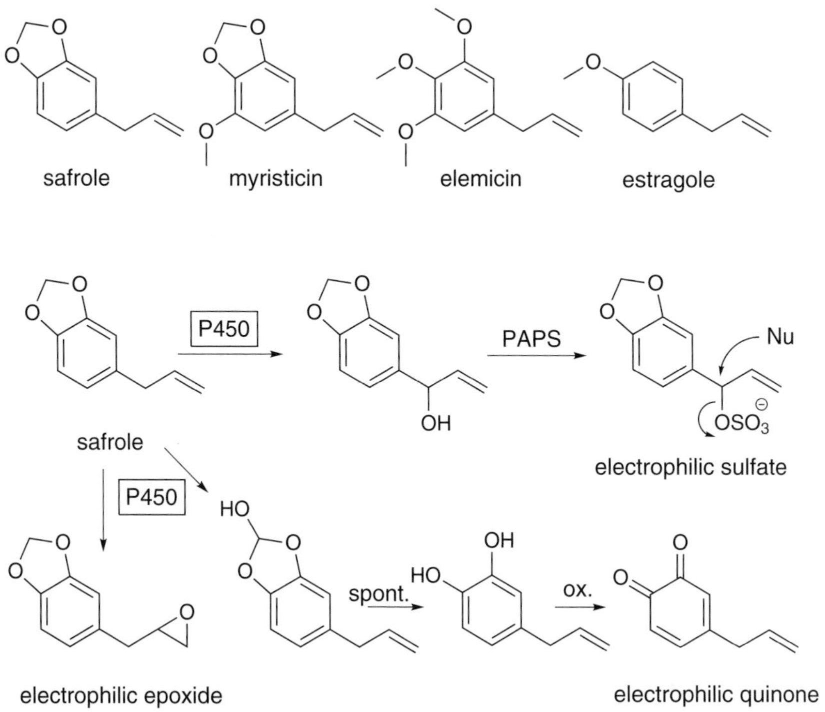

Figure 9.17 Metabolic activation of the aromatic substance safrole.

muscle protein myoglobin and gives a red product that stains the ham (or any meat treated with nitrite). Nitrite can also be formed by the reduction of nitrate, which is naturally present in high concentrations in foods such as spinach. If such food is stored too long at room temperature, bacteria can reduce nitrate to nitrite, although acutely dangerous levels are seldom reached. Bacteria in the oral cavity and especially in the intestines of infants can do the same thing, and one should always be careful to give small children high quality water (well water in agricultural areas which are fertilized with nitrates is not recommended). In addition, nitrite will in an acidic environment (e.g., the stomach) react with amines to form carcinogenic N-nitrosamines, as discussed in Section 4.2.2 and below. In high concentrations, nitrites have a toxic effect on the blood (methemoglobinemia, discussed in Section 7.3.1).

The N-nitrosamines formed from secondary amines are stable compounds that are absorbed into the body and can be metabolized (via α-hydroxylation) to electrophilic and highly carcinogenic metabolites (see Section 5.6.2). They can be formed in foods treated with nitrites as preservatives, in smoked meat and fish (nitrogen oxides in the smoke then act as a nitrosating agent), as well as food that is dried by combustion gases containing nitrogen oxides (such as malt for beer). Particularly rich in N-nitroso compounds is fried bacon, which consequently should be enjoyed in moderation. For a smoker, however, the amounts of N-

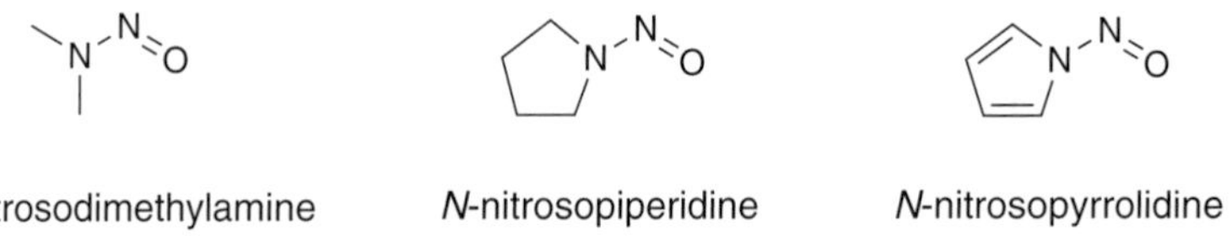

N-nitrosodimethylamine   *N*-nitrosopiperidine   *N*-nitrosopyrrolidine

Figure 9.18 *N*-nitrosamines present in food.

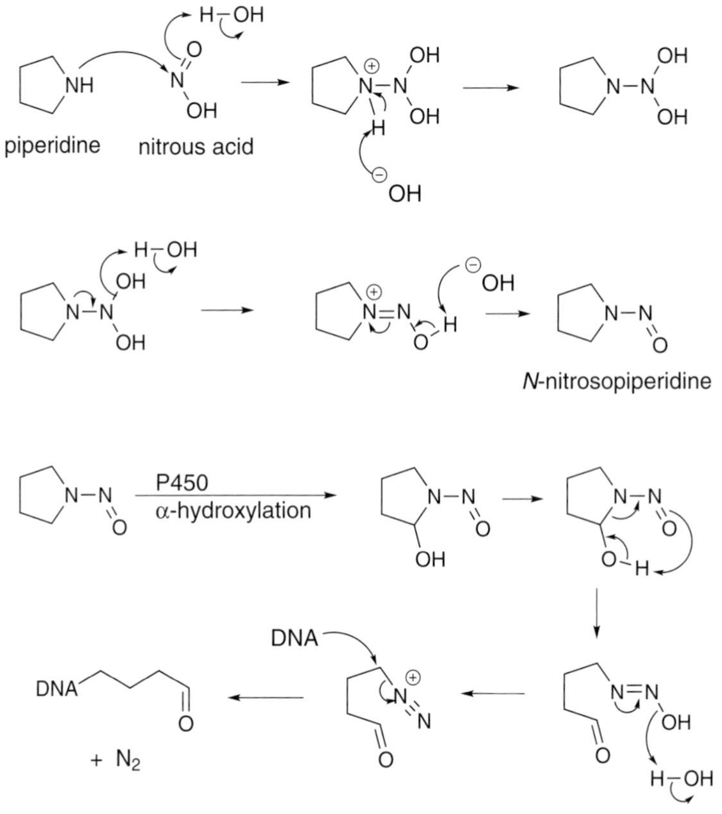

piperidine   nitrous acid

N-nitrosopiperidine

P450
α-hydroxylation

DNA

+ N₂

Figure 9.19 The metabolic activation of *N*-nitrosopiperidine.

nitroso compounds in food are relatively insignificant compared with those found in cigarette smoke, but they all add to the carcinogenic potential of cigarette smoke. Three examples of *N*-nitroso compounds that are present in foods are shown in Figure 9.18. All these are potent carcinogens.

The metabolic activation of dimethyl-*N*-nitrosoamine was discussed in Chapter 5, while that of *N*-nitrosopiperidine is shown in Figure 9.19. The latter has an extra twist, as the final reactive diazonium species also has an 'sticky' aldehyde function in it that will make the compound stick to nucleophilic environments like proteins and DNA.

9.3.3
Polycyclic Aromatic Hydrocarbons and Heterocyclic Aromatic Compounds

Polycyclic aromatic hydrocarbons (PAHs) and heterocyclic aromatic compounds (having one or more heteroatoms, usually nitrogen and/or oxygen, in any of the aromatic rings) are compounds formed during incomplete combustion of organic matter. The PAHs are found in emissions from hydrocarbons such as coal and petroleum products (see above), but they are formed naturally at high temperatures or on incomplete combustion of any organic materials. If the material also contains amino acids and carbohydrates, complex mixtures of heteroaromatics are also formed. A good example of how many new compounds are formed under such conditions is the pyrolysis of tobacco, which will be discussed below. Hundreds of products have been identified, of which at least 70 are carcinogenic. Polycyclic aromatic hydrocarbons may end up in food simply because cereals, vegetables, fruits, and seafood absorb them from contaminated air and water and lack the ability to degrade them, or they are formed together with heterocyclic aromatic hydrocarbons in foods exposed to high temperatures (above 400 °C) during roasting and grilling. One important source during such processes is the fat from the food that drips down onto a hot surface and is pyrolyzed. The polyaromatics formed will condense on the surface of the food that has a relative cold surface. If you enjoy grilled meat but want to avoid exposure to polycyclic aromatic hydrocarbons, you should grill meat without fat. The smoking of food is chemically a complex procedure, there being many compounds formed in the smoke, and their individual levels depend strongly on the wood used to produce the smoke and how the smoking is carried out. Of the polycyclic aromatic hydrocarbons, mainly the smaller ones, anthracene, phenanthrene, and pyrene, are formed, the larger (and more dangerous) ones occurring only in low concentrations. Smoking devices in which the smoke is purified by cooling it, thereby condensing out the larger and less volatile polycyclic aromatics before the food is exposed to the smoke, have been developed. In total, it is estimated that the avarage person is exposed to 1 mg polycyclic aromatics per year from food, although indidual variations are considerable. Of this, on average, one third comes from the cooking processes.

Of the heterocyclic aromatics, it is primarily the heterocyclic amines which have attracted attention. They can be divided into four main groups, the indoles, the quinolines, the quinoxalines and the pyridines. The heterocyclic amines are formed during the heating of organic materials containing different amino acids, this being part of the complex and partly unknown chemistry called the Maillard reactions. Figure 9.20 shows some examples of heterocyclic amines, which are, or are suspected to be, toxic. They have all been shown to be potent genotoxic compounds in bacteria, while the genotoxic and carcinogenic activities in animals appear to be more limited. Industrially used aromatic amines are also interesting in this respect, and have on several occasions been shown to be potent carcinogens in humans. The aromatic amines have already been discussed in Sections 5.6.2 and 7.2.4.2.

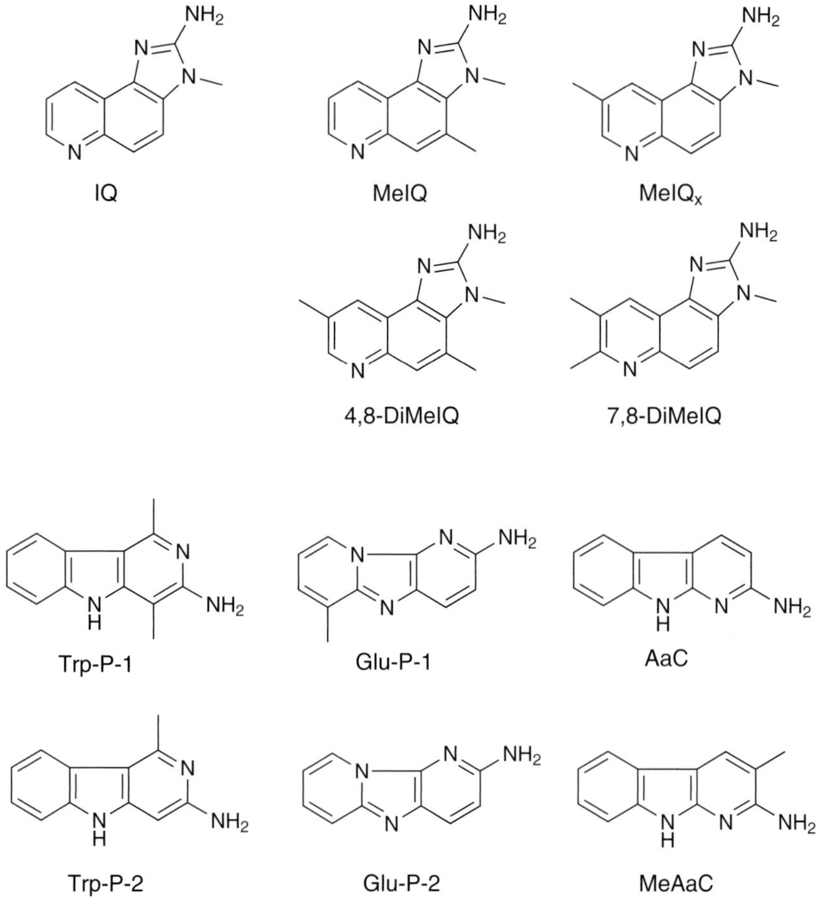

Figure 9.20 Example of heterocyclic amines formed during cooking.

9.3.4
Food Colorants

Every day we are confronted with food and other edible and drinkable items that have all colors of the rainbow, and one may ask oneself whether they are as harmless as claimed. We now know that some of the food colorants used some time ago have later been shown to be quite toxic. An example is the compound called butter yellow, which was used for coloring margarine to give it the same color as butter, but which turned out to be a potent carcinogen. The colorants used today are used in a more restricted way, and some types of food, for example, baby food, are not allowed to contain them. All colorants used today have been tested, and we know at least something about their toxicity. Some natural compounds are used, such as β-carotene (E160a), which is naturally present in carrots and is converted to vitamin A in our bodies. Carotene, which gives a yellow to orange color, is, unlike most modern food colorants, lipophilic, and can consequently be used in lipophilic products such as margarine and salmon. Another example of a natural compound is carminic acid, which, as a calcium-aluminum complex, gives a beautiful carmine red food colorant (E120, not approved in all countries). Carminic acid is extracted from the dried bodies of females of the insect *Coccus cacti*, but this may be unwelcome information for consumers of carmine-colored products. Frequently used synthetic colorants are the aromatic azo compounds, which share the azo functionality with butter yellow but have other substituents and properties that make them less hazardous. Examples (see Figure 9.21) are the red

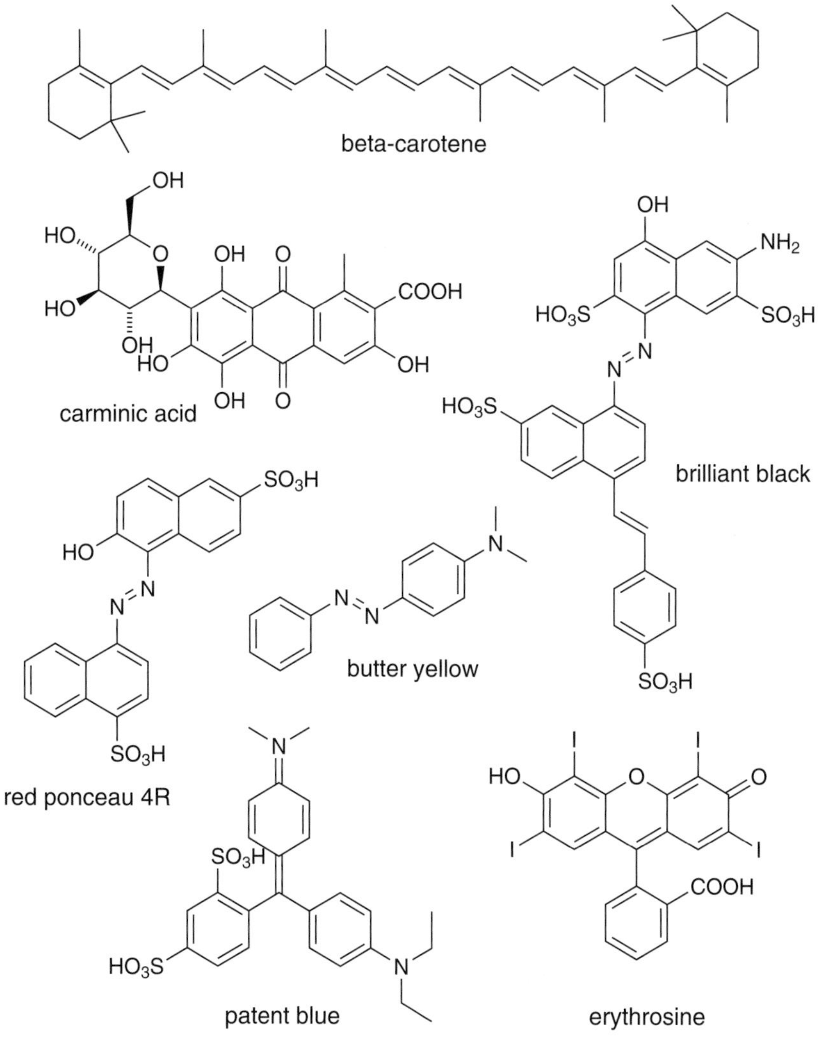

Figure 9.21 Some food colorants.

ponceau 4R (E124), which is commonly used in candy, soft drinks, and ice cream, and brilliant black (E151), used to color fish eggs. Aromatic azo compounds can be reduced by metabolism to the corresponding hydrazine, which in turn can generate two aromatic amines that are potentially carcinogenic. However, the azo colorants used today have ionized sulfonic acid groups as substituents, which make them highly water soluble and difficult to absorb. If that happens, they are easily excreted via the kidneys and therefore unlikely to be metabolized at all. Patent Blue (E131) is used to color blue ice cream, blue soft drinks, and blue candy, while erythrosine (E127) gives us the deliciously red cherries.

9.3.5
Penicillin

Penicillin is the name of a group of mold metabolites (and their derivatives) which have antibacterial or bactericidal effect. They are also called β-lactams, because of the four-membered ring containing nitrogen and an adjacent carbonyl group, and they target bacteria but not fungi, plants, or mammals, making them selective agents. They are usually administered orally, although a penicillin injection may be necessary to treat severe acute infections, and although a substantial part of a dose will be destroyed in the acidic environment in the stomach, enough will pass through to the intestines and be absorbed into the blood stream. Because of their relatively high water solubility, the penicillins can be excreted in the urine via the kidneys, both by regular filtration in the glomerulus and by active transport from blood into the primary urine. The chemical instability and the efficient excretion of the penicillins were a great problem in the days when these were exclusive and expensive drugs, but the problem no longer exists as the losses can be compensated by administrating a higher dose. However, resistance is a rapidly growing problem.

The molecular mechanism by which the penicillins function is by interfering with the bacteria's ability to construct a strong cell membrane that prevents them from bursting open because of their high internal osmotic pressure. This bacterial membrane, which also gives the cells an internal mechanical support for the macromolecules they need to attach, is composed of linear polysaccharides that are cross-linked with short peptides. The penicillins destroy the enzyme that catalyzes the cross-linking, by binding to and reacting covalently to the enzyme. This renders the bacteria less viable, and those that do not perish by themselves become easy prey for the immune system. A very serious problem is that the bacteria are able to develop resistance to the penicillins (as well as to any other antibiotic). They are able to pick up the gene for an enzyme called penicillinase from other microorganisms, and this inactivates the penicillin by hydrolyzing the lactam of the four-membered ring (see Figure 9.22). The problem of resistance to the penicillins is largely caused by the tendency to prescribe these antibiotics too liberally. The solution to the problem is simple: Invent novel antibiotics that target other specific molecular events in the biochemistry of bacteria. That may sound easy, but it is not.

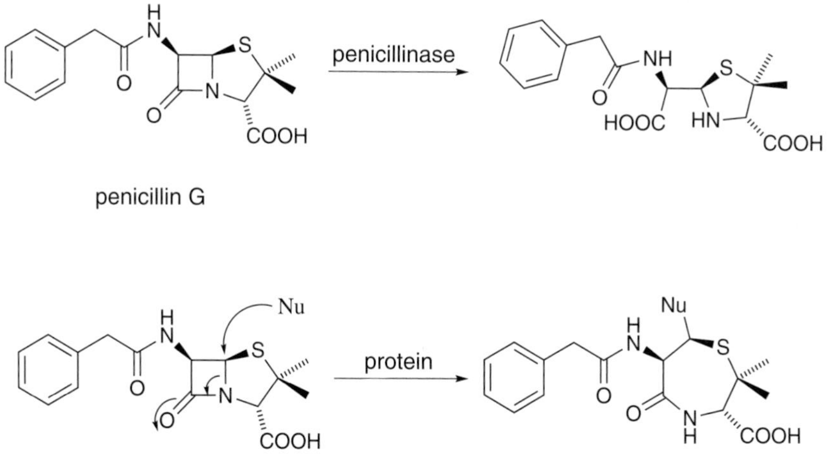

Figure 9.22 The penicillins, hydrolysis and reactivity.

The penicillins are normally safe drugs, and most of us can tolerate high doses without suffering any side effects. The major problem is allergy, which affects a minority (about 1%) of those treated, but it is a huge problem for public health care. It is estimated that one hundred persons die every year from allergic shock caused by the penicillins, and although this may appear as a small number on a global scale it is nevertheless of concern. The reason for the allergic reactions is that the β-lactams are somewhat chemically reactive (see Figure 9.22). One can compare with β-propiolactone, the corresponding four-membered ring lactone (see Section 3.2.3.1). The penicillins will slowly react with our proteins, which will thereby be transformed into macromolecules that are alien and that stimulate the immune system to attack them. One may ask whether the same reaction takes place with DNA, but there is no indication that this is the case.

The chemical reactivity of the penicillins hampered their discovery, and before the English scientist Sir Alexander Fleming (who first discovered that the microrganisms actually produce antibacterial agents) managed to isolate the active ingredient he had to seek help from a chemist.

9.3.6
Acetaminophen

An aromatic amide that is used in large amounts as a medicament (an analgesic and antipyretic) and available without prescription is acetaminophen, or paracetamol (N-acetyl-4-aminophenol). It is considered to be safe, but misuse (typically several grams per day for a long period) can result in liver damage. In a normal situation the half-life of acetaminophen in our bodies is rather short, because it is efficiently conjugated at the phenolic function with sulfate and glucuronic

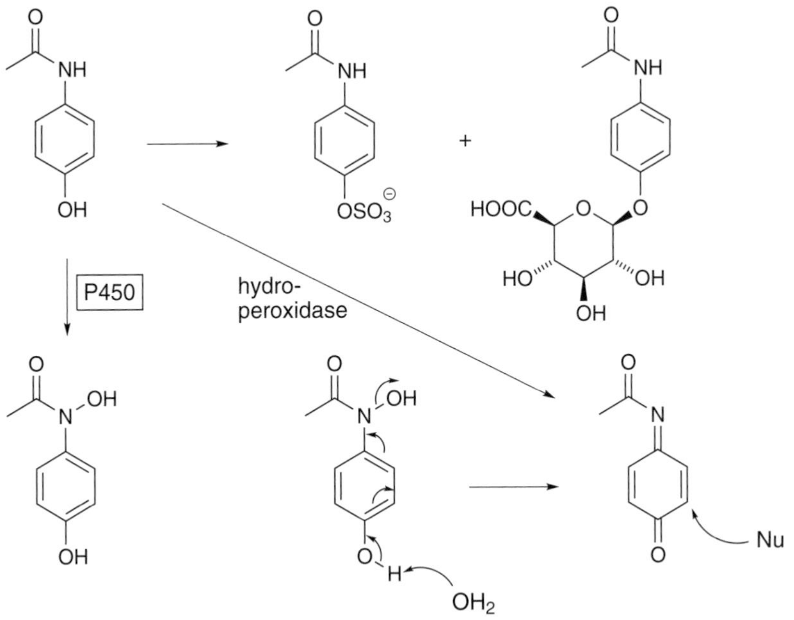

Figure 9.23 The metabolism of acetaminophen.

acid. However, as already noted, the amounts of PAPS (3′-phosphoadenosine 5′-phosphosulfate) available for sulfate conjugation are limited, and if too much of a compound like acetaminophen is taken daily the PAPS stock will be depleted. In such cases only conjugation with glucuronic acid remains, and, as it is considerably slower, the possibility of N-hydroxylation of acetaminophen by cytochrome P450 increases. The N-acetyl-N-hydroxyl derivative can eliminate water and form a conjugated N-acetylquinone imine that is electrophilic (see Figure 9.23) and believed to be responsible for the liver toxicity of acetaminophen by affecting the enzyme glutamate dehydrogenase, which results in the inhibition of mitochondrial respiration. The direct oxidation of acetaminophen to the reactive quinone imine has also been observed, apparently by a peroxidase, but it has been suggested that cytochrome P450 is responsible for the one-electron abstraction from the phenolic oxygen that initiates this conversion.

As well as the amines that we come into contact with, for example, in the working environment, there are also a number of endogenous amines that are not necessarily completely harmless. The DNA base adenine has attracted special attention, and studies have shown that even adenine can be oxidized by cytochrome P450 to the corresponding N-hydroxyadenine which turns out to have genotoxic properties *in vitro*. However, it has so far not been possible to determine whether this activation takes place also *in vivo*, and in that case if it has any biological significance.

9.3.7
Thalidomide

The pharmaceutical agent thalidomide was originally developed in the 1950s by a German pharmaceutical company called Grünenthal. It was marketed in Germany as a tranquilizer and painkiller under the name Contergan and rapidly became very popular among pregnant women because of its inhibitory effect on 'morning sickness' caused by the hormone imbalance during the first trimester of pregnancy. It was sold in many countries, and in those days the only toxicological tests required for registration were, in general, those that had already been carried out by Grünenthal. These tests were perfectly up to the standards of the time, and this is important to remember. However, in some countries, notably the United States, the risks of using the substance were considered to be poorly investigated, and the drug was never sold. After some years of use, it was discovered that the compound is strongly teratogenic in humans between the 23rd and 38th day after fertilization, and in very small doses (a single dose of 100 mg would be sufficient). The most notable effects are malformations of the limbs, but damage to the internal organs was frequent and often resulted in the death of the fetus. The major reason why such a very toxic chemical came to be approved for use as a drug in Europe was the fact that its effect on humans was selective, other mammals being insensitive. In total, thalidomide was the cause of a huge number of injuries, causing an unknown number of deaths and approximately 10 000 malformed babies in the 1950s and 1960s. A remarkable feature of this disaster is that it took such a long time before the link between thalidomide and the malformations could be established, although they are extremely rare in fetuses not exposed to thalidomide and despite the fact that so many pregnant women were exposed. In spite of clear indications that there was a link, the decision to withdraw the drug from the market was not made for some time, and this aggravated the damage. However, it is easy to condemn errors made by others when you know the answers, but this is typical of epidemiological studies of the effects of chemical exposure. It is indeed extremely difficult to prove any links between an exposure and the appearance of toxic effect in humans. It is certain that if thalidomide had given rise to a less spectacular birth defect it might still have been in use.

The molecule of thalidomide can exist as two different enantiomers (mirror images) differing only in the configuration of the carbon indicated in Figure 9.24. While two enantiomers have identical chemical properties, they may interact differently with other enantiomeric molecules, for example, proteins. It has been suggested that one enantiomer of thalidomide, (R)-thalidomide, is harmless and a good tranquilizer, while only the second, (S)-thalidomide is teratogenic. Consequently, as the thalidomide used in the pills consisted of a 1:1 mixture of the two enantiomers one should expect both the positive (tranquilizing) and the negative (teratogenic) effect. However, the two enantiomers have a strong tendency to interconvert (i.e. one can spontaneously turn into the other under physiological conditions), and vice versa, making both enantiomers potentially equally dangerous as drugs. This issue has, however, not been completely clarified.

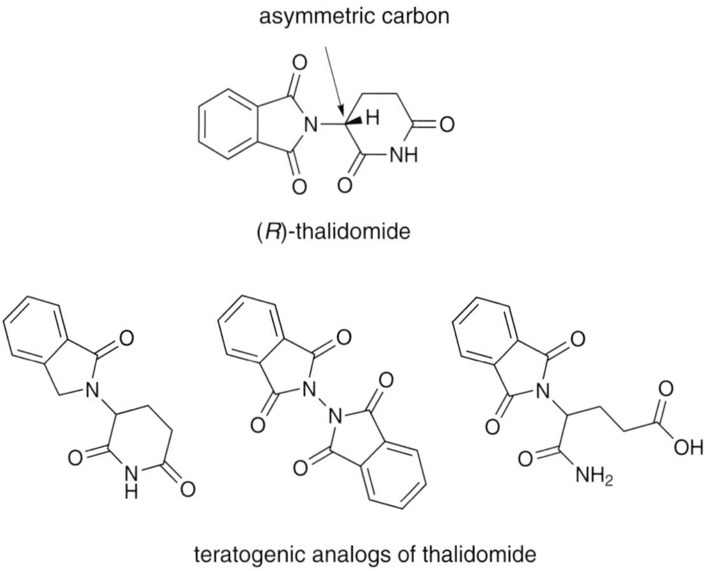

asymmetric carbon

(*R*)-thalidomide

teratogenic analogs of thalidomide

Figure 9.24 Thalidomide, and some analogues that retain the teratogenic activity.

Despite extensive research to find out the molecular mechanism for the teratogenic activity of thalidomide, this is still not understood. Because of the variations observed in different strains of the same species it has been suggested that it is less likely that metabolic activation is responsible for the species selectivity (see below), but this possibility cannot be ruled out. Among the many analogs and derivatives of thalidomide prepared and tested, only a few have the same effect as the parent compound (see Figure 9.24). It is nevertheless believed that hydroxylated metabolites of thalidomide are involved, but this remains to be verified. Recently, it was discovered that thalidomide also has other, and potentially useful, effects, which of course may be linked to its toxicity. For example, it reduces the production of some endogenous signal transmitters (e.g., TNF-α) involved in inflammation and helping tumor cells to divide. Another potentially interesting effect is that it inhibits the development of HIV infection to AIDS, again probably associated with the reduction of TNF-α. In spite of its appalling history, thalidomide therefore continues to intrigue science. Regardless of the unwise use of a chemical substance and the suffering that its adverse effects have caused, the experience and the knowledge will eventually be of use. In the future we need to understand more about how the development of human embryos and fetuses are regulated at the molecular level, and this can at least partially be achieved by studying the effects of thalidomide and its toxic metabolites. Thalidomide has in the last decade been investigated for use against a large number of conditions, and especially interesting may be leprosy, which still is a huge

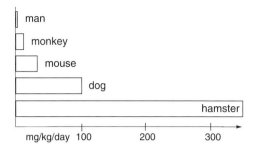

Figure 9.25 The daily doses it takes for thalidomide to provoke a teratogenic effect in various mammals.

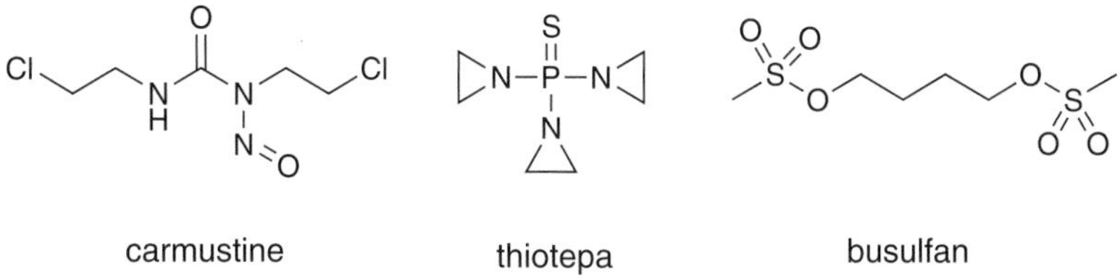

carmustine   thiotepa   busulfan

Figure 9.26 Cytostatic agents that also are good electrophiles.

problem. Time will tell whether the compound will ever again be registered as a pharmaceutical agent.

Surprisingly, some mammals (e.g., rats) are more or less resistant to the teratogenic effect of thalidomide, and for some animals (e.g., rabbit), different strains show highly variable sensitivity. Figure 9.25 gives an impression of how the sensitivity toward the toxicity of thalidomide varies among different species.

9.3.8
Chemotherapy

Drugs that inhibit cell growth and cell division are regarded as cytotoxic or cytostatic agents, and in some instances these are used, often in combination with radiation therapy and surgery, to treat tumors. If such compounds are simply cytotoxic they might be expected to damage both normal cells and cancer cells at the same rate, but cancer cells are considerably more sensitive as they are constantly dividing. There are several different types of antitumor drugs on the market. In general they aim at inhibiting cell division and the formation of new DNA in the cells, and an important group of these is the electrophiles reacting directly with DNA and other macromolecules. The structures of some alkylators that have been or are currently used as antitumour agents are shown in Figure 9.26. These include compounds having classic electrophilic functionalities that we

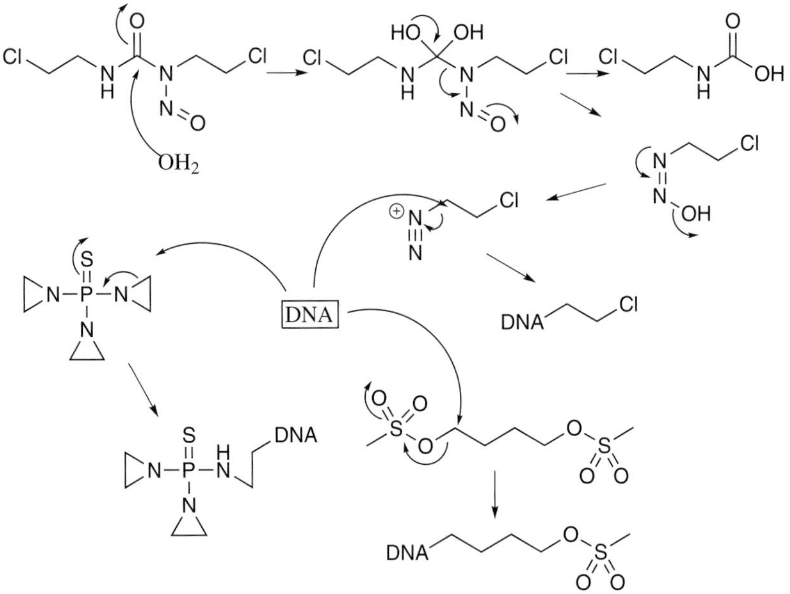

Figure 9.27 Drugs may actually be designed to react with DNA.

are trying to warn everybody about, and it is perhaps surprising that these compounds are actually used as drugs. However, this essentially demonstrates how poor our weapons to combat cancer are and how important it is that we develop more efficient antitumor agents.

Figure 9.26 show some examples of classical but still used drugs used to treat cancer. Carmustine has an LD_{50} value of approximately 20 mg/kg (oral, rat), thiotepa 15 mg/kg (intravenous, rat), and busulfan 1.8 mg/kg (intravenous, rat). All are consequently relatively toxic. Obviously, these compounds are not only electrophilic and ready to react with any nucleophile, but they are also bi- or trifunctional and may give rise to cross-linkages between different parts of the DNA and/or the proteins.

Figure 9.27 shows how carmustine, which at a first glance shows some resemblance to sulfur mustard (see Section 9.2.11 and below), reacts. It reacts with water in a reaction that can be called a hydrolysis, and this is something we can expect with all *N*-alkyl-*N*-nitrosoureas. The leaving group during this hydrolysis is initially the *N*-nitroso derivative, from which the hydroxyl group is displaced in order to form a diazonium ion that is an extremely efficient electrophile having N_2 (nitrogen gas) as a leaving group. Thiotepa contains three aziridine rings, three-membered rings with nitrogen, and these react as epoxides with any reasonable nucleophile. Finally we have the compound busulfan, which contains two sulfonate groups that are good leaving groups that can be displaced by the approach of a good nucleophile (such as DNA).

9.3.9
The Botulinum Toxins

The botulinum toxins are extremely toxic proteins produced by a bacterium, *Clostridium botulinum*. Altogether there are seven variants (named A, B, C, D, E, F, and G), all having a molecular weight around 150 000 and an elemental composition close to $C_{6760}H_{10447}N_{1743}O_{2010}S_{32}$ (botulinum toxin A). The botulinum toxins are the cause of botulism, an illness typically associated with food poisoning. This can be caused by ingesting poorly preserved food containing the bacterial spores that have been stored for a some time at a suitable temperature. In fact, the word botulism comes from the Latin botulus which means sausage. *C. botulinum* is widespread and very heat tolerant. To kill these bacteria food must be heated to 120 °C for 30 min, and the risk of botulism must always be borne in mind in the food industry. The toxins, which by themselves are more heat sensitive and can be destroyed at 60 °C, are extremely dangerous. They are the most toxic compounds known, and 1 μg could be enough to kill a human. They block the transmission of signals between motor neurons and muscle cells by inhibiting the release of the neurotransmitter acetylcholine.

In normal signal transmission (see Figure 9.28), vesicles containing acetylcholine in the signaling nerve cell merge with the cell membrane, and the acetylcholine is released into the synapse (see also Section 7.3.2). The release is made

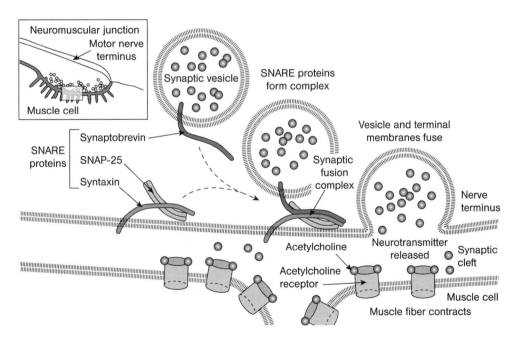

Figure 9.28 The transmission of a signal between motor neurons and muscle cells. (Modified after The Journal of the American Medical Association, 2001, 285: 1059–1070.)

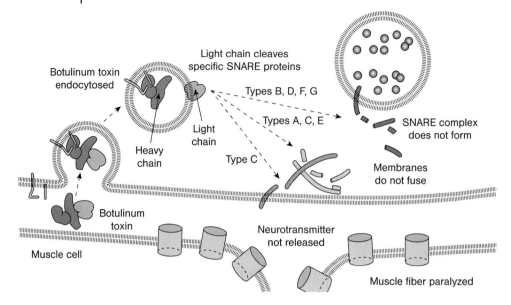

Figure 9.29 The effect of the botulinum toxins. (Modified after The Journal of the American Medical Association, 2001, 285: 1059–1070.)

possible by the three SNARE proteins (synaptobrevin, SNAP-25, and syntaxin) which fuse the vesicle with the cell membrane, and it is these proteins that are the target of botulinum toxins (see Figure 9.29). They inactivate the SNARE proteins, which results in a lack of acetylcholine in the synapse. The muscle cells will not contract, and the intoxication leads to paralysis.

The symptoms, which may appear several days after an exposure, is initially weakness and dizziness, followed by gradual paralysis of the muscles, starting in the face and progressing from there. When the respiratory muscles are affected the victim will die from lack of oxygen.

Botulinum toxin A is actually used clinically, the commercial product being known as Botox and is used to treat various spasms and cramps. Strabismus, for example, a condition in which the eyes are not properly aligned with each other, can be corrected if a small amount of botulinum toxin A is injected in the right place so that the muscles that pull the eye in the wrong direction are paralyzed. The effect will last for several months, after which the treatment has to be repeated. It is also used to reduce wrinkles on the forehead and around the eyes. The amounts used for cosmetic purposes are of course minute.

9.3.10
The Tetanus Toxins

Tetanus is caused by two proteins, tetanolysin and tetanospasmin, produced by the bacterium *Clostridium tetani*. The latter is almost as toxic as the botulinum toxins, but differs in several respects. For example, the tetanus toxins do not tolerate the acidic conditions encountered in the gastrointestinal tract and will not cause food poisoning. Tetanus can only affect us if we get the bacteria directly into the bloodstream via a wound, and the infection can thus be caused by a dog bite or a cut from a garden tool. Most people receive basic protection against tetanus from a vaccination, although this should be repeated regularly in order to be effective. In the blood, the bacteria will multiply and produce sufficient amounts of the toxins. Tetanospasmin is a neurotoxin that blocks inhibitory impulses by interfering with the release of the neurotransmitters glycine and γ-aminobutyric acid (GABA). The result is the opposite of the paralysis that the botulinum toxins cause, tetanus being characterized by severe cramps and convulsions.

9.3.11
Aflatoxins

The aflatoxins are a group of mycotoxins, which in addition to being strongly toxic to the liver are also potent carcinogens. They are common contaminants in food, for example nuts and figs, that is stored in an improper way, but they may also occur in animal feed and from thence, for example, in the milk of cows. Fungal growth on stored food is always a major problem, especially in poor areas with a hot and humid climate. It is likely that exposure to aflatoxins is the cause of a significant proportion of the tumors that affect people such areas. The most toxic and carcinogenic of the aflatoxins is aflatoxin B1, which is epoxidized efficiently in the liver to form a toxic and carcinogenic metabolite (see Figure 9.30).

In spite of their toxic potency, the aflatoxins were discovered by accident. A product based on peanuts was being used as feed in a turkey farm in England in the 1960s, and the peanuts happened to be heavily contaminated with aflatoxins. A hundred thousand turkeys died suddenly, and after a scientific investigation it was found that the cause was not an infection but rather a poisoning. The toxin was isolated from the feed and was shown to have been produced by the fungus *Aspergillus flavus*. The biosynthesis of the aflatoxins, which from a chemical point of view are classified as polyketides, depends on a number of factors: heat, humidity, substrate, genetic makeup of the fungus, and so on. Other *Aspergillus* species can also produce these toxins, which are surprisingly common at low levels in a

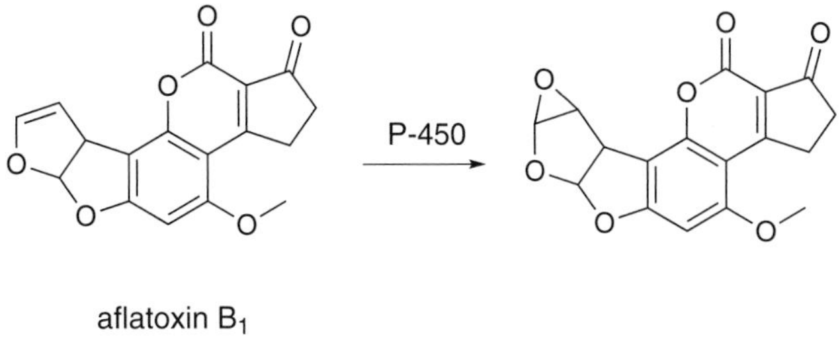

aflatoxin B$_1$

Figure 9.30 The activation of aflatoxin B1 to a toxic epoxide.

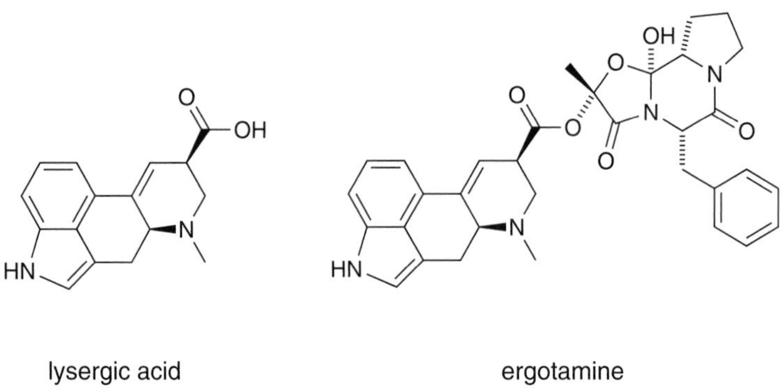

lysergic acid   ergotamine

Figure 9.31   The ergot alkaloids.

wide range of foods. The aflatoxins are chemically stable and do not decompose even after prolonged boiling.

9.3.12
The Ergot Alkaloids

The ergot alkaloids are responsible for a type of poisoning called ergotism, and are produced by the ergot fungi (e.g., *Claviceps purpurea*) that infect rye. Ergotism, caused by the ingestion of bread baked with infected rye, was very common during the middle ages and affected more or less everybody. Today it is quite rare. The ergot alkaloids are a cocktail of alkaloids of which the most important are amides of lysergic acid (for example ergotamine, see Figure 9.31). Lysergic acid is also a constituent of other substances that have an extremely strong effect on the nervous system, the most well known being LSD (see Section 9.5.7). The ergot alkaloids have a complex effect on the nervous system, constricting the blood vessels and decreasing the blood flow by blocking the receptors that respond to epinephrine. Epinephrine is a stress hormone that has similar physiological effects to those of the neurotransmitter norepinephrine, which increases blood flow to the muscles.

   The physiological effects of the ergot alkaloids are potentially useful, and ergotamine has been used to inhibit bleeding from the uterus after childbirth as well as to relieve migraine pains. The headache pain of a migraine attack is caused by an increased pressure inside the brain, and if the blood vessels constrict and allows less blood to reach the brain the pressure will be reduced and the migraine attack will calm down.

   There are also a number of other toxic mycotoxins that we come into contact with in different ways (see Figure 9.32). The ochratoxins are stable toxins produced, for example, by *Aspergillus ochraceus* and *Penicillium viridicatum*. They attack cereals, coffee beans, and fruits (including grapes), and can be present in, for example, bread, coffee, grape juice, wine, and fruit juices. They are able to pass along food chains, and can also be present in meat. The ochratoxins are quite toxic to the kidneys and the liver, and as they also are potent kidney carcinogens their

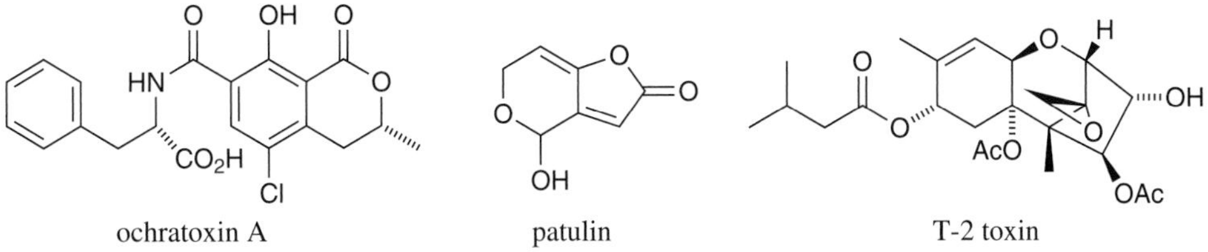

Figure 9.32 Other potentially dangerous mycotoxins.

possible presence in food is of concern. Another mycotoxin of interest is patulin, produced by molds that attack fruit and berries, which has been shown to be geno-toxic. It is chemically less stable but survives long enough in more acidic food products such as fruit juices. Especially apples and apple juice may contain patulin in concentrations of concern, and for these products there are contamination limits determined by the authorities. Acute poisoning primarily affects the gas-trointestinal tract, and patulin is believed to be the cause of some outbreaks of diarrhea among children. The T-2 toxin, which chemically belongs to the group of trichothecenes, are primarily produced by a *Fusarium* species that attack cereals. The effect of the poison is to inhibit the synthesis of proteins in the cells, but they also disrupt the mucosa of the stomach and the intestines, which, among other things, leads to bleeding.

The T-2 mycotoxin and other toxins produced by various *Fusarium* species caused a series of massive poisonings of an epidemic nature of Russia during the Stalin era, when flour from infected corn was used to bake bread.

9.3.13
Muscarine

Fly agaric (*Amanita muscaria*) is a red mushroom with white dots that is familiar to all and generally known to be toxic. Most associate its toxicity with the substance muscarine. However, there are several other biologically active substances present in the mushroom and involved in poisonings with fly agaric (which by the way are almost never fatal). While muscarine is a highly toxic compound, it is only present in low concentrations in the mushroom. It acts on the nervous system as an agonist and activates a particular type of acetylcholine receptors that are sensi-tive to muscarine, and was long considered to be the component that is responsible for the alleged hallucinogenic properties of the mushroom. Instead it provokes nausea and sweating, which definitely are not desirable effects. In addition, the mushroom produces ibotenic acid and muscimol, the former is an agonist of glutamate receptors while the latter is an agonist of the GABA$_A$ receptor. (Can you see the chemical resemblance between muscarine and acetylcholine, between ibotenic acid and glutamate/glutamic acid, and between muscimol and GABA in Figure 9.33) Ibotenic acid and muscimol are consequently psychoactive and responsible for the hallucinogenic effects. The 'fly' in the English name fly agaric

muscarine   ibotenic acid   muscimol

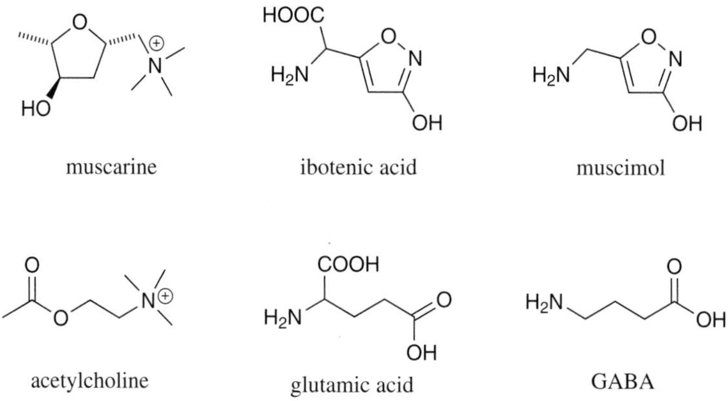

acetylcholine   glutamic acid   GABA

Figure 9.33 Active metabolites present in the fly agaric.

has, by the way, nothing to do with the psychoactivity (as in getting high and flying): it comes from the traditional use of the mushroom as an insecticide to kill flies. The ibotenic acid is responsible for this effect.

Fly agaric is believed to have been used in traditional rites in different cultures. For example, early Indian cultures (about 1000 BC) used a special drink called Soma that was important in the ritual ceremonies to make contact with the gods, and it is believed that fly agaric was an important part of it. Also, the Vikings were believed to consume the mushroom in order to get into the right mood before a fight, although this has not been established. However, it is known that the shamans of Siberian cultures used fly agaric to open the door to the spiritual world. They valued the mushroom highly (a fruiting body could be worth the price of a reindeer), and they even discovered that by drinking the urine of the primary consumer, a secondary consumer could reuse the psychoactive components of the mushroom for his pleasure. However, it appears that the effects obtained from the use of fly agaric vary from one culture to another, which is strange as we know today that consumption of the mushroom results in excessive sweating and vomiting. Probably the explanation is that the relative concentrations of the three psychoactive metabolites in Figure 9.33 vary among different subspecies, the fungus having slightly different genetic properties when grown under different conditions in different parts of the world. Actually it is quite reasonable to assume that fly agraric growing in Siberia will contain different levels of the metabolites from that growing in India.

9.3.14
Summary

The total amount of food and drink that a human being ingests over a lifetime is colossal. It is also clear that very low concentrations of contaminations (so low in some cases that they are difficult to detect, even for an analytical chemist) could over time become significant quantities. Chemical carcinogens and substances that modulate the development of cancer cells are obviously also present in food, either as natural products or added and/or formed during cooking. Many of our prescribed drugs are quite toxic, being designed to have a biological effect, and have to be used with care, for example, by ensuring that the dose is correct. Also, over-the-counter drugs, although in general considered safe, can give rise to toxic effects if misused. Natural products are produced by all organisms, including humans, but in most cases we have no idea about why they are formed. The mycotoxins and the antibiotics produced by microorganisms, for example, may be thought to be used to kill competitors for food, but we really do not know. They may taste or smell good, and we can choose to use them as spices or perfumes, but they can also be horribly toxic and highly carcinogenic.

9.4
Polymers, Adhesives, and Other Materials

Measured by volume of products used, this section covers probably the most significant use of chemicals. The construction of something big, such as a skyscraper or a bridge, normally consumes many thousands of tonnes of concrete, which consequently is an example of an extremely high volume chemical product. Indeed, concrete was at one time a huge environmental problem and was often the cause of contact allergy among builders. This was caused by the presence in concrete of small amounts of hexavalent chromium, which is highly allergenic. When the cause of the allergies was identified, it was eliminated by the addition of divalent iron to the concrete. This reduces the chromium to the trivalent state, which is not allergenic, at the same time being oxidized to trivalent iron.

9.4.1
Vinyl Chloride

Vinyl chloride (chloroethene, chloroethylene) is used in large quantities as the monomer for the production of the plastic polyvinyl chloride (PVC) and in organic synthesis. It became famous in the end of the 1960s, up to which point it had been regarded as an almost harmless solvent that could be used without any special precautions. However, it was shown that heavily exposed workers are more likely than nonexposed workers to get cancer. Vinyl chloride causes a very unusual form of liver cancer, which is the reason why it was possible to establish the relationship between the exposure and its effect. Its potency as a carcinogen is quite low, and the effect would not have been detected if the tumor had been of a more common type, only relatively few individuals having been affected. Nevertheless, as a consequence of these findings its use was severely restricted. It is metabolized by epoxidation, and the epoxide is long-lived enough to be hydrolyzed as well as conjugated with glutathione (see Figure 9.34). However, the epoxide has a tendency to rearrange spontaneously to form the electrophile

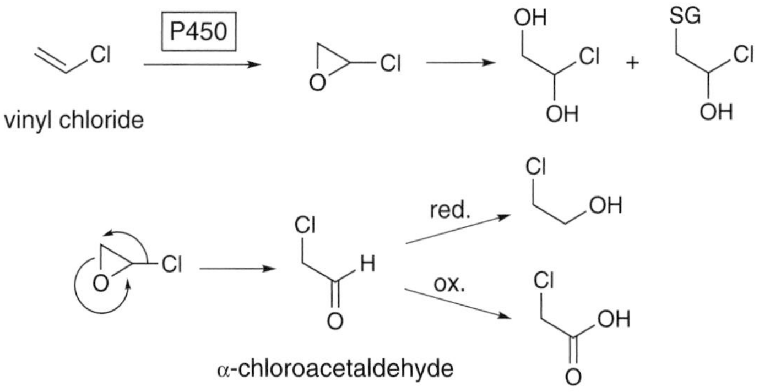

Figure 9.34 Metabolic activation of vinyl chloride to electrophilic species.

α-chloroacetaldehyde, which can be either reduced to the alcohol or oxidized to the carboxylic acid. The toxic effect is assumed to be caused by the reactivity of the epoxide and α-chloroacetaldehyde.

9.4.2
Acrylamide

Acrylamide is a small and seemingly insignificant molecule, like most polymer monomers, but it is used in enormous quantities not only to produce polyacrylamide but also for many other purposes. Polyacrylamide is a polymer used, for example, as a gel in scientific research, as a flocculant in the paper industry, and as a sealant in tunnel constructions. It is also used to purify drinking water and waste water, and consequently acrylamide is often present at low concentrations in drinking water (the limit is commonly 0.5 mg/L). It is not extremely toxic, the LD_{50} value for rats after oral exposure being 120 mg/kg. Acrylamide is easily absorbed through the skin and in the gastrointestinal tract and the lungs, and it is metabolized relatively quickly to water-soluble products that are excreted via the kidneys in the urine. The main metabolic route is conjugation with glutathione, not hydrolysis to acetic acid and subsequent oxidation of this in the citric acid cycle, as one perhaps would have expected. The acute toxicity affects the nervous system, an exposed individual experiencing numbness in the hands and the feet, muscle weakness, and balance problems. Acrylamide affects both the central and peripheral nervous systems, although the exact molecular mechanism of what it does is not known yet. Furthermore, and perhaps more seriously, acrylamide is genotoxic and carcinogenic, at least to animals. The fact that acrylamide is conjugated with glutathione in the metabolism suggests that it is a reasonably good electrophile, reacting as a Michael acceptor. However, it is believed that the genotoxic effects of acrylamide are due to its epoxidation by cytochrome P450 to form the corresponding glycidamide (see Figure 9.35), which reacts with DNA.

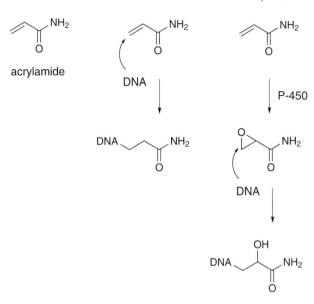

Figure 9.35 The turnover of acrylamide in organisms.

When road and train tunnels are drilled, acrylamide is used in enormous quantities as an underground seal that stops the natural flow of water in the soil. The product (Rhoca Gil) used is actually composed of a mixture of acrylamide and N-methylol acrylamide, which is the addition product between acrylamide and formaldehyde. The latter generates acrylamide for the polymerization reaction acrylamide to polyacrylamide, but also releases formaldehyde, which is a gas at room temperature and quite hazardous, that acts as a cross-linker and reinforces the polymer. The polymerization of Rhoca Gil depends on its dilution, and there have been several accidents in which the water flow has been too strong and has simply diluted and swept away the acrylamide/N-methylol acrylamide mixture. That water would then be heavily polluted and dangerous to use, both because it contains acrylamide but also because of its formaldehyde content.

Recently, there have been reports that acrylamide is also present at alarming levels in fried (either in a frying pan or in oil) starch-rich products such as potato chips, which is worrying because these are popular products consumed by most people. Exactly how acrylamide is formed here is not known, but it is reasonable to assume that the nitrogen in the acrylamide orginates from amino acids in some protein present, and that the high temperature during the frying is important for the transformation. The starch, which is the main component of potatoes, is probably not directly involved in the reactions, but may be important in adsorbing the acrylamide formed so that it does not evaporate. Boiled potatoes do not contain acrylamide, so the temperature is critical. It is also of interest that acrylamide is a component of cigarette smoke (see Section 9.5).

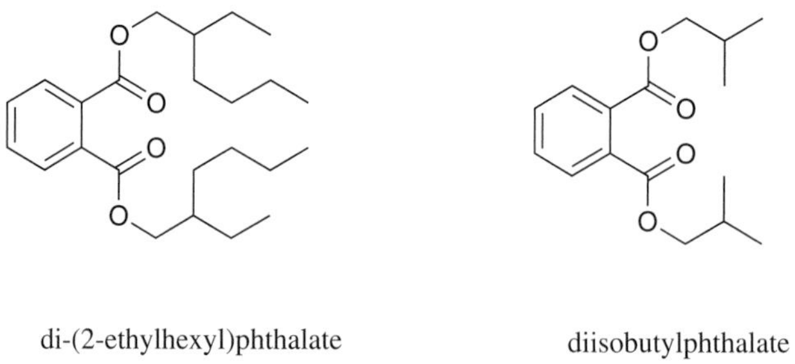

di-(2-ethylhexyl)phthalate   diisobutylphthalate

Figure 9.36 Dialkyl phthalates of concern.

9.4.3
Dialkyl Phthalates

Dialkyl phthalates are diesters of phthalic acid (1,2-benzene dicarboxylic acid), the most commonly used being di-(2-ethylhexyl) phthalate (also known as dioctyl phthalate or DOP) and di-isobutyl phthalate (see Figure 9.36). Esters of phthalic acid are found practically everywhere on earth as they are produced on a very large scale, most of the production eventually ending up in the environment, and they are relatively stable beause of their lipophilicity. They are used in many processes and products, such as paints, adhesives, and plasticizers in a range of plastics, and as a result they are exposed to and present in the environment everywhere. Their chemical stability and high lipophilicity enable them to accumulate in all organisms. DOP, with a log P value of 7.64, which is quite impressive, will be concentrated from sea water 28 000 times in algae and 24 000 times in invertebrates. The bioconcentration in mammals is somewhat lower, because these are able to metabolize the compounds by enzymatic hydrolysis, but it is still very high. As a consequence, the dialkyl phthalates are also concentrated in the sediments of lakes and seas, and in the Baltic Sea the sediments contain up to 100 times more phthalates than, for example, PCBs or PAHs. Their chemical stability to hydrolysis in neutral water (half-life) is between 3.2 years (dimethyl phthalate) and 2000 years (diethyl-hexyl phthalate). Photochemical degradation is less important because of their low volatility. There is a general degradation by micro-organisms, which hydrolyze esters and oxidize the ring, but this is limited to the dissolved material, which is a very small amount. Humans are primarily exposed to these phthalates by consuming food that is contaminated by them. As they are extremely lipophilic, they are rapidly absorbed and distributed in the body, and they pass through both the placenta and the blood-brain barrier, with the result that both the fetus and the CNS are exposed. Eventually, they will be distributed to the adipose tissue, where they can stay for a long time. In a water-soluble state they would be relatively easily metabolized to phthalic acid and the corresponding alcohol, but the problem is that they are not in solution. Phthalates are generally considered to be teratogenic, although not potent, and carcinogenic effects have been reported (although at very high doses).

Another type of plasticizer that we should worry about is chlorinated paraffins. These are used as technical mixtures of unspecifically chlorinated alkanes having between 10 and 30 carbon atoms and a chlorine content of 30–70%. The number of individual components in such a mixture is gigantic, and the characterization of its toxic properties is practically impossible. Chlorinated paraffins are primarily used as plasticizers in PVC, the rate of use exceeding 100 000 tonnes per year. They are not considered to be volatile, but they accumulate efficiently, the log P value of the component with 25 carbons and 42% chlorine being, on average, 6.2. The chlorinated paraffins that are small (C_{10}) and have a high chlorine content (70%) are very toxic to aquatic life. However, the main concern about chlorinated paraffins is that they yield chlorinated products such as PCBs and dioxins when they end up in waste-fed power plants and are combusted.

9.4.4
Brominated Flame Retardants

It is probably not well known, but an important use for chemicals is as flame retardants. We use many products that are potentially dangerous because they may catch fire and give off hazardous emissions if they burn. An example is a TV set, which is a major source of fires in the home. Flame retardants are chemicals used to make such products safer. These will hopefully prevent them from catching fire (or at least retard the flames once they appear), and TVs have to have chemical flame retardants in their components. Brominated flame retardants are a specifically important class of these chemicals. They act as radical scavengers, reducing the concentration of radicals and thereby slowing the radical reactions that fire is composed of. A comparable product is the halon fire extinguishing agent, composed of low-molecular weight brominated alkanes and used to extinguish fires in an very efficient way in places like chemical stores. Flame retardants that are intended to protect a specific product can either be chemically bonded to the material they are designed to protect, for example, by a covalent ether bond, or only adsorbed/absorbed by the material. If the latter method is used, the flame retardant can evaporate (if volatile) or be washed away (if exposed to humidity), and sooner or later it ends up in the environment.

Tetrabromobisphenol A (TBBPA) is probably the commonest brominated flame retardant (see Figure 9.37), several thousands of tonnes of this substance having been used, for example, to protect printed circuit boards in TVs and computers. Polybrominated diphenyl ethers (PBDEs) are also used, essentially for the same purpose. Of the PBDEs, there are 209 structurally different variants, making (again) PBDE a product that is difficult to characterize when it comes to properties. The fully brominated version, decabromodiphenyl ether, is the commonest single component of the PBDEs. Hexabromocyclododecane (HBCD) is used, for example, in the foam used to fill the seats of furniture and cars, and also in fabrics. Polybrominated biphenyls (PBBs) are currently being phased out because they are similar to the PCBs (see above) and have similar effects. The exposure of humans to PBBs today, which is still significant, is therefore mainly due to PBBs

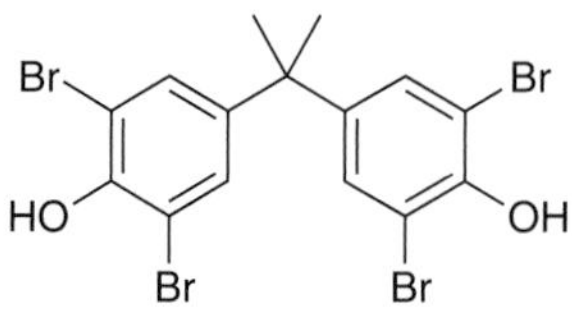

tetrabromobisphenol A

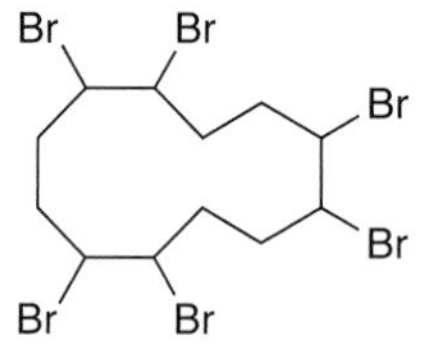

hexabromocyclododecane

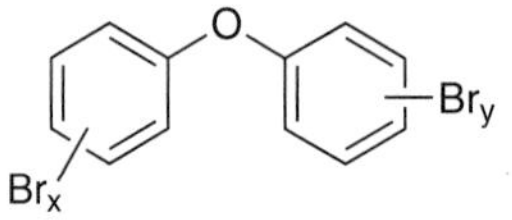

polybromodiphenylethers
(x + y = 1-10)

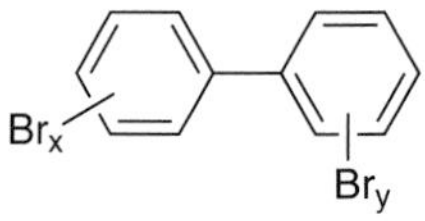

polybromobiphenyls
(x + y = 1-10)

Figure 9.37 Frequently used brominated flame retardants.

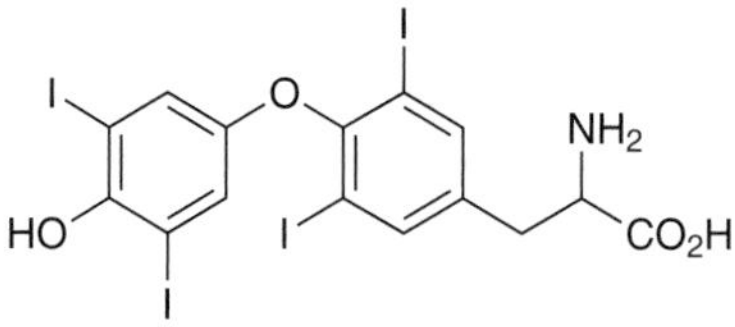

thyroxine

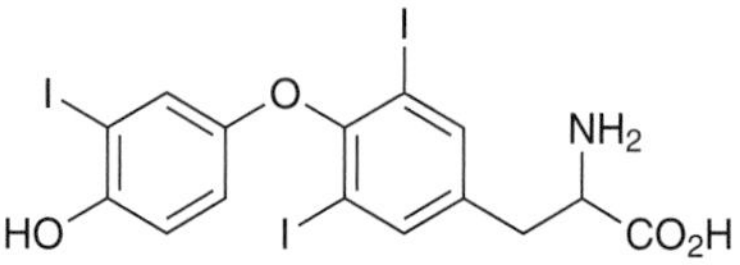

triiodotyronine

Figure 9.38 The thyroid hormones.

used some time ago. These compounds are certainly persistent, and will be with us for a long time.

Mammals easily absorb brominated flame retardants from the environment, as they are all lipophilic compounds. However, they are not compounds that are easily metabolized, something that can be imagined when one considers the chemical structures. The rate of metabolism depends strongly on the degree of bromination, and those with few bromines can be expected to be oxidized, like other aromatics, to phenols, which are conjugated and excreted. However, the bulk will, at least temporarily, end up in the fat tissue. Some PBDEs interact with the Ah receptor (also called the dioxin receptor) for which they compete with dioxin and the PCBs (see Section 10.6). Such PBDEs can consequently be mitogenic. TBBPA, and also the PBDEs hydroxylated metabolically, bind to and block the protein that transports the thyroid hormones thyroxine and triiodotyronin (see Figure 9.38) as well as their receptors. If their chemical structures are compared, it is possible to understand why. In addition, brominated flame retardants are suspected to be both immunotoxic and neurotoxic. However, there is little hard evidence for the effects of these compounds in these respects.

9.4.5
Cyanoacrylates (Superglue)

This type of glue, which was introduced in the late 1960s, has become popular both in industry and in the home. As a glue, it is relatively easy to handle because it consists of only one component which hardens quickly to give a very strong joint. Compared to the epoxy and the polyurethane glues, the cyanoacrylates offer several advantages. The single component is an acrylic acid ester substituted at the α-position (see Figure 9.39) with a cyano group. The alkyl portion of the ester varies, some commonly used being shown in Figure 9.39, and each, of course, has slightly different properties.

It is obvious that these compounds are excellent electrophiles and will react readily as Michael acceptors with nucleophiles. The carbon–carbon double bond is doubly activated by two electron-withdrawing groups, and the partial positive charge on the b-carbon of the double bond is relatively big. Cyanoacrylates are therefore not only superglues but also superelectrophiles, and it is this property that makes them function so well as glue and, astonishingly, also limits their toxicity. Because they are so reactive, they will not survive the transport into the body because they will react as soon as they encounter the first nucleophile. They are by no means harmless, as they will give local effects on the part of the body that first comes into contact with them, which normally is the skin. Many using this product have found themselves with two finger glued together, but although it is an annoying situation it will solve itself with time. Contact allergies to cyanocrylates can develop after repeated dermal exposure. In addition, cyanocrylates will certainly affect the respiratory system and eyes if these are exposed to the vapor. The cyanocrylates are not very volatile, but during industrial use it happens that heat is applied to accelerate the curing. Then it is important to make sure that the ventilation works properly.

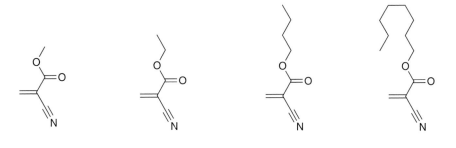

methyl cyanoacrylate   ethyl cyanoacrylate   butyl cyanoacrylate   octyl cyanoacrylate

Figure 9.39 Frequently acrylates used in superglue.

The curing of cyanocrylates takes place by a process known as anionic polymerization. An anion, any one, will serve as a nucleophile and react with the β-carbon of a cyanoacrylate molecule, resulting in a negative charge on the a-carbon in that molecule, which thereby is transformed into an anion and can react with the next cyanoacrylate molecule, and so on. The glue does not contain any nucleophiles that are strong enough to initiate the polymerization, so it is stable at room temperature. Once outside the bottle, it comes into contact with moisture containing nucleophiles, and the curing starts.

9.4.6
Asbestos

Asbestos, which takes its name from the Greek word for inextinguishable, is an inorganic fibrous silicate mineral that exists in a variety of forms. Examples are white asbestos, or chrysotile $[Mg_3(Si_2O_5)(OH)_4]$, blue asbestos or crocidolite $[Na_2(Fe^{3+})_2 (Fe^{2+}Mg)_3Si_8O_{22}(OH)_2]$, and brown asbestos or amosite $[(Fe^{2+}Mg)_7Si_8O_{22}(OH)_2]$. All types of asbestos fibers have high heat resistance, conduct heat poorly, and have a large surface area. These properties have made asbestos a very useful material, and, as it is cheap, it has been used on a large scale, initially mainly as an insulating material in chimneys, stoves, and ovens, but with time also as a construction material because of its ability to resist fire. Mixed with cement, plastics, and other materials, new products with fantastic properties were invented that resisted weather, fire, and microrganisms. Eternit boards, for example, were originally made from asbestos and cement, but today the asbestos has been replaced with other, safer materials. Brake pads and shoes in cars and trucks used to contain asbestos, and their use resulted in a fine asbestos-containing dust that was spread in the air of all circulated areas. The large surface of asbestos fibers also made it useful in technical systems for exchanging moisture and heat. Some health hazards associated with exposure to asbestos were known at the beginning of the twentieth century (the fact that it causes asbestosis, for example), while the connection to lung cancer was revealed in the 1950s when the relationship between smoking and lung cancer was established. Nevertheless, asbestos played a major role in the construction boom in the 1960s and 1970s, and the restrictions that were eventually imposed came too late for those who were exposed during this period.

Asbestos is a natural material, and is consequently found in low concentrations in most waters. It is also naturally present in the soil in many areas. Asbestos is known to be carcinogenic, and in areas with high natural concentrations of asbestos it has been shown that pulmonary changes (so-called pleural plaque) is more common than in areas with little natural asbestos. The toxic effects of asbestos depend on the asbestos fibers being inhaled, and the smallest fibers, with a length of less than 1 mm, are the worst. This is because they are able to penetrate all the way down to the alveoli. In the alveoli, they will adhere to the walls and cause a fibrosis, an inflammation that will result in excess fibrous connective tissue. Fibrosis caused by asbestos is called asbestosis. The additional tissue increases the distance between the air and the blood in the alveoli, which hampers the oxygenation of the blood. Typical symptoms in induviduals suffering from asbestosis is consequently shortness of breath and lack of endurance. In addition, asbestos causes lung cancer, but the mechanism for this is unclear. One possibility that has been suggested is that metal ions complexed in the asbestos catalyze redox reactions in the lung cells that generate reactive oxygen species (ROSs). The risk of lung cancer due to exposure to asbestos is strongly linked to smoking, and smokers working with asbestos are at high risk. Possibly, the mitogenic properties of asbestos, which are also important for the development of asbestosis, are critical in this respect.

9.4.7
Dental Amalgam

The word amalgam comes from the Greek 'malagma', which means soft pulp, and refers to a solution of various metals in liquid mercury. The amalgam used in dentistry consists of mercury, silver, tin, and small amounts of copper and zinc (typically 50% Hg, 35% Ag, 13% Sn, 1.5% Cu and 0.5% Zn). To arrive at this composition as a suitable filling material in teeth took quite some time, and many other materials were tested in parallel with varying results. For example, extensive experiments using molten lead to fill holes in teeth were made, but this technique was abandoned because the lead had a tendency to crack the tooth when it solidified. This obviously took place during a time when the environmental hazards of heavy metals were more or less ignored. However, it was known that mercury was a toxic metal, and this drawback of using dental amalgam has actually been debated since the 1850s. Relatively recently it has become possible to measure the release of mercury from fillings, and thereby to quantify the problem.

From dental amalgam, mercury will continuously be released in the form of elemental mercury, Hg (0), or as mercury vapor. The amount depends on a variety of conditions, how many fillings you have and how big the surface is, what you eat and drink and how often (weakly acidic food and hot beverages increase the release), whether or not you chew gum (chewing gum and using a toothbrush polish the amalgam surface and increase the release), and so forth. The release is normally between 10 and 200 µg/day. In general, if you breathe through your mouth the released mercury will be absorbed in the lungs where it has entered, but if you breathe through your nose the mercuty will dissolve in the saliva and be swallowed. The latter route reduces the uptake because the mercury will be oxidized on its way through the intestinal tract and a substantial proportion will react with proteins to form adducts that are not absorbed. In addition, mercury can be absorbed directly from the tooth to the brain via the nerves that communicate sensory impulses from the teeth and the olfactory nerves.

Individuals who suffer from mercury poisoning can have their dental amalgam removed and exchanged for composite fillings that are mercury-free and mechanically as good. The composite material consists of fine glass or silica grains that are glued together by a plastic material polymerized with UV light. Most people who undergo dental amalgam removal and replacement find that the initial discomfort largely disappears. Mercury will also be discussed in Section 10.7.6.

9.4.8
Concrete Containing Radon

Radioactive chemicals are dangerous because they emit ionizing radiation that can damage the macromolecules of our cells. Radiation as a health hazard is more physics than chemistry, but as the emitter is a chemical we will include this specific example in the text. Ionizing radiation can be either particle radiation, that is, α-radiation, which consists of helium nuclei, each of which consists of two protons and two neutrons, β-radiation, which consists of electrons which are left over when a neutron is transformed into a proton, high energy protons or neutrons released in nuclear reactions, or electromagnetic radiation (γ-radiation), which is radiation like light but with considerably more energy. α-radiation has a very short range and cannot penetrate the skin from the outside. It is therefore dangerous only when an α-radiation emitter has been absorbed into a human and emits its radiation from within the tissue/cell. β-radiation can easily penetrate into the body and is consequently more dangerous if the emitter is an external source. Dangerous electromagnetic radiation, X-rays and γ-rays, have a much shorter wavelength and higher energy content compared to the electromagnetic radiation we are constructed to withstand (visible light) and is potentially very dangerous. However, most of this kind of radiation goes straight through our bodies without affecting them, and we have made use of it for medical examinations. Nevertheless, ionizing electromagnetic radiation must always be used with care.

Ionizing radiation in high doses is lethal to all organisms, it is a true antibiotic. Massive exposure, however, is unusual under normal circumstances, but we have to recognize that also lower levels of this radiation may give rise to toxic effects. Of special concern is the risk of genotoxic effects of ionizing radiation, and this is essentially why it is so important that any use of radioactive materials must be regulated. However, it is important to remember that we are all continuously exposed to low levels of radiation, from natural radioactive sources on the earth as well as from the cosmos. This is what we call background radiation, and it is something that evolution has in principle constructed us to withstand. Nevertheless, it is estimated to cause a few percent of all the 'spontaneous' mutations in our cells, as well as a few percent of the tumors that affect mankind. The mutations are the driving force for evolution, so that we cannot complain about that, but we could do without the tumors. Unfortunately, from an evolutionary point of view tumous are not that important, especially as they normally affect individuals who are no longer reproducing. Evolution is rather cruel, making sure that abilities that promote successful reproduction are carried on to future generations, but having little interest in the individuals of the present generation if they are not reproducing. When it comes to background radiation it is difficult to prove the relationship between it and any biological effect. Even for the relatively large groups of individuals who were exposed to ionizing radiation in Hiroshima and Nagasaki during World War II, or those who were exposed during the nuclear power plant meltdown in Chernobyl, it has not been possible to prove statistically that anyone has had any hereditary defects due to the radiation. This is definitely not to say that no hereditary defects are due to man-made or background radiation, only that it is difficult to show the connections. In addition, we also need to remember that ionizing radiation has a number of important uses, for example, as a theurapeutic agent against tumors. Cancer cells are killed by the radiation more efficiently than normal cells are, because the former divide constantly and are more sensitive to genotoxic events.

Radon is a major health problem that affects all parts of the world, and is linked to the presence of the chemical element uranium. Often it does not receive the attention due to it, but it is a major contributor to what we call the background radiation and therefore contributes to the background incidence of human cancer. Radon is a noble gas, chemically unreactive but radioactive, and is formed in different amounts in the bedrock beneath us. As it is a gas that we breathe, the lungs are the primary target for its toxic effects. Radon is the second most important cause of lung cancer, after smoking, and is estimated to cause about as many deaths as road traffic in affected areas. It is part of the decay chain of uranium-238, which through thorium-234 and uranium-234 gives radium-226. This radium slowly (half-life of four days) decays to produce radon. There are altogether 36 isotopes of radon, all radioactive, with a range of atomic masses from 193 to 228. The health effects of radon were noted as early as the sixteenth century, when miners in certain areas experienced a dramatic increase in the frequency of lung diseases. However, as a chemical element, radon was not characterized until Mr and Mrs Curie isolated radium and observed that radon is a product of its decay.

Radon is therefore produced naturally and constantly in all areas containing uranium, eventually departing as a gas to the atmosphere if the ground is gas permeable. In areas where significant amounts of radon are given off, hazardous levels can be found in houses. How much depends on the construction of the house and how well it is ventilated. Radon is also present in some tap water, especially if it comes directly from wells. Before using such water one should let it aerate so that the radon is lost to the atmosphere. The main source of radon, however, is stone-based building materials, which always contain a little radium. The radon eventually formed will leak into the house at a steady rate. Especially dangerous is the blue porous concrete, a special form of concrete based on alum shale. The problem received attention in the 1970s, and since then all concrete produced is monitored and has low radioactivity. In older houses built with blue porous concrete, there is not much else to do but to increase the ventilation to reduce radon levels. It is easy to measure radon levels in the air. People living in especially bad areas or in houses constructed with blue porous concrete can monitor radon levels inside and ensure that they are acceptable.

9.4.9
Summary

In this section we have highlighted some 'bad guys', from a chemical point of view, and the story in general has been that these chemicals have been used and have caused problems, and now we have replaced them. However, we should not exclude the possibility that our children will be able to tell similar stories about the replacements. In addition, we should never forget that it is not primarily the chemicals themselves that are hazardous: it is the way we handle them. Some of the chemicals discussed in this section are still used, but our experiences have made us wiser and we have adopted safer handling procedures.

9.5
Legal and Illegal Chemical Pleasures

A chemical pleasure could be a fragance, a sweet, or some other innocent product, but what we will focus on here are those that have an effect on the central nervous system (CNS). It appears that humans have always been interested in products that affect the mind, and there are always good reasons to use a drug – as a pain-reliever, an anti-depressant, a door into the world of gods and spirits, or simply as a road to temporary happiness. The CNS is fantastically complex, there are obviously a large number of substances that can affect its functions, and the science of psychopharmacology together with the pharmaceutical industry are constantly developing increasingly selective drugs that can be used to treat mental conditions and diseases. Many of these are used both legally and illegally, and the story of the development and use of a psychopharmaceutical agent is fascinating, but here we will focus on what can be called the traditional drugs. These, in most cases, come from nature, and often have a long history of use. We should also note that the interests of society do not always go hand in hand with the desires of individuals to use drugs. In many cases, the drugs are so dangerous and destructive that society has forbidden them. Others are almost forbidden but can be used in special situations. Yet others are legal and sold more or less openly, although often heavily taxed. For society, this is a delicate and difficult balance. If it is not handled correctly, serious criminal activities will be encouraged that in the long run pose a serious threat to society. For example, the American Mafia was established thanks to Prohibition between 1919 to 1933, when the manufacture, transportation and sale of alcohol for consumption were banned nationally. Alcohol and tobacco are still legal chemical pleasures in most parts of the world, and the taxes on alcoholic beverages and tobacco products are adjusted to the black market.

9.5.1
Caffeine

Caffeine (see Figure 9.40) is a common natural product, being present in the leaves, seeds, and fruits of around 100 different plant species. The main sources

caffeine   theophylline

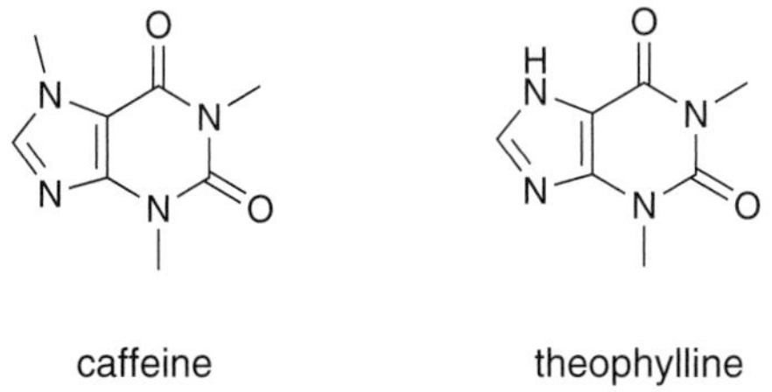

Figure 9.40   Bioactive purines in coffee.

are coffee beans (e.g., from *Coffea arabica*), tea leaves (e.g., from *Camellia sinensis*), nuts of the cola tree (*Cola nitida*), and cocoa beans (*Theobroma cacao*). Humans are exposed primarily by drinking coffee, which contains about 90 mg per cup, but also tea (about 50 mg per cup, plus 1 mg of the analog theophylline, which relaxes smooth muscle), cola soft drinks prepared from cola nuts (about 100 mg per liter), and cocoa beverage (about 5 mg per cup). In addition, it is often an important ingredient in the so-called energy drinks. Chemically, caffeine is a purine derivative, and it has some chemical similarities to the DNA bases adenine and guanine.

Caffeine is water soluble, but it is still absorbed rapidly in the intestines and distributed throughout the body. The stimulating effect of coffee is attributed to the ability of caffeine to increase the secretion of the hormone epinephrine in the body, as well as to block the receptor that responds to elevated levels of the nucleoside adenosine.

1) Epinephrine, a catecholamine, is released by specific neurons when the blood sugar level is low, and it stimulates the degradation of glycogen and triglycerides by inducing the formation of cyclic AMP in cells. This increases the levels of glucose and fatty acids, resulting in an increased production of the energy molecule ATP.

2) When ATP is used as a fuel in the body's chemical machinery, adenosine is formed, and the more adenosine that is formed the lower is the level in the fuel tank. We have specific receptors that detect the concentration of adenosine, and if they are activated the cells are ordered to slow down and consume less energy. Caffeine, as an antagonist, blocks this receptor, and we do not feel as tired as we actually are.

Caffeine is metabolized primarily by hydroxylation of methyl groups (see Figure 9.41), resulting in oxidative dealkylation, as well as by epoxidation of the carbon–nitrogen double bond in the five-membered ring. The main metabolites are demethylcaffeine, didemethylcaffeine, and trimethyl uric acid, all of which are water soluble and excreted in the urine. The half-life of caffeine in the body is approximately 3 h.

It has long been debated whether the consumption of coffee is associated with an increased risks of cancer, cardiovascular disease, miscarriage, and so forth.

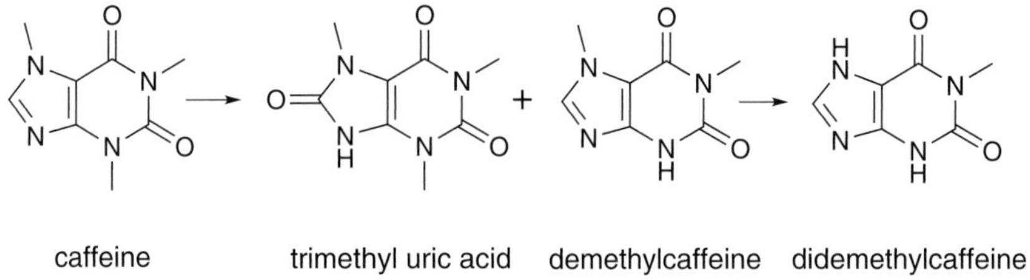

caffeine   trimethyl uric acid   demethylcaffeine   didemethylcaffeine

Figure 9.41 Metabolites of caffeine formed in man.

After all, the beverage is an aqueous extract of roasted coffee beans, and we have already seen what can be formed from organic materials that are heated strongly. In fact, not only does coffee look like a complicated chemical soup, it is. Over 1000 chemical components have been found in roasted coffee, 28 of these have been tested for carcinogenic activity in rodents, and not less than 19 of these have tested positively. However, extensive epidemiological studies of coffee drinkers have not provided any indication that moderate consumption of coffee is harmful. Caffeine is, of course, fairly toxic, the LD_{50} value (rats, oral administration) being 300 mg/kg, which corresponds to about 200 cups of coffee.

9.5.2
Tobacco Smoke

The use of tobacco, especially cigarette smoking, is the single greatest risk factor for cancer. It is today believed that around one third of all tumors are caused by smoking. In particular, smoking causes lung cancer. Of all lung cancer cases, 85–90% are attributed to smoking. However, several other types of tumor (in the esophagus, trachea, oral cavity, kidney, bladder, pancreas, and cervix) have also been linked to smoking. Tragically, but also interestingly, although cigarette smoke is a relatively potent carcinogen and there are so many people who smoke, the link between smoking and lung cancer was not scientifically established until the 1950s, and the conclusion was questioned (especially by the tobacco industry!) for another decade. Since then, great efforts have been made to understand why cigarette smoke is carcinogenic and to develop safer cigarettes. Approximately 5000 chemicals have been identified as components in cigarette smoke, which is reasonable if you consider how cigarette smoke is formed. The plant tobacco, either *Nicotiana tabacum* or *N. rustica*, is very rich in secondary metabolites, which obviously is important for tobacco flavor. In addition, some chemicals, such as glycerol, are added to the tobacco in order to prevent the products from drying. A cigarette is pyrolyzed, that is, it is subjected to heat without free access to oxygen, which is a fertile environment for the formation of new compounds. Not less than 69 carcinogenic chemicals have been identified in cigarette smoke, which consists primarily of polycyclic aromatic hydrocarbons, heterocyclic aromatics, *N*-nitrosamines, aromatic amines, aldehydes, phenols, epoxides, and

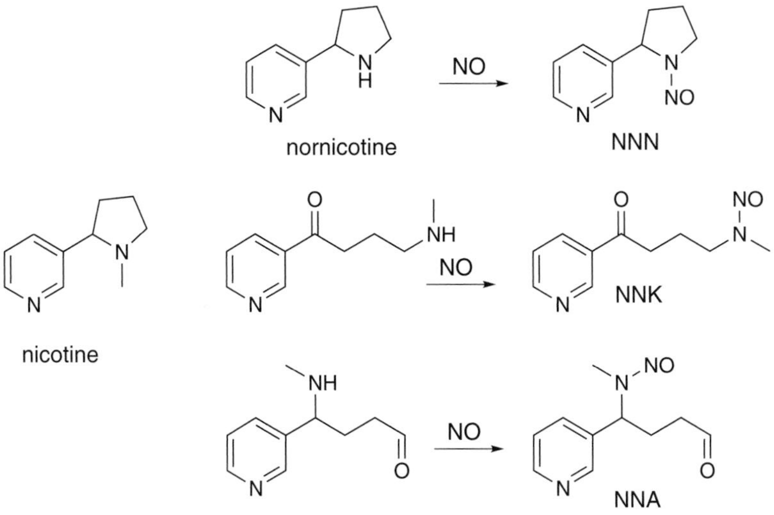

Figure 9.42 The nitrosation of nicotinoids.

inorganic substances such as heavy metals and radioactive polonium-210. Cigarette smoke causes lung cancer primarily because it contains polycyclic aromatic hydrocarbons such as benzo[a]pyrene (see Section 9.1) and *N*-nitrosamines. The latter are formed by the nitrosation of nicotine derivatives during the drying of tobacco leaves (by bacteria) and by smoking (from NO, formed from nitrate, and amino acids). Especially NNN and NNK (see Figure 9.42) are considered important for the carcinogenic effect.

The metabolic activation of *N*-nitrosamines to the final electrophilic form is described in Section 5.6.2. Although cancer, and especially lung cancer, is the illness primarily associated with cigarette smoking, one should not forget that smoking is also linked to a number of other illnesses. For example, heavy smokers will, with time, even if they are lucky enough to escape lung cancer, almost certainly get emphysema and/or COPD (chronic obstructive pulmonary disease), resulting in severely impaired lung capacity. In addition, cardiovascular diseases occur more frequently among smokers.

Nicotine is the most important secondary metabolite in tobacco, and the dried plant contains 0.5–3% nicotine. Tobacco was initially used as a medicine and to control insects (nicotine is still used as an insecticide). Nicotine affects both the central and the peripheral nervous systems, by binding to a specific type of acetylcholine receptors (which are therefore called nicotinic receptors), and is curiously enough both relaxing and stimulating in small doses. At toxic doses, palpitations and seizures occur, the lethal dose for a nonsmoker being about 60 mg. Heavy smokers are less sensitive because nicotine is metabolized/excreted relatively quickly because of enzyme induction, and they may actually be exposed to more than 60 mg per day.

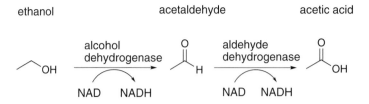

Figure 9.43 The metabolism of ethanol.

9.5.3
Ethanol

More than 95% of the ethanol ingested or otherwise absorbed into the body is metabolized, less than 5% being excreted unchanged with the exhaled air or the urine. Hyperventilation may increase the part that is excreted unchanged to 10%, but not more. It is oxidized by alcohol dehydrogenase to acetaldehyde, which in turn is oxidized by aldehyde dehydrogenase to acetic acid (see Figure 9.43). The acetic acid fuels the citric acid cycle and will be used to form ATP, and alcoholic beverages therefore have quite a high calorie content. Both enzymes (alcohol dehydrogenase and aldehyde dehydrogenase) use NAD as a co-factor, which is reduced during the oxidation to NADH. NADH is also produced in the citric acid cycle and can be used in the respiratory chain to produce ATP. At high exposure levels (which is equal to high consumption), ethanol will also be oxidized by cytochrome P450 in the endoplasmatic reticulum as well as by catalase in the peroxisomes. As has been discussed (Section 5.2.2), cytochrome P450 can be induced by substances it oxidizes, and also by ethanol. Alcoholics will therefore metabolize ethanol faster than normal consumers, and this creates a sort of tolerance to ethanol that is not really appreciated.

When the cytochrome P450, which uses oxygen for its oxidations, is induced, this may have serious consequences in other respects. More cytochrome P450 means that the ability to metabolize other chemicals increases, which can lead to problems for a person taking medication. (The degradation can be so rapid that the drug does not work, or there may be an efficient production of toxic metabolites.) It also increases the oxidative stress in the cells. Catalase normally catalyzes the disproportionation of two molecules of hydrogen peroxide to one of oxygen and two of water, but it can also use methanol or ethanol to replace one hydrogen peroxide molecule. The remaining hydrogen peroxide will have more time to participate in ROS reactions. However, in humans the capacity is low, not more than a few percent of the alcohol consumed being oxidized by catalase. Ethanol is rapidly and efficiently absorbed into the blood in the lungs and the gut. Typically, the ingestion of 1 mL pure ethanol per kg body weight will result in a blood concentration of ethanol of 1 mg/mL after one hour. For a man weighing 75 kg, this means that consumption of 15–20 cL of a strong alcoholic drink (containing

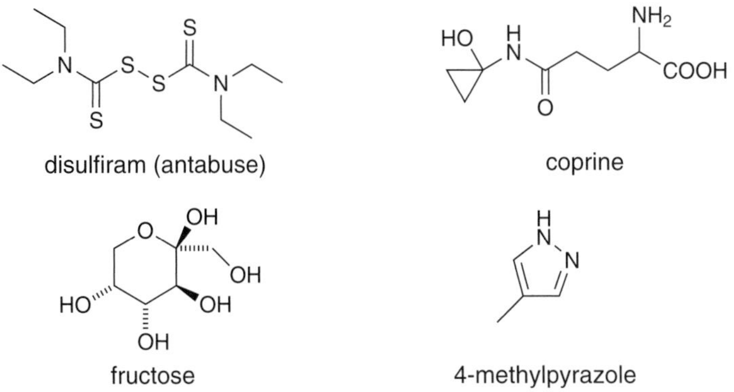

Figure 9.44 Compounds that affect the metabolism of alcohols to carboxylic acids.

40% ethanol) will give a blood level of 1 mg/mL. Approximately 10 g of ethanol is typically metabolized per hour, meaning that the amounts consumed in the previous example will be gone in some 8 hours. The metabolic rate is slightly increased by fructose, but it can also be decreased by some compounds. As mentioned above, ethanol will be metabolized via acetaldehyde, but as the activity of aldehyde dehydrogenasis several times higher in the liver and kidneys, where most of the metabolism of ethanol takes place, the acetaldehyde is a transient intermediate that goes unnoticed. However, individuals that are undergoing treatment with antabuse (disulfiram) or a similar drug will not be able to convert acetaldehyde, at least not as efficiently, and will suffer badly from the effects of acetaldehyde if exposed to ethanol. Besides the drug antabuse, which also is used as a rubber accelerator and vulcanizer in the rubber industry, coprine, a fungal metabolite isolated from the fruit bodies of *Coprinus atramentarius*, also inhibits aldehyde dehydrogenase. Fructose has been shown to increase the conversion of ethanol, while 4-methylpyrazole decreases the conversion by inhibiting alcohol dehydrogenase (see Figure 9.44 for structures). The latter may also be used to prevent the generation of formaldehyde during methanol intoxication.

The effects of ethanol on the CNS are complex. As relatively large amounts of ethanol are required to produce intoxication, it can be assumed that part of the effect is caused by non-specific disturbance of the neurons. However, ethanol has also been shown to affect some CNS receptors specifically. The neurotransmitter GABA (γ-aminobutyric acid) has attracted special interest because it is the major inhibitory neurotransmitter in the CNS. Compounds that are known to stimulate the receptor for GABA, enforcing the effect of GABA, produce similar physical effects to those of ethanol, and it is believed that ethanol has a specific effect on the $GABA_A$ receptor. Too much ethanol will be lethal, and this is due to a general inhibition of the respiratory center in the CNS. Normally, 0.5–1% ethanol in the blood is required for this effect, which corresponds to an adult consuming at least a full bottle (75 cL) of a strong alcoholic drink (40% ethanol) or more in a short time.

People who consume larger amounts of alcohol over a longer time may suffer from various liver-related diseases, such as fatty liver (which in principle can be result from one evening of heavy drinking), alcohol-related hepatitis, and cirrhosis of the liver, which leads to necrosis (tissue death). At this point, only a liver transplant can save the life of the drinker. Fatty liver is a condition in which the fat cells normally produced in the liver from a surplus of nutrients cannot be transported from the liver to adipose tissue where it should be stored. Instead, the fat remains in the liver as small fat globules. This leads to an expansion of the liver, which assumes a yellowish color. The reason for this effect is that the intensive oxidation of ethanol in the endoplastic reticulum causes oxidative stress, which eventually affects the protein synthesis in the liver cells. To transport fat from a liver cell requires a specific transport protein, and if this is not produced the fat remains where it is. Luckily, fatty liver is a reversible condition that normally does not cause any other effects. If the consumption of alcohol stops, or at least is moderated, the protein production will go back to normal, the fat will be transported from the liver cells, and the liver will be normalized. However, it should be recognized as the first step in a process that ends with cirrhosis and liver collapse.

Jaundice is a symptom, not a disease in itself, and is caused when bilirubin (a reddish-yellow pigment formed when old red blood cells are degraded and normally excreted in the feces) accumulates in the blood and circulates in all tissues. It causes the yellow color of the skin and the whites of the eyes that is typical of jaundice. The processing of bilirubin takes place in the liver, and the final product is excreted with the bile into the small intestines. There are many conditions that will cause jaundice, but, after a period of particularly excessive consumption, alcoholics may have it because of serious liver failure. There is a connection between the appearance of jaundice in alcoholics and liver cirrhosis, but jaundice is a condition that after some time can disappear if the consumption stops.

Cirrhosis of the liver is a final stage in which the liver tissue is degrading. Connective tissue grows into the liver, which hampers the blood flow and which in turn hampers the liver function. As it develops, cirrhosis of the liver becomes increasingly more difficult to treat.

9.5.4
Amphetamines and Ecstasy

Amphetamine and ecstasy (also called MDMA) are synthetic central stimulants, which belong to the group of phenylethylamines (see Figure 9.45). They function by stimulating the release of the neurotransmitters dopamine (associated with well-being, aggressiveness, and motor as well as mental activity) and serotonin (which among other things regulates impulse control, anxiety, sexuality, and aggression), and the combination of the two gives an effect in which significantly increased activity predominates. Chemically they are similar to the natural product ephedrine, which has been an important drug against asthma as it, among other things, imitates the neurotransmitter epinephrine. The effect of amphetamine and

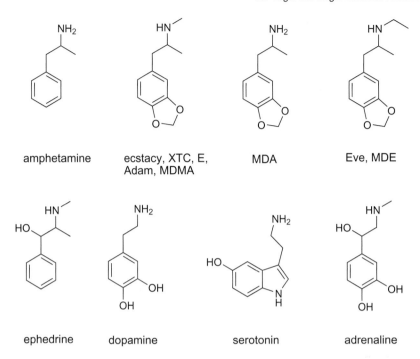

Figure 9.45 Amphetamine and related compounds.

ecstasy is described as an experience of empathy and openness, and is compared by many to what one feels when falling in love. However, the experience is individual and depends on both the current mood as well as what expectations one has. In addition, differences between one person's metabolism and another's will influence the experience. Both compounds will also increase the sensitivity to sound and touch, and reduce such pain that is caused by fear. Furthermore, they reduce our sense of hunger, and ecstasy was patented in 1914 in Germany as an appetite inhibitor. Consumers of amphetamine or ecstasy will increase their activity (work, dance, etc.) while they no longer get hungry and will quickly lose weight. Amphetamine is a drug that is used to treat obesity, as well as some other conditions like narcolepsy (characterized by an abnormal tendency to fall asleep). It is also used to counteract fatigue and increase performance and was formerly popular among students at examination time to enable them to study round the clock. The down side, as you may have guessed, is that there is a price to pay when the amphetamine disappears from the body. This type of drug has, of course, also been used for doping in different sports (e.g., cycling), as they make you feel that you are not tired. There is nothing stopping you from working yourself to death, and this happened frequently. Most of the amphetamine or ecstasy consumed will be excreted unchanged via the kidneys, as they are water soluble in the protonated form. The remainder is metabolized, mainly by oxidative dealkylation to cleave off a methyl group from the nitrogen, which produces MDA (see Figure 9.45). The turnover is relatively fast, only a few percent is left in the body after one day. Ecstasy is often considered to be a harmless and innocent drug, appropriate for use at parties, but this not the case. Together with its metabolites it will slowly degrade the serotonin-producing cells in the brain, and the important balance between dopamine and serotonin changes in time with the abuse. When it eventually is mostly dopamine that is released, the intoxication will be unpleasant experience producing aggressive behavior. The LD_{50} values for amphetamine and ecstasy are between 50 and 100 mg/kg (oral, rat).

There are a number of compounds that are chemically similar to amphetamine and ecstasy and have similar effects, and there is a tendency to produce new versions (for example MDE, see Figure 9.45) as the authorities ban the old ones. The substances are relatively simple and can be synthesized in large quantities in primitive laboratories hidden in the countryside. Therefore, certain chemicals which are needed for the synthesis of phenylethylamines are blacklisted and difficult for private persons to purchase.

9.5.5
Cannabis

Cannabis preparations, especially hashish and marijuana, are probably the most commonly used drugs, after alcohol. They are based on products from the hemp plant, *Cannabis sativa*, and while marijuana is simply dried plant parts of the plant, hashish is the resin of the plant. Hashish and marijuana contain a number of so-called cannabinoids, some of which have a strong effect on the central nervous system. In marijuana, the cannabinoids constitute one or a few percent, while hashish contains around 10% cannabinoids. In addition there are extracts of the plant that can be much more concentrated. The most important of the cannabinoids is Δ9-tetrahydrocannabinol (Δ9-THC) (see Figure 9.46), because it occurs in relatively high concentrations. Δ9-THC is lipophilic and easily absorbed. It can be oxidized to 11-hydroxy-Δ9-THC, which is much more potent and readily crosses the blood-brain barrier.

11-Hydroxy-Δ9-THC affects the brain in a complex way, which is only partially understood. It binds to CNS receptors (cannabinoid binding receptors) that exist

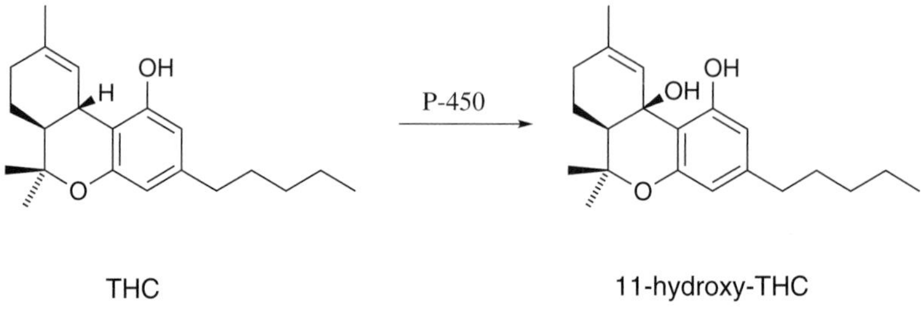

Figure 9.46 The conversion of Δ9-THC to 11-hydroxy-Δ9-THC.

in many parts of the brain and consequently gives rise to a number of effects: euphoria, moudlated perception of visual and audial input, relieve of pain, and memory as well as concentration problems. Eventually, 11-hydroxy-Δ9-THC is oxidized to the less active and inactive metabolites, mainly in the liver, and these degradation products present in blood and urine can be used to detect cannabis use several weeks after the consumption of hashish or marijuana. Chronic exposure may lead to psychosis and other mental illnesses, as well as permanent memory problems. The cannabinoids will also gives effects on the endocrine system, resulting in fewer and poorer sperm in men and irregular ovulation in women.

9.5.6
GHB

γ-Hydroxybutyric acid (GHB) (see Figure 9.47), can give a feeling of euphoria, and as it is very easy and cheap to produce it is misused. It is actually an endogenous neurotransmitter in the CNS, where it acts together with GABA by affecting the levels of dopamine (another neurotransmitter). This will, in a short term, result in a calming effect, probably by inhibiting the release of dopamine, and GHB has been used as a calmative during childbirths in some countries. When the effect of GHB decreases (it is rapidly metabolized by the primary metabolism and its half-life is approximately 20 min), the release of dopamine increases above normal, which makes a person feel alert and euphoric.

9.5.7
LSD

LSD is one of the most potent hallucinogenic compounds known. It is chemically related to the ergot alkaloids and also causes ergotism (see Section 9.3.12), being a lysergic acid derivative. LSD (see Figure 9.48) was discovered during the research

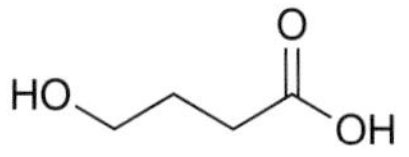

Figure 9.47 GHB.

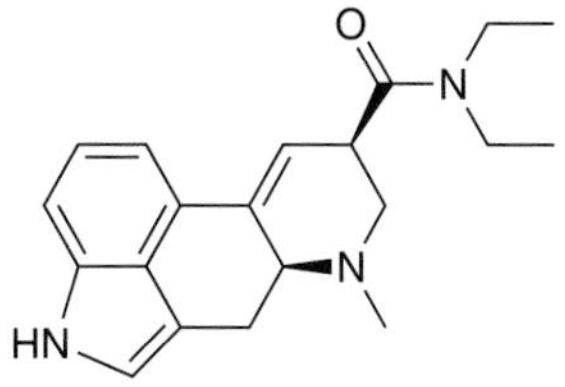

Figure 9.48 LSD.

on ergot alkaloids, conducted in the 1930s with the aim of developing new drugs. LSD was initially named LSD-25 because it was the 25th derivative of lysergic acid produced from ergot.

It became famous during the hippie period in the 1960s and 70s, when the testing and use of drugs was considerably more liberal than it is today. However, it never really left the scene, and has gained renewed popularity as a rave party drug. A dose of 100 mg of LSD gives a trip that lasts for several hours, during which tones, smells, and colors are perceived more intensely than usual. The hallucinations are considered to be similar to those experienced by schizophrenics. At the molecular level, LSD is an antagonist of the neurotransmitter serotonin, which for that reason can be suspected to be involved in schizophrenia. The acute toxicity of LSD is high, but perhaps more dangerous is the mental state of the exposed person and the risk that he or she will be involved in an accident. It is also believed that LSD intoxications may trigger latent mental illness or aggravate one that already exists.

9.5.8
Opium, Morphine, and Heroin

Opium contains a score of biologically active alkaloids, of which morphine is the most important (see Figure 9.49). Others are include codeine, thebaine and papaverine. It is a product obtained from the opium poppy seed capsules, which are cut with a knife producing a white latex, or milk, that seeps out. The product opium is the dried latex, which appears as a viscous brown mass. Opium poppy (*Papaver somniferum*) is just one of many species of poppy, but it contains most of the active alkaloids. It is one of the oldest medicinal plants, and has been reported to cure everything from snakebites to deafness. There is no sharp line between use and abuse, but it is evident that opium all the time has been used as a recreational drug as well as a cure. With time the abuse grew, especially in Asia but also in Europe, and peaked during the nineteenths century. It is still significant, but because of competition from other drugs the importance of opium has declined. Already very early, in the childhood of chemistry, attempts were made to understand what in the opium gives rise to its effect, and in 1803 a German

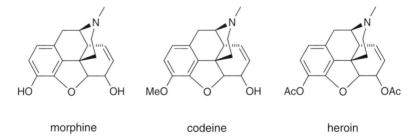

Figure 9.49 Morphine and its derivatives, codeine and heroin.

chemist isolated pure morphine (named after the Greek God of sleep, Morpheus). At the end of the nineteenth century, when modern pharmaceutical industry was born, efforts were made to chemically modify morphine to give products with retained pharmacological effects but without the addictive properties of morphine. A product that at first appeared to have such qualities was diacetylmorphine, which was obtained by a simple acetylation of morphine with acetic anhydride. The compound was named heroin, from the German word 'heroisch' (meaning heroic) and seemed to be an ideal drug with which to treat bronchitis and asthma until it became clear that heroin is also highly addictive.

All three opioids, morphine, codeine, and heroin, have the same effects on humans, but the less polar codeine and heroin are absorbed more easily and cross the blood-brain barrier faster. The excretion, which mainly takes place via a conjugate of morphine (for example, glucuronic acid), is slower for codeine and heroin. The opioids act primarily on the CNS, although there are receptors for them throughout the body. They are efficient pain relievers and are generally calming while at the same time producing a strong feeling of euphoria. Another important and useful effect is to reduce the cough reflex, and especially codeine is used in cough medicines. The receptors in the CNS that the opioids bind to, called the opioid receptors, have been identified, and these are involved in the experience of pain. This has inspired research that has led to the discovery of endogenous compounds that bind to the same receptors. Such compounds are called endorphins and are produced by the body when they are needed. It is worth noting that morphine is still a very important pharmaceutical agent used for pain relief.

9.5.9
Summary

The eternal human need for escapism has led to the discovery of a wide range of drugs that have effects on the CNS. Most of these were originally natural products, and, by using them as scientific tools, great advances in our understanding of how the mind works have been made. Many conventional drugs have the drawback of being highly addictive, because the nervous system quickly adapts to their presence and the changes they provoke, and this is often the reason for why drug misuse often leads to mental as well as physical collapse. Other intoxicants, such as alcoholic beverages and tobacco, have become more accepted and integrated in different civilizations.

10
Environmental Effects of Chemicals

The environmental effects of pollutants are often very complicated and difficult to foresee, as sensitive balances in ecosystems are affected. An example is, again, the use of DDT to kill mosquitoes that spread malaria, which as well as the long-term effects also resulted in dramatic local effects in the areas where it was used. In Borneo, for example, the insects killed by DDT were eaten by lizards, which were intoxicated and became easy prey for cats. The cats in turn became ill and died, the absence of cats resulted in an explosion of the rat population, and the rats consumed the local crops and spread dangerous diseases. The government of Borneo eventually had to reintroduce cats to cope with the problems! Chemicals that we consider to pollute the environment may be anthropogenous (resulting from human activities) or natural. This chapter will mainly deal with anthropogenous chemicals, which are those we can do something about, but it is important to relate these to the natural emissions, which are substantial in many instances and include many classes of chemicals.

10.1
Natural Emission of Chemicals

10.1.1
Inorganic Chemicals

Volcanoes are the major polluters of our environment. Since the beginning of time, volcanic activity has more or less continuously belched out huge amounts of, for example, sulfur compounds and ash into the atmosphere, where they are distributed worldwide by the weather systems. A massive volcanic eruption is extremely dramatic and dangerous. Several of those that happened during the most recent centuries have instantly killed more than 10 000 persons, and under such circumstances the emission of chemicals could be regarded as a secondary problem. The eruption of 'El Chichon' in Mexico in 1982, which did not cause too much damage in terms of lives lost, is nevertheless estimated to have brought $500\,000\,000\,m^3$ of ash, pumice, and dust into the atmosphere, and 40 000 tonnes of HCl into the stratosphere (accounting for an immediate 40% increase in the

Chemistry, Health, and Environment. Olov Sterner
© 2010 WILEY-VCH Verlag GmbH & Co. KGaA, Weinheim
ISBN: 978-3-527-32582-5

HCl concentration). The local effects of major eruptions are, of course, massive, but global effects can also be observed. During the Mount Pinatubo eruption in the Philippines in 1991, approximately 10 million tonnes of sulfur in the form of sulfur dioxide were blasted into the stratosphere. The sulfur dioxide is oxidized by molecular oxygen (via reactive oxygen species such as the hydroxyl radical and hydrogen peroxide) to sulfur trioxide, which reacts with water to form sulfuric acid. The sulfuric acid condenses to small droplets with a diameter of around 1 μm, which are spread around the globe and can stay in the stratosphere for a long time (more than a year). Such aerosols not only reflect part of the radiation from the sun, but they will also act as cloud condensation nuclei. Clouds will also scatter the sun's rays, but will in addition absorb and to some extent reflect the heat radiation from the earth, the overall effect of a major volcanic eruption on the temperature being global cooling.

Lightning releases large amounts of energy into the atmosphere, and this transforms molecular oxygen and nitrogen into nitrogen oxides, especially nitric oxide (NO), nitrogen dioxide (NO_2), and ozone (O_3). Nitric oxide, which is also formed during the combustion process in, for example, car engines, is easily oxidized in air to nitrogen dioxide (see below).

Winds will transport dust, and huge quantities will be moved, especially in desert areas. Finer particles (with a diameter of less than 50 μm) will form an aerosol that can be transported even between continents. A few hundred million tonnes of mineral dust are transported to the atmosphere as an aerosol yearly.

The continuous movement of the water in the oceans produces small water droplets that are carried away by the wind and dried. The remaining salt particles (mainly chlorides but also sulfates) that result from this process form an aerosol that transports approximately 1 billion tonnes of sea salts per year to the atmosphere. Most of it comes down again with the rain close to the spot where it was produced, but the wind may of course transport it for substantial distances, and part of this ends up on land. The inorganic gases carbon monoxide, carbon dioxide, nitric oxide, and nitrogen dioxide are the most significant of these

10.1.2
Organic Chemicals

More complex and highly hazardous organic compounds are emitted in forest fires, for example, mixtures of polyaromatic hydrocarbons, polychlorinated dioxins, and polychlorinated dibenzofurans. Yes, the famous 'doomsday chemical' TCDD (dioxin) is also formed naturally and was present on earth long before humans arrived. The amounts formed in forest fires are difficult to estimate, but several investigations have suggested that they are in fact comparable to the amounts formed anthropogenously. It is likely that polychlorinated dioxins and polychlorinated dibenzofurans are also formed *in vivo* by water and soil microorganisms, as it has been shown that chlorophenols (e.g., 1,4,5-trichlorophenol) can be converted into polychlorinated dioxins and polychlorinated dibenzofurans by peroxidase enzymes, and halogenated phenols are formed by organisms, although this remains to be demonstrated. Also, polychlorinated biphenyls (PCBs) are formed naturally, as demonstrated by a study of the ash from the 1980 eruption of Mount St. Helens. The dioxins and the PCBs will be further discussed in Section 10.6.

Living organisms produce large amounts of organic as well as inorganic chemicals, which are released into the environment. Some of these compounds, natural products of which we have encountered several examples throughout the text, can be highly toxic. Others possess less biological activity but are more interesting because of the amounts emitted, but in general one can say that natural products are relatively easily biodegradable.

10.1.2.1 Hydrocarbons
Simple hydrocarbons (also emitted from petroleum sources) are important because they participate in several of the effects discussed in this chapter. The simplest, methane, is mainly produced by anaerobic microbial degradation of organic matter in, for example, beds of lakes, landfills, and the stomachs of ruminant animals. Other hydrocarbons, for example, the terpenoids, are emitted in large amounts (estimated amounts 1 billion tonnes yearly) from plants, especially in coniferous forests. A few examples of some common hemiterpenes (with 5 carbons), monoterpenes (with 10 carbons), and sesquiterpenes (with 15 carbons) are shown in Figure 10.1.

10.1.2.2 Sulfur Compounds
Sulfur is an important element for organisms, and is especially important for some microorganisms (e.g., the chemoautotrophic sulfur bacteria which produce energy by oxidizing sulfur to sulfate). Volatile sulfur compounds, mainly hydrogen sulfide, carbon disulfide (formed by anaerobic fermentation), carbonyl sulfide (from volcanic emissions), mercaptans, sulfides, and disulfides, are released by microorganisms as well as by plants, and it is estimated that in total 65 million tonnes are emitted yearly. The major contributors are the plankton of the oceans, producing approximately 43 million tonnes of dimethylsulfide. All volatile sulfur compounds will be oxidized to sulfur dioxide in the atmosphere, and in turn will be transformed into sulfuric acid (as described above) and will contribute to acid rain (see Section 10.3).

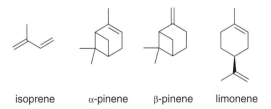

isoprene   α-pinene   β-pinene   limonene

Figure 10.1 Some of the many volatile terpenes that are produced by plants.

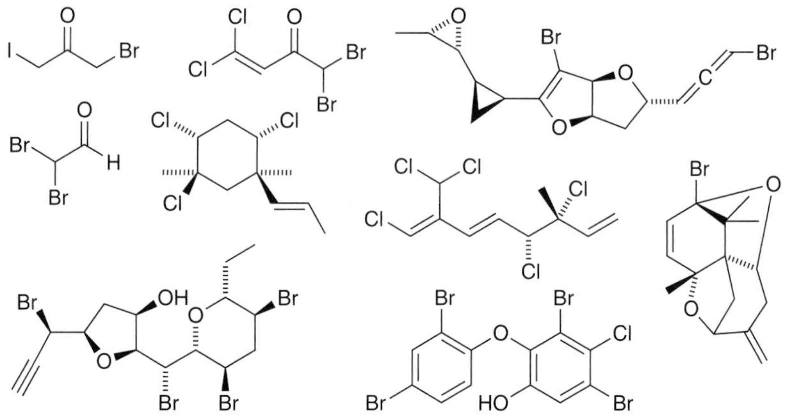

Figure 10.2 Some halogenated organic compounds produced by marine organisms.

10.1.2.3 Halogenated Compounds

Although they are generally considered to be typical synthetic products, halogenated compounds are produced in enormous quantities in nature. We have already encountered a few: some halogenated fungal metabolites that resemble synthetic pesticides (in Chapter 2) and two toxic fluorinated plant metabolites (in Chapter 7). Figure 10.2 shows some examples of halogenated compounds produced by marine organisms, believed to be the major producers of organic compounds in nature. The quantitatively most important halogenated compounds are of course the most simple, and it is estimated that the natural production of chloromethane and iodomethane is in the order of several million tonnes per year, which far exceeds the emission from anthropogenic sources. As these compounds are volatile, they are important for the global transport of halogens, especially from the oceans to land.

As well as fungi and marine organisms, all other organisms produce and excrete halogenated compounds, and large amounts are also emitted by volcanoes, as mentioned above.

10.2
Smog and other Air Pollution

Smog (smoke and fog) is a typical local problem, affecting an area within 100 km of the place of emission. It is a mixed aerosol, formed as a result of the heavy pollution of sulfur dioxide and soot mainly due to the burning of coal together with an inversion of the temperature layers (a weather condition characterized by cold air close to the ground and warmer air above it that prevents the air pollutants and water vapor escaping). It was formerly a big problem in densely populated areas and industrial regions. The smog that reigned for a week in London in 1952, with sulfur dioxide concentrations of approximately 0.7 ppm and 1.5 mg soot per m³, killed approximately 4000 persons. Happily, as the use of coal has decreased, the incidence of smog has as well.

However, inversions regularly occur over cities, and although modern heating systems no longer emit soot and particles, there is still a lot of air pollution in the form of, for example, nitrogen oxides and hydrocarbons from traffic, and this may give rise to what is called photochemical smog. This is a process for which sunlight is essential and which generates photochemical oxidants like ozone (see below).

10.2.1
Degradation of Trace Gases in the Troposphere

Most of the UV radiation from the sun is filtered out by the ozone in the stratosphere, but some amounts pass though and come down to the troposphere, where it can transform some of the few molecules of ozone present into molecular oxygen and an oxygen atom. The oxygen atom thus generated is in an electronically excited state (if the wavelength is below 310 nm), and it can react with a molecule of water to form two hydroxyl radicals. This reaction is the major generator of the very important hydroxyl radicals in the atmosphere. Other examples are the reaction of superoxide with ozone or nitrogen monoxide, or the homolytical splitting of hydrogen peroxide by light. In the troposphere, although it is present in very low amounts (100–1000 times less than in the stratosphere), the hydroxyl radical will act as a cleaner and react with trace gases that are present and transform them into more hydrophilic forms that are washed away by the rain. Examples are methane and carbon monoxide, which together are responsible for the degradation of 90% of the hydroxyl radicals in the troposphere in reactions that generate ozone (see Figure 10.3).

Both methane and carbon monoxide are oxidized to carbon dioxide, methane via formaldehyde and carbon monoxide (see Figure 10.3), while the hydroxyl radical transforms other (minor) air pollutants into water (molecular hydrogen), sulfuric acid (hydrogen sulfide and sulfur dioxide), hydrogen chloride (chlorinated organic compounds), and nitric acid (nitrogen dioxide and ammonia). The products formed after the reaction of these gases with the hydroxyl radical are all water soluble and washed away by the rain. More complex organic compounds present in the atmosphere, for example, terpenes emitted by trees, will also be attacked by the hydroxyl radical and transformed by similar reactions into more water-soluble products.

10.2.2
Photochemical Oxidants

The mechanism for the oxidation of methane and carbon monoxide by the hydroxyl radical in the troposphere is shown in Figure 10.3. The oxidation is coupled with the oxidation of nitric oxide to nitrogen dioxide, which is split by light back to

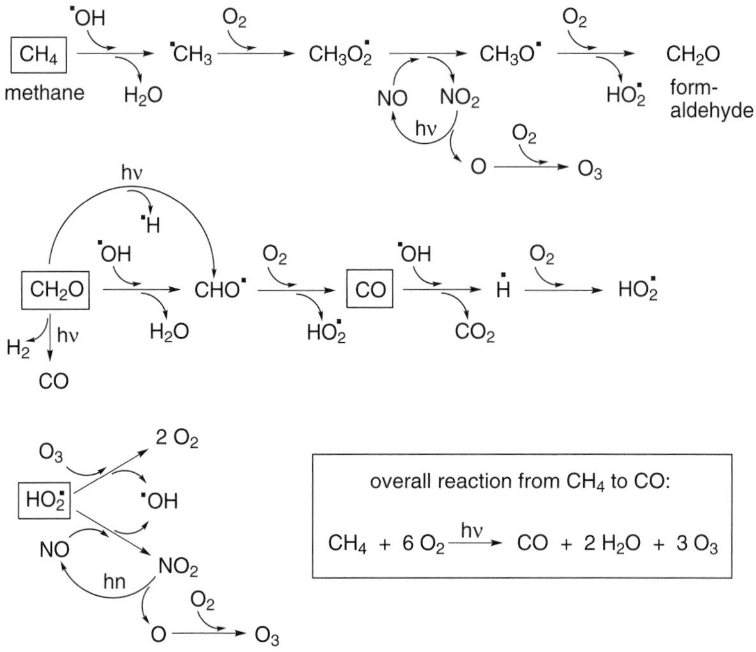

Figure 10.3 Degradation of methane and carbon monoxide in the troposphere.

nitric oxide and an oxygen atom, which forms ozone with molecular oxygen. Normally, the ozone formed by this photochemical reaction will be consumed by the oxidation of nitric oxide to nitrogen dioxide, but in the presence of oxygen radicals, for example, superoxide or alkylperoxy radicals, nitrogen dioxide will be formed anyway and there is a net formation of ozone. Ozone is the most important photochemical oxidant, but also for example peroxyacetyl nitrate (PAN), peroxypropionyl nitrate (PPN) and peroxybenzoyl nitrate (PBN) (see Figure 10.4), formed by oxidation of the corresponding aldehydes as shown in Figure 10.4, are important (although their concentrations are much lower).

Ozone is a strong oxidizing agent of which most (90%) of the total amount in the atmosphere is present in the stratosphere, where it plays an important role in the absorption of ultraviolet radiation from the sun (the subject of the discussion in Section 10.4). The remaining 10% is found in the troposphere, where it is more of a nuisance, but most unfortunately the good ozone (in the stratosphere) is decreasing while the bad (in the troposphere) is increasing (by approximately 0.7% per year in the lower layer of air). The effects of photochemical smog are essentially due to ozone, and this smog can be highly irritating, especially for individuals suffering from bronchitis or asthma, although it is not as acutely lethal as the London smog. Plants are affected by ozone, and many polymers, as well as

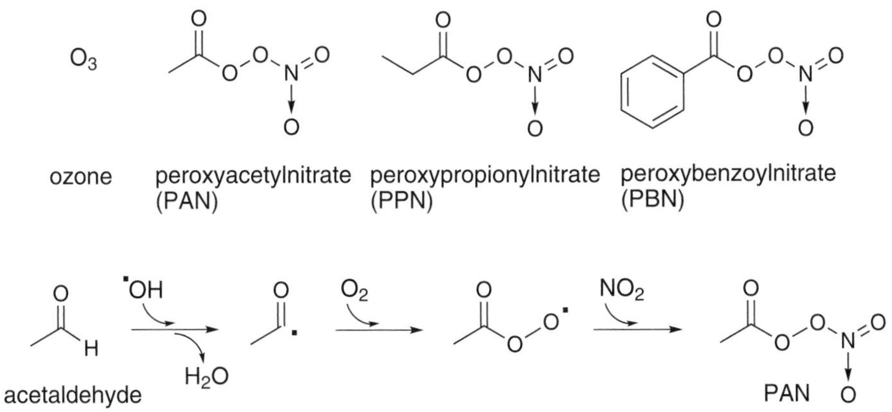

Figure 10.4 Major photochemical oxidants and the formation of PAN.

natural materials, are attacked. In addition, ozone is a greenhouse gas (discussed in Section 10.5).

10.3
Acid Pollution

As well as the biotic production of volatile sulfur compounds (see Section 10.1), even bigger amounts (approximately 100 million tonnes per year) are produced by human activities. The main contribution comes in the form of sulfur dioxide from the burning of fossil fuels, especially coal, but the smelting of sulfide minerals, for example, is also an important source. The sulfuric acid eventually formed from the anthropogenous emissions adds to those formed from the natural emission of volatile sulfur compounds (all forms of sulfur, organic and inorganic, reduced and oxidized, will be transformed to sulfuric acid in the atmosphere). Although the formation of sulfuric acid from various sulfur species can take a number of routes, these routes all involve an initial formation of sulfur dioxide that is oxidized by reactive oxygen species (hydroxyl radical, ozone, superoxide ...) in either the gas phase or in water droplets to sulfur trioxide, which reacts with water to sulfuric acid. The sulfuric acid will be dissolved by the water in rain and clouds and increase the concentration of cloud condensation nuclei, and will eventually come down with the rain as acid rain. Rain in completely unpolluted areas will also be weakly acidic (pH 5.0–5.5), because carbon dioxide will be present, and this forms carbonic acid in contact with water. However, rain in polluted areas can have pH values down to 4. Because the sulfuric acid is formed quickly in the atmosphere and is washed out by rain, the acid rain will essentially come down close (within 1000 km) to the emission source, and this is consequently a typical regional

$$3 \ NO_2 \ + \ H_2O \longrightarrow 2 \ HNO_3 \ + \ NO$$

$$2 \ NO_2 \ + \ 0.5 \ O_2 \ + \ H_2O \longrightarrow 2 \ HNO_3$$

Figure 10.5   The formation of nitric acid from nitrogen dioxide.

problem. As well as sulfuric acid, which is responsible for 80–85% of the acidity in acid rain, nitric acid (from nitrogen oxides formed, for example, during combustion) and hydrochloric acid (from hydrogen chloride emitted, for example, by volcanoes) also contribute.

Nitric acid is formed from the nitrogen oxides NO and NO_2 (often called NO_x when they are together). NO is oxidized in the atmosphere by reactive forms of oxygen to NO_2, which, in the presence of water (and oxygen), generates nitric acid by the reactions shown in Figure 10.5.

The nitrogen oxides NO and NO_2, both of which are toxic gases that generate nitric acid in the tissues they contact, are formed naturally in the stratosphere [by the oxidation of nitrous oxide (N_2O), see Section 10.4], by lightning (see above), by oxidation of ammonia (NH_3) present in the atmosphere, and by microorganisms in the soil. However, the natural production is exceeded by the anthropogenic. The main anthropogenic sources are the burning of fossil fuels (traffic, heating, and generation of electricity), during which molecular nitrogen of the air is oxidized, and the burning of biological material which contains organic nitrogen. Minor contributions come from the conversion of fertilizers and urea by microorganisms. Although nitric oxide is quickly oxidized to nitrogen dioxide, this is rapidly degraded by photolysis (wavelength below 420 nm) to nitric oxide and oxygen (which may combine with molecular oxygen to form ozone), and the concentrations of NO and NO_2 in the atmosphere are therefore comparable. On average, the concentrations are 30–50 ppb. In remote maritime areas far from land the concentrations are very low (1 ppt), but in large cities with heavy traffic values around 1 ppm have been reported.

The effects of the increased acidity due to the acidic gases (sulfur dioxide, the nitrogen oxides, and hydrogen chloride) are in some cases obvious and in others a matter for discussion. Some ecosystems are more sensitive than others, for example, lakes in areas where the pH buffer capacity is limited due to low amounts of dissolvable carbonates in the ground. Such lakes will rapidly have their flora and fauna changed, and especially fish will be among the first living things to disappear. The direct effects of acid rain on forests are not very clear. The 'Waldsterben' that was believed to be wiping out forests and received much attention during the 1980s turned out to be an exaggeration. It is clear that a low pH will affect plants directly and even kill them if it is low enough, as shown by the devastating effects on the forests in the vicinity of industrial complexes in former Eastern European countries, but the effects observed in, for example, western Germany are not so clearly correlated with emission of acid gases. It is a complex situation where several effects (e.g., eutrophication, acid rain, other pollution, soil quality, parasites, etc.) have to be considered together. Perhaps most important are the effects on the soil, from which the calcium and potassium ions, which are crucial for plants, are leached by acid rain.

An economically important effect of acid rain is the damage it causes to materials used in buildings, for example, metals, polymers, protective paint, and so forth. This costs society enormous sums every year, money that could have been used on more constructive projects.

10.4
Depletion of Stratospheric Ozone

Life on earth is sensitive to UV radiation and depends on the protection that the stratospheric ozone affords. Ozone is a reactive and toxic gas whose formation and effects in the troposhere have been discussed in Section 9.1.10, but it is at the same time the only compound in the stratosphere that absorbs UV radiation with wavelengths between 200 and 300 nm. In the stratosphere it is formed from molecular oxygen, which is continuously degraded by photolysis (wavelength <242 nm) to two oxygen atoms that either combine back to molecular oxygen or react with molecular oxygen to form ozone (see Figure 10.6). Ozone is also photolyzed to molecular oxygen and an oxygen atom, and can in addition react with an oxygen atom to form two oxygen molecules. There is consequently an equilibrium between molecular oxygen and ozone in the stratosphere, and this, under normal conditions, maintains a steady concentration of protective ozone.

The concentration of ozone is not affected by chemicals in the stratosphere that react with ozone and are destroyed. Instead, the species that seriously disturb the equilibrium between molecular oxygen and ozone act as catalysts and establish new equilibria. Another criterion is that such chemicals, or their immediate precursors, must be fairly long-lived in order to be able to reach the stratosphere. The chlorine atom and nitric oxide are examples of species that will degrade ozone catalytically, as shown in Figure 10.7.

In addition, other species are also involved, an example being the hydroxyl radical. These three, the chlorine atom, nitric oxide, and the hydroxyl radical are active at different heights. The concentration of ozone is highest at approximately 30 km above sea level, and this is also where nitric oxide is responsible for around 70% of the degradation of ozone. The NO/NO_2 cycle is consequently the most

$$O_2 \xrightarrow{h\nu < 242\ nm} 2\ O$$

$$O + O_2 \longrightarrow O_3$$

$$O_3 \xrightarrow{h\nu < 1180\ nm} O + O_2$$

$$O + O_3 \longrightarrow 2\ O_2$$

Figure 10.6 The natural formation and degradation of ozone in the stratosphere.

$$O_3 \longrightarrow O + O_2 \qquad\qquad O_3 \longrightarrow O + O_2$$

$$Cl + O_3 \longrightarrow ClO + O_2 \qquad NO + O_3 \longrightarrow NO_2 + O_2$$

$$ClO + O \longrightarrow Cl + O_2 \qquad NO_2 + O \longrightarrow NO + O_2$$

$$\text{overall: } 2\,O_3 \longrightarrow 3\,O_2 \qquad \text{overall: } 2\,O_3 \longrightarrow 3\,O_2$$

$$\text{byreaction: } \quad ClO + NO \longrightarrow Cl + NO_2$$

Figure 10.7 The catalytic degradation of ozone by chlorine atoms and nitric oxide.

$$N_2O \xrightarrow{h\nu} N_2 + O$$

$$N_2O + O \longrightarrow 2\,NO$$

$$N_2O + O \longrightarrow N_2 + O_2$$

$$CCl_2F_2 \xrightarrow[\text{(l < 220 nm)}]{h\nu} CF_2 + 2\,Cl$$

R 12

Figure 10.8 The generation of nitric oxide and chlorine atoms in the stratosphere.

important for the degradation of ozone. The chlorine atom contributes most to the degradation at 45 km, while the hydroxyl radical is most important at the extreme limits (10 and 70 km above sea level) of the stratosphere.

NO/NO$_2$ formed at sea level (see Figure 10.7) is too unstable to be transported to the stratosphere, but two major sources of stratospheric nitric oxide are the exhausts of airplanes and nitrous oxide (N$_2$O). Nitrous oxide is a nontoxic and (in the troposphere) very stable (half-life 150 years) gas that is used as an anesthetic (laughing gas). It is produced naturally by organisms in both the soil and the oceans, but is also formed in substantial amounts by the degradation of fertilizers and by burning fossil fuels. It can be transported to the stratosphere, where it will generate nitric oxide (see Figure 10.8). (Another major environmental effect of nitrous oxide is its contribution to the anthropogenic greenhouse effect, and this is discussed in the next section.) Chlorine atoms (and bromine atoms, which are even more potent ozone degrading species) are mainly generated by photolysis of fully halogenated hydrocarbons that are stable enough to be transported to the stratosphere. In the form of chlorofluorocarbons (CFCs) and halons (containing bromine), they have been used in enormous quantities during the last decades as propellants, solvents, refrigerants, in fire-extinguishers, and as blowing agents for plastic foams. A major advantage of CFCs when they were introduced was their very low toxicity, due to their stability, especially compared to the chemicals that they replaced (e.g., ammonia in refrigerators).

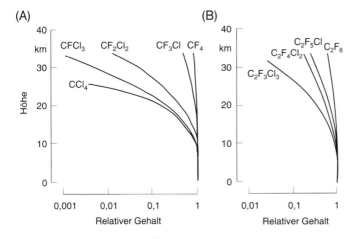

Figure 10.9 The proportions of halogenated methanes and ethanes that reach various altitudes of the atmosphere.

The CFCs are named 'Rxyz', where R stands for refrigerant, x the number of carbons – 1, y the number of hydrogens + 1, and z the number of fluorine atoms. R 11 is $CFCl_3$ (it should really be R 011), R 12 is CF_2Cl_2, R 13 is CF_3Cl, and R 14 is CF_4. The halons have a similar nomenclature, halon xyzst, where x gives the number of carbons, y the number of fluorines, z the chlorines, s the bromines, and t the iodines. Halon 1211 is consequently CF_2ClBr, while halon 2402 is $C_2F_4Br_2$.

Fully halogenated hydrocarbons are in general long-lived (half-life > 100 years in the atmosphere), and many of them are efficiently transported to the stratosphere (see Figure 10.9). As can be seen, there is a strong tendency to improved stability the more fluorine they contain. While carbon tetrachloride, with a half-life of approximately 50 years, will not reach an altitude of 30 km in significant amounts, almost all the carbon tetrafluoride emitted (half-life in the atmosphere 50 000 years) will end up in the stratosphere.

The halons are less stable compared to the CFCs but degrade ozone more efficiently because they generate bromine atoms. Halon 1301 (CF_3Br), for example, has a half-life 3 times shorter that that of R 13 (CF_3Cl), but it degrades stratospheric ozone 10 times more efficiently (1 tonne halon 1310 emitted into the atmosphere degrades as much ozone as 10 tonnes R 13). In 1985, the CFCs were responsible for approximately 80% of the ozone degradation by halogenated hydrocarbons, while halons and chlorinated hydrocarbons were responsible for approximately 10% each. The compounds that have been developed as direct substitutes for the CFCs and halons are not fully halogenated, and the presence of a hydrogen–carbon bond makes them more prone to oxidation in the atmosphere (and to attack by organisms which make them more toxic!). In addition, saturated hydrocarbons such as pentane and cyclopentane have replaced the CFCs as blowing agents for plastic foams.

The use of CFCs and halons has been restricted since their effect on the stratospheric ozone became generally known, and the production and emission of these chemicals in the industrialized countries have dropped from approximately 600 000 tonnes R 11 and R 12 per year in 1985 to close to zero. However, it will be a long time before the amounts already emitted become degraded and disappear from the atmosphere, and we will therefore have to live with the higher levels of UV radiation let through by the lower concentrations of stratospheric ozone for another couple of generations.

10.5
Greenhouse Effect

The energy that the earth receives from the sun is mainly dissipated back into space by thermal radiation, but the atmosphere reflects some of this radiation back to the earth. Thanks to this, the temperature at the surface of the earth is, on average, approximately more than 30 °C higher than it would have been without the atmosphere. The gases in the atmosphere, for example, water vapor and carbon dioxide, act like the panes in a greenhouse, letting the radiation from the sun pass through but absorbing the infrared radiation (heat radiation) from the earth and then emitting it in all directions including back to earth (see also Figure 10.10). Water is the most important greenhouse gas, responsible for over 60% of the natural greenhouse effect, but its concentration in the atmosphere is not changing significantly due to human activities.

While the natural greenhouse effect is a prerequisite for life on earth in its present form, it is not difficult to imagine that the balances that have been established between the incoming radiation from the sun, the chemical and biological processes that affect the atmosphere, and the heat radiation from earth

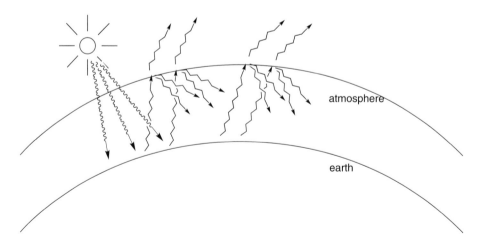

Figure 10.10 Heat radiation from earth without (left) and with (right) an atmosphere.

are sensitive. Changes in the sun's activity, for example, or the composition of the atmosphere, can be imagined to have an significant impact on the temperature on earth, leading to changes in, for example, the weather systems that could have a dramatic impact on life on earth. However, although the greenhouse effect is a reality that we know exists, we know relatively little about how sensitive the systems regulating the greenhouse effect are and exactly what would be the results of a temperature increase of, for example, 1 °C. The weather and the climate are constantly changing, and if we look back a couple of centuries it is evident that the earth has experienced both cold and warm periods that have no apparent connection with changes in the atmosphere. What has been happening over the last decades is that the emission of greenhouse gases, for example carbon dioxide, methane, nitrous oxide, ozone, and the freons, has increased so much that an anthropogenic greenhouse effect is possible and even to be expected. The greenhouse gases differ in several ways, for example, in which frequencies of infrared radiation they absorb (see Figure 10.11), and hence in their respective abilities to contribute to the greenhouse effect.

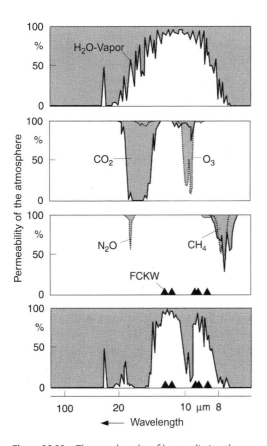

Figure 10.11 The wavelengths of heat radiation that are absorbed by the greenhouse gases.

Carbon dioxide is the end product from carbon compounds after the metabolic processes of humans and many other organisms, and on average every person emits approximately 700 g carbon dioxide per day. It is not really toxic, although accidents are not uncommon, as in some instances it may be present in such high amounts that it supersedes the oxygen necessary for breathing (carbon dioxide is a heavier gas than oxygen), and concentrations exceeding 10% for a long time may be dangerous. On the other hand, carbon dioxide is desirable for plants, which grow stronger and faster in the presence of higher concentrations of it. Carbon dioxide exists as a gas in the atmosphere, as a solute in the hydrosphere, as carbonic acid (hydrogen carbonate) and carbonate ions also in the hydrosphere, and as various solid carbonates in the lithosphere and the pedosphere. The hydrosphere contains approximately 50 times more than the atmosphere.

Carbon dioxide is a relatively poor greenhouse gas, but the amounts present in the atmosphere compensate for this weakness, and it is responsible for approximately 22% of the existing greenhouse effect and 50% of the increase in this effect presently observed. Almost one thousand billion tonnes of carbon dioxide are emitted each year, approximately 50% from biological processes, almost 50% from the oceans, and a few percent from the use of fossil fuels. Most of the anthropogenic carbon dioxide is taken up by the oceans and by organisms, but a significant part (approximately 40%) stays in the atmosphere and is responsible for the annual increase of carbon dioxide of about 0.3% that has lately been registered. Since the eighteenth century, the concentration of carbon dioxide has increased by almost 30%, and the increase has accelerated dramatically during recent decades. A rather frightening picture emerges if the concentration of carbon dioxide (and the other major greenhouse gas methane) in the atmosphere is compared with variations in the temperature over the last 160 000 years (data obtained from drill samples obtained from the permanent ice of Antarctica, see Figure 10.12). There is obviously a strong correlation, although it is impossible to say whether the concentrations of carbon dioxide and methane determine the temperature or *vice versa*. (It can be argued that an increase in temperature will increase the biological processes and thereby increase the amounts of greenhouse gases, especially methane.) The concentrations of the two gases have in recent decades increased to levels that are unsurpassed, and if the small increase in temperature that has been observed (approximately 0.7 °C in 100 years) is in fact due to the increasing concentrations of the greenhouse gases we may be in big trouble.

The concentration of methane in the atmosphere is increasing even faster (approximately 1% per year) than that of carbon dioxide, and although the total contribution of methane to the greenhouse effect is small today (a few percent) it is contributing approximately 13% of the present increase. As mentioned in Section 10.1, anaerobic bacteria that produce methane as an end product of their metabolism and degrade organic matter (especially in wetlands, rice paddies, and landfills, in the fore-stomachs of ruminant animals, and in termites) are the major source of methane, while the use of fossil fuels (especially natural gas) also makes a significant contribution. The degradation of methane in the atmosphere is

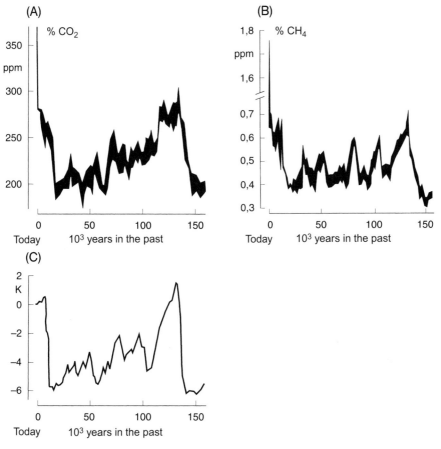

Figure 10.12 The relationship between concentrations of carbon dioxide (A) and methane (B) in the atmosphere and the variation in temperature (C).

almost exclusively due to its oxidation by the hydroxyl radical and molecular oxygen in the troposphere (see Figure 10.3). Less than 10% is transported to the stratosphere, where it is degraded photochemically. Methane is only one of many volatile hydrocarbons that are released in large quantities into the atmosphere. All hydrocarbons are potential greenhouse gases, and methane is the least efficient of these (though approximately 20 times more efficient than carbon dioxide), but, in contrast to all other hydrocarbons, methane is relatively stable, with a half-life of approximately 4 years.

Ozone, perhaps primarily interesting for its role in photochemical oxidation but also a highly efficient greenhouse gas (approximately 2000 times more efficient than carbon dioxide), contributes approximately 5% to the supplementary greenhouse effect due to the activities of man (i.e. in addition to the natural greenhouse effect).

Nitrous oxide is around 200 times more efficient as a greenhouse gas than carbon dioxide. It is also involved in the depletion of the ozone layer in the stratosphere and has been discussed in Section 10.4. Its contribution to the supplementary greenhouse effect due to the activities of man is also approximately 5%.

CFCs (e.g., R11 and R12) are extremely potent greenhouse gases, more than 10000 times more efficient than carbon dioxide, in addition to their effect on the stratospheric ozone. Their present contribution is relatively low, a few percent, but their concentration has increased rapidly during the last three decades of the twentieth century (approximately 5% per year). As discussed in Section 10.4, the use of CFCs has now been restricted, and it is expected that the increase in their concentration will slowly turn into a decrease.

The global warming due to the greenhouse effect is to some extent balanced by the cooling caused by, for example, the emission of sulfur dioxide from both natural and anthropogenous sources (see above). Many other factors are also studied and discussed, for example, the fact that the increase in temperature leads to increased formation of clouds that reflect the sun's radiation, and so forth, and it is still not possible to draw any conclusions about what will happen.

10.6
Pollution by Halogenated Hydrocarbons

Halogenated, in particular chlorinated, hydrocarbons are, in the minds of many people, generally toxic and environmentally unfriendly. It is true that many of the accidents and scandals involving chemicals during recent decades have been caused by halogenated compounds, but it is the ways in which we have used these chemicals that have led to the undesired effects, not the chemicals themselves. Lately we have become aware that many natural products are halogenated, in most cases chlorinated, and chlorine is actually an essential element for life (an adult human contains approximately 100 g chlorine). A number of other halogenated hydrocarbons that are also hazardous to the environment are discussed in Chapter 9.

10.6.1
Dioxins

Polychlorinated dibenzodioxins (PCDDs) and polychlorinated dibenzofurans (PCDFs) are often called dioxins as a group, and the same name is used for the most toxic member 2,3,7,8-tetrachlorodibenzodioxin (TCDD, dioxin). They are formed in very low amounts (typically nanograms per m³ exhaust) as an impurity when organic material containing chlorine (organic or inorganic) is incompletely burned in conventional waste dump combustion plants. Because of the chemical stability of PCDDs and PCDFs, the temperature during combustion has to attain

polychlorinated dibenzodioxins
(PCDDs)

polychlorinated dibenzofurans
(PCDFs)

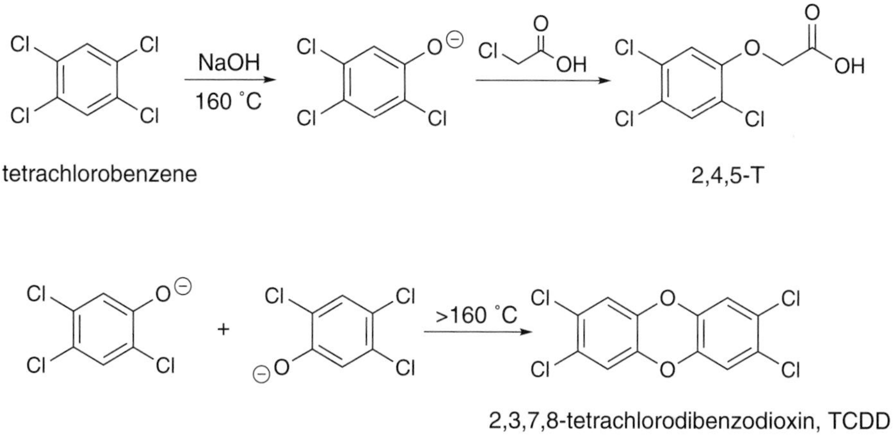

tetrachlorobenzene

2,4,5-T

2,3,7,8-tetrachlorodibenzodioxin, TCDD

Figure 10.13   The formation of TCDD from 2,4,5-trichlorophenol.

1000 °C in order to degrade them as well, and such temperatures are not normally employed. Another source has been the bleaching of paper pulp in the paper industry when this was carried out using chlorine as an oxidant, but new processes that make use of other oxidants have been developed. A third source, which used to be the most important and which focused attention on the dioxins, is the formation of dioxins as a by-product during chemical reactions with chlorophenols. An example is 2,4,5-trichlorophenoxyacetic acid (2,4,5-T), a chlorophenol derivative that has been produced in large quantities and used as a herbicide (a component of the notorious Agent Orange). 2,4,5-T is formed by the reaction of 2,4,5-trichlorophenol (formed from the reaction between 1,2,4,5-tetrachlorobenzene with sodium hydroxide in an alcoholic solvent at 160 °C) with chloroacetic acid (see Figure 10.13). TCDD is a by-product of this process, as can be seen in Figure 10.13, and Agent Orange contained approximately 10 ppm TCDD. Later it was discovered that the formation of TCDD depends on the temperature and the concentration of the trichlorophenoxide anion, and if the reaction was carried out under conditions when these two factors were controlled the amounts of TCDD in the final product was less than 0.1 ppm. The accidental contamination of the surroundings of Seveso in Italy 1976 with TCDD was caused by a chemical reactor in which this reaction was carried out. For some reason it was overheated, TCDD was formed in significant amounts, and when the pressure became to high the contents of the reactor, a few hundred kg of 2,4,5-trichlorophenol and a few kg of TCDD, were blown into the air via a pressure relief valve.

The toxicity of TCDD has already been discussed. It is highly toxic to certain mammals (the ground in the contaminated area in Seveso contained enough TCDD to kill 50 000 guinea pigs per square meter!), although humans do not appear to be the most sensitive. A slight increase in the incidence of cancers in the population of Seveso was detected 20 years after the accident. The acute effects of TCDD are believed to be caused by metabolites, although these have not yet been identified, and it appears that the initial metabolic conversion involves epoxidation of the aromatic ring. However, TCDD is a very poor substrate for the metabolic systems of mammals, and the half-life in humans is in the order of 10 years. The risk of bioaccumulation of TCDD over the years must therefore be considered. Other PCDD derivatives containing more than 4 chlorines are less toxic, and the fully chlorinated octachlorodibenzodioxin is approximately 1000 times less potent than TCDD. Its carcinogenic activity is believed to be caused by its binding to the Ah receptor (Ah because it also binds other aromatic hydrocarbons), which in a very complicated and not completely understood way is involved in cell division.

10.6.2
PCB

In contrast to the dioxins, which were only formed as unwanted by-products, the polychlorinated biphenyls (PCBs) have been produced in large quantities for many years. As the name implies, they are formed by chlorinating biphenyl, a process that cannot be controlled very well. PCBs are therefore always mixtures of many isomers, and the commercial products were simply characterized by the relative amount of chlorine they contained. PCBs are chemically extremely stable, not volatile, and will not burn easily, but are still oily liquids at room temperature. They were used for many things, as coolants in electrical transformers and capacitors (they are good insulators), as hydraulic oils, as plasticizers for various polymers, to impregnate paper and wood, and so forth. The PCBs are less toxic than the more potent dioxins, although the mechanism for the toxicity of the two classes may well be the same. Instead, it is a combination of the amounts of the PCBs produced and emitted to the environment (in total far more than 1 000 000 tonnes) which, together with their chemical stability, make them relatively abundant, and their high lipophilicity, which makes them subject to bioaccumulation. In addition, they may be contaminated by polychlorinated dibenzofurans (PCDFs), as PCBs may be oxidized at high temperatures to PCDFs. This possibility also shows how important it is that products that contain PCBs are taken care of properly, and not simply burned together with other waste (Figure 10.14).

The toxic effects of the PCBs have been observed in species that are at the top of food chains of exposed species, for example sea birds that prey on fish (that eat other marine organisms, etc.). The banning of the PCBs in many countries in the 1970s has resulted in lower concentrations in the species affected, but it will be a long time before the problem has disappeared.

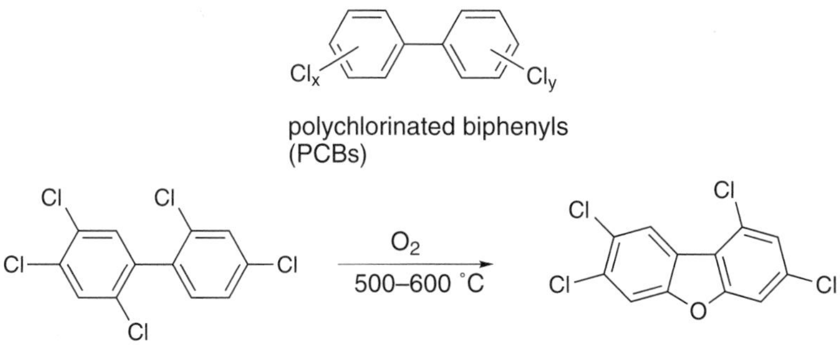

Figure 10.14 The general stucture of PCBs, and their transformation to PCDFs.

10.7
Pollution by Metals

Many metals and metalloids are seriously toxic to both humans and the environment, yet they are produced in large quantities and are commercially very important products. Metals are, of course, also a natural part of the environment, and toxic metals are, for example, emitted by volcanoes, forest fires, and sea spray (see above). However, these concentrations can be increased as a result of anthropogenic emissions, and it is not only the direct discharge from various activities but also secondary effects due to, for example, acid pollution (see Section 10.3) that are responsible. Most metals are mined, and the waste from mining and the refining of metals is a major source of pollution. Another major source is the combustion of fossil fuels (coal, oil, natural gas) and wood, which contain trace amounts of metals that are emitted into the atmosphere with the fumes.

Atmosphilic elements experience most of their mass transport via the atmosphere, while lithophilic elements mainly pass via rivers to the oceans. A characteristic of atmosphilic elements [for example antimony (Sb), arsenic (As), cadmium (Cd), copper (Cu), lead (Pb), mercury (Hg), molybdenum (Mo), selenium (Se), silver (Ag), tin (Sn), and zinc (Zn)] is that they exist in relatively volatile forms as they are, as oxides, or methylated in organometal derivatives, or they are emitted by industrial processes as particles in the atmosphere. In general, atmosphilic metals, which easily bind to bioorganic compounds, for example, to the –NH– or –SH groups of proteins, are the most dangerous to all organisms, in addition to being relatively mobile and readily absorbed by organisms. Such metals may be bioaccumulated in food chains and eventually give toxic effects in certain species.

The effects of metals depend on a number of factors, of which some are related to the organism intoxicated, for example, the ability to metabolize metals, and others on the form in which the metal is present. The oxidation state and the ligands are examples of such factors, and a metal ion is not as reactive as a 'free' ion (free metal ions in water are always hydrated) as it is as a complexed ion. Unicellular organisms, with a large interfacial surface area between themselves and their surroundings, will absorb metals, as well as other substances, in a different way from that of multicellular organisms such as humans, which have specialized organs for this. For obvious reasons, metals in the form of metal ions will not readily pass through cell membranes but will be mainly taken up by individual cells with the help of porter molecules (ligands) and carrier proteins. However, as discussed in Chapter 4, for most metals little is known in detail about the mechanisms of uptake.

Resistance in microorganisms may be the result of the complexing (protein complex, sulfide, etc.) of the metal whereby it is rendered immobile, either inside or outside the cell, or by active exclusion from the cell (e.g., by volatilization by methylation of Pb, Hg, and Sn.

In the following, a summary of the effects of a selection of the metals that are toxic and cause severe environmental effects is given, but it should be stressed that this in no way should be considered to be comprehensive. All metals discussed here are what we call heavy metals, having a density of at least $5\,g\,cm^{-3}$. Often one associates the term 'heavy metal' with 'toxic metal', but it is important to remember that other metals (having a density of less than $5\,g\,cm^{-3}$) can also be very hazardous.

10.7.1
Arsenic (As)

The earth's crust contains approximately 2 ppm arsenic. This is well known by everybody as the classic poison used in detective stories, although its importance has decreased with the development of more potent organic biocides. Arsenic exists in several forms, of which arsine (AsH_3, with the oxidation state −3), arsenic trioxide (As_2O_3, with the oxidation state +3) and arsenic pentoxide (As_2O_5, with the oxidation state +5) are important examples. The consumption of arsenic has decreased during the twentieth century, as arsenic compounds used as insecticides and to treat various illnesses in humans and animals have gradually been replaced by organic compounds, but it is still used for electronic applications and as a component of various alloys.

Arsenic is emitted into the environment from the burning of coal (coal contains approximately 20 g arsenic per tonne and brown coal up to 1.5 kg per tonne), from the smelting of metals (copper and nickel), and from volcanoes. The natural releases are approximately 8000 tonnes per year, which is a small amount compared with the releases caused by human activities. In the atmosphere, arsenic is transported over long distances before it precipitates, and it may then circulate many times in various forms through water, soil, and air before it is eventually bound in the sediments. Plants do not readily absorb arsenic compounds, except if arsenic-containing pesticides have been used for their cultivation or if they are grown in a contaminated area, and the consumption of seafood (arsenic is strongly bioaccumulated by algae) is the main source of the arsenic found in humans. In mammals it is absorbed well, by both the intestines and the lungs, and it is transported by the blood (bound to hemoglobin in the red blood cells) to the liver

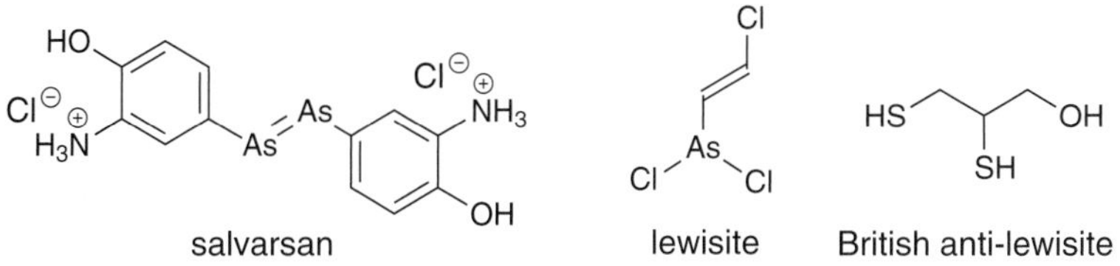

Figure 10.15 Some well-known arsenic compounds.

and the kidneys. In humans, some amounts will be deposited in bones, hair, nails, and skin, but approximately 70% is excreted (mainly with the urine). It is believed to disturb the biochemical reactions either by replacing phosphorus in the phosphate groups of the nucleotides (Section 2.3.6) or by reacting with thiol groups in enzymes, and arsenic is consequently toxic to all organisms. Also of concern is the carcinogenicity and teratogenicity of arsenic, although it is not the most potent of the heavy metals in this respect. Arsine and arsenic trichloride ($AsCl_3$) are considerably more toxic than the arsenic oxides, and are a serious problem in the industries where they are used (e.g., for the manufacture of electronic components), but because of their instability they do not survive long enough to be an environmental hazard. Like several other heavy metals, arsenic oxides may be reduced and methylated by microorganisms to organic forms that are more volatile and are bioaccumulated more efficiently. A few examples of historically important arsenic compounds are shown in Figure 10.15.

Salvarsan was developed (by the German chemist P. Ehrlich) in the beginning of the 19th century, and was successfully used to treat syphilis until the arrival of the penicillins in the 1940s. Lewisite was developed as a war gas (by the American chemist W. Lee Lewis). It has similar reactive and irritant properties to those of mustard gas, but shows in addition a long-term toxicity due to the presence of arsenic. British Anti-Lewisite (BAL) was developed as an antidote to lewisite, it chelates (forms a complex with its three heteroatoms) to arsenic (and other heavy metals) and renders it less toxic.

10.7.2
Cadmium (Cd)

The amount of cadmium in the earth's crust is approximately 0.1 ppm, and cadmium minerals (e.g., CdS, CdO, and $CdCO_3$) are most commonly found as a minor component in zinc minerals. Cadmium compounds have been used in large amounts as a stabilizer in PVC and as pigments. Lately these uses have been restricted with the intention of reducing the consumption of cadmium, but the exploding use of rechargeable batteries (with a nickel-cadmium cell) has instead increased the amounts of the metal in circulation. In addition, cadmium is used for surface treatment of metals (a thin layer of 10 μm on iron will prevent corrosion) and in alloys.

Approximately 8000 tonnes of cadmium is emitted into the environment per year, and almost all of this is due to human activities. Acid rain has also increased the amount in circulation, as the solubility and thereby the amounts absorbed by plants increase with decreasing pH (plants need zinc, and cadmium is absorbed by the same system). Cadmium has a very strong affinity for proteins, which will affect its absorption in mammals (cadmium in the food is not taken up efficiently in the intestines) and excretion (cadmium has perhaps the longest half-life of all chemicals in humans). Many factors will moderate the absorption of cadmium, as well as any other metal, in humans, and zinc deficiency is very important for the efficiency with which ingested cadmium is absorbed. In man, small amounts are deposited in the kidney bound to metallothionein (the presence of cadmium induces the biosynthesis of this protein), while larger amounts are stored in the liver bound to various proteins. It is toxic, causing lung edema if an aerosol containing cadmium is inhaled (5 mg cadmium per m^3 for 8 h may be fatal) or acute toxic effects on the gastrointestinal tract. The chronic effects are severe. For example, the kidneys are irreversibly damaged if the amount of cadmium reaches approximately 200 μg per g kidney tissue (in 'unexposed' individuals the kidney normally contain 20–40 μg/g), and the kidney dysfunction will eventually affect the metabolism of phosphorus and calcium and promote osteoporosis (the elimination of mineral constituents from bone). For several decades an area close to an abandoned lead/zinc mine in the Jinzu valley of Japan was so contaminated with cadmium that the people living there became ill (contaminated water was used to irrigate rice fields). This illness was called the itai-itai disease (itai is an expression for pain in Japanese, indicating how it was experienced by the afflicted) and led via damaged kidneys to porous bones that were deformed and easily collapsed. The low pH (approximately 5) of the soil in the Jinzu valley made cadmium stay in solution and facilitated its absorption by the rice; similar contaminations with cadmium in areas with soil containing calcium carbonate and having a pH above 7 did not lead to any noticeable effects. Although hundreds of people, of whom many died over the years, were afflicted in Japan it took a long time to establish the cause (and admit the guilt!) of the itai-itai disease. Cadmium is considered to be a carcinogen, although not a very potent one. It does not pass through placenta very efficiently and does not appear to be teratogenic.

10.7.3
Chromium (Cr)

Chromium is a relatively common metal, the earth's crust containing approximately 100 ppm. It exists in a number of different oxidation states, but only the oxidation states 0 (metallic chromium), +3 (e.g., chromic oxide, Cr_2O_3), +4 (e.g., chromium dioxide, CrO_2) and +6 (e.g., chromium trioxide, CrO_3) are of practical importance. Most of the chromium produced is, of course, used for the production of stainless steel, and chromium plating of various metals is another important application that consumes the metal. Chromic oxide is used as a (green) pigment and as a catalyst, chromium dioxide is magnetic and used in audio, video, and data storage tapes, and chromium trioxide is an important oxidizing agent in organic chemistry. Chromium is an essential chemical (Section 4.2.1) that is needed for the biosynthesis of the glucose tolerance factor, and the total absence of chromium in man results in diabetes. Together with cobalt and molybdenum, it is the chemical that (so far) we need the lowest amounts of: an adult contains approximately 5 mg chromium, which is around 600 000 atoms per cell.

While chromium in the oxidation state +3 is very poorly take up by cells, chromate anions (CrO_4^{2-}, oxidation state +6) are readily transported by phosphate–sulfate carrier, and chromium toxicity is closely linked to the higher oxidation state. However, inside the cells, chromium (+6) is reduced to chromium (+3) and thereby trapped, and it may well be the reduced form that causes the toxic effects on the molecular level. Chromium in the oxidation state +6 is toxic to all organisms, microorganisms, plants, and fish. In man it has been shown to be a potent allergen, causing a delayed-type hypersensitivity that results in eczemas, and it is also considered to be a carcinogen. Like other metals, chromium ions may bind to proteins and DNA, and the allergic effect is caused by the transformation, by chromium, of endogenous proteins to forms that are no longer recognized by the immune system.

10.7.4
Copper (Cu)

Copper, of which the earth's crust consists of 50 ppm, is found naturally both in its metallic form and as various compounds with the oxidation state +1 or +2. As it is easily available in its metallic form (either directly or from the ores), copper was one of the first metals to be exploited by man. It is hard yet formable, it is relatively resistant to corrosion, and it is an excellent conductor of electricity, and man has consequently found a number of uses for it. Together with other metals it forms valuable alloys such as brass (with zinc) and bronze (with tin), and it is an essential chemical (Section 2.3) of which an adult contains approximately 100 mg. It forms a part of several enzymes, for example, cytochrome oxidase and superoxide dismutase, and severe copper deficiency will consequently lead to, among other things, the blocking of cell respiration and the destruction of cells by superoxide.

In spite of the fact that copper is essential, it is a very toxic metal. This means that it must in principle be present in any organism but only within a narrow range of concentrations. In mammals, evolution has solved this dilemma in an elegant way, by making sure that excess copper is eliminated rapidly and that the copper that the cells make use of is tightly bound to specific copper proteins and not available for other reactions. As a result of this, we are less sensitive to intoxication by copper that one might fear. However, excess copper is highly toxic to microorganisms (copper sulfate, $CuSO_4$, and copper oxide, Cu_2O, have been used extensively as pesticides), and humans exposed to excessive copper in their diet will initially notice it from the effect that it has on the bacterial flora in the guts.

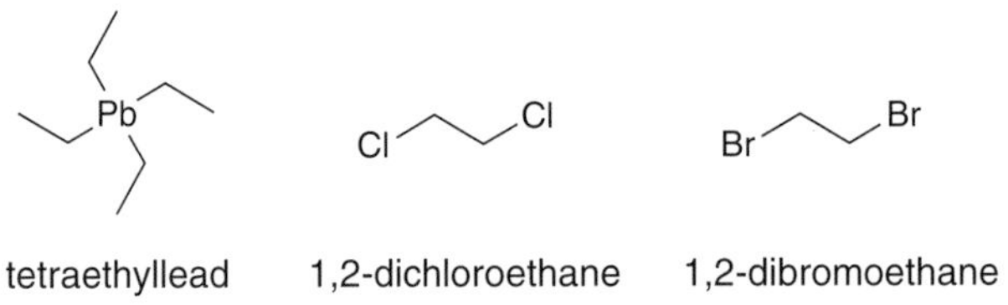

tetraethyllead   1,2-dichloroethane   1,2-dibromoethane

Figure 10.16 Important components in leaded petrol.

10.7.5
Lead (Pb)

Lead, 13 ppm of the earth's crust, is not found in nature in its metallic form, and its inorganic compounds are mainly in oxidation state +2 (e.g., lead oxide, PbO, and lead sulfide, PbS). The major use for metallic lead is in lead–acid batteries, the kind used in all normal cars. In addition, metallic lead is used for cable sheathing, in shot, and in alloys (solder, for example, contains approximately 50% lead). Inorganic lead compounds are used as pigments, as protective agents in paints, and in glassware. Organic lead (which is absorbed very efficiently), especially in the form of tetraethyllead, has been used in enormous quantities (hundreds of thousands of tonnes, often together with similar amounts of 1,2-dichloroethane and/or 1,2-dibromoethane) as antiknock agents in petrol (see Figure 10.16).

During the combustion process, tetraethyllead is heated and dissociates to a lead atom (which is oxidized to lead oxide) and four ethyl radicals. These will quench any radicals formed too early (before the sparking plug has given the ignition spark) and thereby prevent self-ignition (knocking) at inappropriate times during the ignition cycle in a car engine. Lead oxide reacts with dichloroethane (or dibromoethane) to give volatile (inorganic) lead compounds that are emitted with the exhausts. In addition, the lead compounds formed in this process act as a lubricant for the valves of the engine.

However, with the introduction of catalytic converters in cars and our awareness of the hazards of emitting such large amounts of lead into the atmosphere, the use of tetraethyllead has very much decreased and will (hopefully) soon disappear. Instead, *tert*-butyl methyl ether, which is combusted to water and carbon dioxide, is used as an antiknock agent in petrol.

The first to experience lead poisoning were the Greeks, who realized that drinking acidic beverages from lead containers was harmful and should be avoided. The Romans did not learn from the Greeks, of course, and used lead for their water pipes (the word 'plumber' has its roots in the Latin name for lead, plumbum) and were consequently chronically exposed to lead. Although it has not been proven in any way, the hypothesis that this lead poisoning played an important role in the decline of the Roman empire has been put forward and is not completely inconceivable! In our bodies, lead is able to take the place of calcium in bone tissue, which by itself is not harmful but prevents its excretion. In mammals, lead gives rise to a multitude of toxic effects. Perhaps the most well-known effect of lead is on the biosynthesis of hemoglobin, resulting in the formation of zinc-porphyrin instead of iron-porphyrin (lead interferes with the enzyme ferrochelatase which inserts iron into the porphyrin). Because the spectroscopic properties of zinc-porphyrin are different from those of iron-porphyrin it is easy to measure and quantify. High exposure to lead will eventually lead to anemia because hemoglobin with zinc instead of iron will not be able to transport oxygen. In addition, considerably lower concentrations of lead appear to be associated with the degeneration of nerve cells of children, resulting in the impairment of growth, hearing, and mental development. Again, as far as we understand it, the molecular mechanism of lead is that it will bind to sulfur and nitrogen groups in proteins and other macromolecules. Lead poisoning can be treated by the administration of a chelator (such as BAL, see above) that will bind to and inactivate the metal.

The well-known use of lead to make ammunition, for example, for shotguns leads to the deaths of birds and animals not only by shooting but also by poisoning when they eat the lead shot that is spread around in significant quantities during the hunting season Especially the waterfowl and the animals that prey on waterfowl are the victims of this, and in several countries lead shot has been banned.

10.7.6
Mercury (Hg)

Mercury (0.08 ppm in the earth's crust) is mainly obtained from mined cinnabar (mercury sulfide, HgS) by roasting the ore at 600 °C and condensing the metallic mercury, which is a liquid that boils at 356 °C. At room temperature, mercury has a vapor pressure that corresponds to 14 mg mercury per m³ air, which is considerably higher that the concentrations tolerated at workplaces (typically 0.05 mg/m³) and is lethal after a couple of hours. The evaporation rate is not high, approximately 6 μg/h/cm² at 20 °C, but if rooms where mercury is handled are not properly ventilated they will soon become dangerous to stay in. In contrast to other metals, mercury is to some extent soluble in both water and organic solvents. It has a high surface tension, and its physical and chemical properties have made it useful for a number of technical applications (in for example thermometers, barometers, relays, and fluorescent tubes). The ability of mercury to form alloys (known as amalgams) with other metals has been used in metallurgy, in chemical processes (e.g., the production of chlorine) and for dental amalgams (composed of mercury, silver, tin, and small amounts of copper or zinc). Organic mercury compounds were prepared at an early date. Phenyl mercury chloride, for example, was launched as a fungicide in 1915, although the use of this and other similar mercury-containing pesticides (alkyl-, alkoxy- and arylmercury compounds) did not peak until the 1960s.

In water organisms mercury is readily absorbed and bioconcentrated in food chains. As fish and shellfish is food for many different organisms, including humans, the pollution of waters with mercury can lead to catastrophes (as discussed below). In certain areas where lakes have been severely polluted, there are recommendations that pregnant women in particular should not eat the fish. We have already discussed the absorption of metallic mercury in humans, which is poor for liquid mercury in the intestines but efficient for mercury vapor in the lungs. Mercury salts, especially those with mercury in the oxidation state +2, are absorbed in the intestines, as, of course, are organic mercury compounds. The primary target organs for inorganic mercury poisoning are the CNS (chronic exposure) and the kidneys (acute and chronic exposure). Organic mercury will primarily affect the CNS, resulting in anything from hearing defects to tremors and memory loss, and poisonings of large numbers of individuals have been reported on several occasions. Many have been poisoned after consuming bread prepared from wheat that had been treated with an alkylmercury fungicide and that was intended to be used as seed for planting. The worst accident took place in Iraq in 1971 when approximately 60 000 people were exposed and over 2000 died. In the 1950s, a chemical company producing vinyl chloride and using mercury oxide as a catalyst discharged mercury wastes into the Minamata bay in Japan. The mercury was converted by microorganisms in the sediments (see below) into methylmercury, which was absorbed by the sea organisms and ended up in the shellfish and fish that were consumed by the local inhabitants. The result was a massive outbreak of methylmercury poisoning, the Minamata disease, afflicting in total more than 1000 people and killing approximately 100. The cause of the disease was soon identified, and in 1958 it became illegal to *sell* fish caught in the Minamata bay, but the inhabitants continued to catch fish for household use, which compounded the catastrophe. Interestingly, the chemical plant was not closed until 1968, and then for other reasons, and during the last 10 years the pollution of the Minamata bay continued as if nothing had happened.

Mercury is circulated more efficiently than other metals, and the amounts that are released as a result of human activities will 'hang around' for a long time. The major reason for this is that microorganisms, present everywhere and also in the sediments, are able to methylate inorganic mercury to monomethyl- and dimethylmercury as well as convert organic mercury compounds to atomic mercury. For the microorganisms, such conversions can be regarded as an excretion mechanism because the result is that they get rid of mercury, but for other organisms it will mean that mercury is mobilized. Monomethylmercury, CH_3Hg^+, will be absorbed by organisms, for example, algae and fish, while dimethylmercury, $(CH_3)_2Hg$, and atomic mercury are volatile and will be transported to the atmosphere. In this way mercury can be transported long distances before it is brought into its original form or an oxidized form.

10.7.7
Tin (Sn)

Tin (2–3 ppm in the earth's crust) is an essential chemical for humans, being a part of the hormone gastrine, which is released from the stomach into the bloodstream after a meal. If it is absent from our bodies we will suffer from impairment of digestion and growth. Tin compounds have many different uses in modern society, but especially interesting are the organotin compounds that are used, for example, as biocides in various situations. Tributyltin oxide is used as an antifouling agent in marine paints (to prevent algae and other sea-living organisms from attaching themselves to the hull of a boat), as a preservative in cooling liquids, paint and wood, and as a molluscicide. Other organotin derivatives are used as catalysts and stabilizers for plastics, but the trialkyltin derivatives are significantly toxic (trimethyl- and triethyltin oxide are considerably more toxic than tributyltin oxide). Tributyltin oxide, which is used in large amounts, is toxic to a number of algae, molluscs, and fish. In mammals they interfere with the oxidative phosphorylation in the mitochondria as well as giving rise to cerebral edema.

Glossary

Abiotic	Devoid of life.
Abortion	Expulsion of an embryo or a fetus from the uterus before it is viable.
Acetyl coenzyme A	A metabolic intermediate that transfers acetyl groups to the citric acid cycle and other metabolic pathways.
Acetyl group	Derived from acetic acid by the removal of the hydroxyl group in the carboxylic acid functionality, $CH_3(C=O)-$.
Acetylcholine	A nervous system transmitter that is inactivated by ester hydrolysis by the enzyme acetylcholine esterase.
Achiral	Not superposable on its mirror image.
Acid rain	Rain with a pH below 5.0, formed because the atmosphere is polluted by anthropogenic acids.
Acidosis	The condition in which the pH of the blood is lower than normal.
Activation energy	The energy necessary to make a chemical reaction pass through the state with the highest energy (transition state) and proceed to the formation of products.
Active site	The region of an enzyme or other protein to which the substrate (ligand) binds (before it, for example, is converted), also called the binding site.
Active transport	A protein-assisted transport system that transports molecules across a membrane from a region with low concentration to a region with high concentration at the expense of energy in the form of ATP.
Acyl group	A substituent with the general formula $R(C=O)-$, derived from the corresponding

Chemistry, Health, and Environment. Olov Sterner
© 2010 WILEY-VCH Verlag GmbH & Co. KGaA, Weinheim
ISBN: 978-3-527-32582-5

	carboxylic acid by removal of the hydroxyl group.
Addition reaction	Any reaction in which a molecule or parts of two molecules add to a double or triple bond.
Adipose	Fatty. The adipose tissue is composed of fat-storing cells and holds most of the body fat.
ADP	See ATP.
Aerobic	In the presence of molecular oxygen.
Affinity	The degree of attraction between the binding site of an enzyme or a receptor and a ligand.
Agonist	A ligand that binds to and stimulates a receptor.
Albumin	A protein found in relatively high concentrations in the blood and an important component during the transport of chemicals in the blood.
Alkaloid	A class of basic nitrogen-containing natural products that are produced by plants, many of which have pharmacological activity.
Alkalosis	The condition in which the pH of the blood is higher than normal.
Alkane	A saturated hydrocarbon, also called an aliphatic hydrocarbon.
Alkene	A hydrocarbon containing one or more carbon–carbon double bonds.
Alkoxy group	An R–O– substituent that is derived from an aliphatic alcohol by removal of the hydroxyl group hydrogen.
Alkyl group	A substituent that is derived from an alkane by removal of a hydrogen atom.
Alkyne	A hydrocarbon containing one or more carbon–carbon triple bonds.
Allergenic	The ability of something, for example, a chemical, to provoke an allergic response by the immune system.
Allylic position	An sp^3 carbon attached to a carbon–carbon double bond.
Alveolus	A thin-walled air-filled bubble in the lungs where the exchange of molecular oxygen and carbon dioxide between the air and the blood takes place.
Ames test	A quick, cheap, and reasonably accurate mutagenicity assay that detects the ability of

a compound to provoke point mutations in the bacterium *Salmonella thyphimurium*.

Amidases	Enzymes that hydrolyze carboxylic acid amides to the corresponding carboxylic acid and an amine.
Amino acids	The building blocks of proteins. Human proteins contain 20 different amino acids, although many more are known.
Amphipathic compound	A compound whose molecules have both a polar or ionized part and a nonpolar part that preferentially take a position at surfaces and lowers the surface tension. In larger amounts they may form, for example, lipid bilayers.
Anaerobic	In the absence of molecular oxygen.
Anemia	A decreased concentration of hemoglobin in the blood.
Anion radical	A negatively charged ion that has an unpaired electron and therefore also will react as a radical.
Antagonist	A ligand that binds to the active site of a receptor without stimulating it, but blocks it from agonists.
Anthropogenic	Caused, created, or changed by the influence of human activities.
Antibiotic	Literally a compound that is 'against life', that is, a toxic compound, but today a term for compounds that are used to treat bacterial infections.
Antibody	A protein that binds to the antigen that induced its synthesis (by the B cells of the immune system).
Antigen	A foreign chemical structure of certain complexity (e.g., with a molecular weight exceeding 5000) that provokes a response by the immune system.
Arene	An aromatic hydrocarbon.
Aromaticity	A highly stabilizing electronic configuration occurring in rings of atoms with p orbitals, when the number of electrons is 2, 6, 10, 14, etc. Note that the number of p orbitals does not have to be the same as the number of electrons.
Artery	A larger, thick-walled vessel that carries blood away from the lungs and heart. In the

	tissues, it is divided into finer vessels, arterioles, that regulate the blood flow from arteries to the capillaries.
Aryl group	A substituent that is derived from an arene by removal of a hydrogen atom, often shown as Ar–.
Atmosphere	The air stratum surrounding the earth, consisting of the troposphere, the stratosphere, the mesosphere, and the thermosphere.
ATP	Adenosine triphosphate, the primary repository of energy that can be used in chemical reactions in cells. Energy is released by the hydrolysis of ATP to ADP (adenosine diphosphate) and P_i (inorganic phosphate).
B cells	Lymphocytes capable of becoming antibody-secreting plasma cells.
Base pair	Two nitrogenous bases in opposite strands in double-stranded DNA that pair by hydrogen bonding.
Benign tumor	A mass of tumor cells that divide more rapidly than normal cells. It only grows to a certain size and does not invade surrounding tissues.
Benzylic position	An sp^3 carbon attached to a benzene ring.
Bile	A yellowish fluid containing bile salts, lecithin, cholesterol, hemoglobin derivatives, and other end products of the metabolism, also certain metals that are being excreted.
Binding site	See active site.
Bioaccumulation	The term used for the mixture of bioconcentration and biomagnification in ecosystems.
Bioconcentration	The extraction by an organism of a chemical from the abiotic environment, resulting in a higher concentration inside the organism than outside.
Biodegradation	Degradation of compounds or materials by organisms, which either use them as a food source or co-metabolize them.
Biomacromolecules	Big biomolecules like proteins and DNA.
Biomagnification	The concentration of a chemical in a food chain.

Biomolecules	Compounds that take part in the normal biochemical reactions that take place in the majority of cells.
Biosphere	The parts of the world where living organisms are found.
Biotic	Consisting of organisms or the products of organisms.
Biotransformation	The metabolism of an exogenous compound in an organism.
Blood-brain barrier	A barrier that controls the kinds of compounds (and their rates of transfer) that enter the brain from the blood.
Botulin toxins	Extremely toxic proteins (seven have been isolated so far), with molecular weights between 200 000 and 400 000, produced by the bacterium *Clostridium botulinum*, which may infect food causing botulism. The toxins are heat sensitive and are destroyed if heated to 100 °C.
Bowman's capsule	The sac in the beginning of the kidney nephron that accepts the fluid produced in the glomerulus.
Bronchi	The parts of the air passage that enter the lungs and eventually are divided into fine tubes called bronchioles that reach the alveoli.
Capillary	The smallest blood vessel, surrounded by a single layer of endothelial cells through which many compounds can readily diffuse.
Carboanion	An organic anion in which a carbon atom has an unshared pair of electrons and a negative charge.
Carbocation	An organic cation in which a carbon atom is surrounded by only 6 valence electrons.
Carbonate group	Derived from carbonic acid, which is produced by the addition of water to carbon dioxide. Organic carbonates are esters of carbonic acid.
Carbonyl group	A functionality containing a carbon and an oxygen connected by a double bond, present in, for example, ketones, esters, and amides.
Carcinogenicity	The potency of something, for example, a chemical, to increase the number of tumors or to make the tumors appear at an earlier

stage compared to those that arise spontaneously.

Catalyst
A chemical or material that accelerates a chemical reaction by lowering the activation energy without undergoing any net chemical change itself.

Cell
The basic structural and functional unit into which an organism can be divided and still retain the characteristics that are associated with life.

CFCs
See freons.

Chemophobia
The abnormal fear of chemicals.

Chiral
Compounds are said to be chiral when their mirror image cannot be superimposed on the molecule itself. A chiral compound may exist as one of the two mirror images (enantiomers) or as a mixture of the two.

Chromatid
Half (one arm and one leg) of the 'chromosome man' when the genetic material is duplicated and condensed prior to a cell division. If the two chromatids of a chromosome are identical they are called sister chromatids.

Chromatin
The genetic material, consisting of DNA and proteins (histone and nonhistone proteins).

Chromosome
A threadlike structure containing the genes. Prokaryotic cells contain one chromosome, a DNA molecule, while eukaryotic cells contain several chromosomes (in a nucleus) which consist of DNA molecules complexed with proteins.

Chromosome aberration
A change in the number of chromosomes or the structure of a chromosome, also called chromosome abnormality or chromosome mutation.

Chyme
A solution of partly digested food in the stomach and the intestines.

Cilia
Hairlike projections from the surface of specialized epithelial cells that sweep back and forth in such a way as to propel material along the surface.

Citric acid cycle
A series of enzymatic conversions that convert acetic acid (as acetyl coenzyme A) to carbon dioxide and reduced cofactors, which

	later are used to produce ATP. Also called Krebs cycle.
CNS	Central nervous system: the brain plus the spinal cord.
Coenzyme	A small organic molecule that is associated with an enzyme and essential for its activity by serving as a carrier molecule which transfers atoms or small groups during the enzymatic conversion without being consumed itself.
Coenzyme A	A coenzyme essential for the enzymatic transfer of acetyl and other acyl groups.
Cofactor	A coenzyme or a metal ion that is important for the activity of an enzyme, for example, by maintaining the shape of an active site, without being consumed itself.
Colloids	Particles that are so small ($<1\,\mu m$) and therefore so light that they behave as solutes in a solution and do not precipitate as heavier particles would do.
Conformation	Any three-dimensional arrangement of the atoms in a molecule that is the result of rotation about single bonds.
Conjugated addition	Nucleophilic addition to the double or triple bond in alkenes or alkynes where the unsaturation is conjugated with an electron-withdrawing group (e.g., a carbonyl group); also called a Michael addition.
Conjugated unsaturations	Double or triple bonds that are separated by exactly one single bond.
Covalent bond	A chemical bond formed between two atoms that share one or more electron pairs.
Cytochromes	A group of iron-containing proteins that take part in redox reactions by transferring electrons.
Cytoplasm	The region in a cell that is outside the nucleus.
Cytosol	The water solution of a cell that is outside the organelles.
Cytotoxic	Toxic to cells. All chemicals will in principle be cytotoxic if the dose is high enough, but the term is used to describe potent toxicity.
Deamination	The removal of an amino group from a molecule.

Dehydrohalogenation	The removal of a hydrogen atom and a halogen atom from adjacent positions, a type of β-elimination.
Deletion	A chromosome aberration that is characterized by the loss of a chromosome segment, also called a deficiency.
Density	The specific weight: the weight of a certain volume of a chemical at a specified temperature and air pressure. Normally given in $kg\,m^{-3}$, $g\,cm^{-3}$ or $g\,mL^{-1}$.
Deoxyhemoglobin	Hemoglobin that is not combined with molecular oxygen.
Dielectric constant	Indicates the ability of a compound to be polarized by an electric field that is applied over it.
Differentiation	The process in which cells develop and acquire specialized structures and properties that enable them to perform special functions.
Diffusion	The transport of a chemical from one place in a solution or a gas to another due to random thermal molecular motion.
Dipole moment	The sum of the individual bond moments in a molecule, given in Debye units (D).
Dispersion forces	See van der Waals forces.
Disulfide group	A persulfide, with a sulfur–sulfur single bond (–S–S–).
Diuretic	A compound that causes an increase in the amount of urine excreted.
DNA	Deoxyribonucleic acid, a polymer of nucleotides in which the sugar is deoxyribose and the bases adenine, cytosine, guanine, or thymine. The carrier of genetic information.
Dominant	A gene, or its corresponding trait, is dominant if it is manifest in an organism that has different genes for the same trait (a heterozygote). A mutation is called dominant if the effects are immediately observable.
Dose-response relationships	The way in which the magnitude of a biological effect caused by an exogenous agent, for example, a chemical, on an individual or a group depends on the dose. Often assumed to be linear, but frequently found to be nonlinear.

Down's syndrome	A set of symptoms in humans caused by a chromosome aberration such that there is an extra copy of chromosome 21 in all cells.
Duplication	A chromosome aberration that is characterized by the presence of an additional copy of a chromosome segment in the same chromosome or in another.
Ecology	The study of the relationships between various organisms and their interaction with the physical environment.
Ecosystem	An ecological system – a limited area consisting of organisms and the environment they live in that is more or less self-supporting. A lake, a forest, or a greenhouse may be regarded as an ecosystem, although it is important to remember that there are no closed systems on earth.
Electromagnetic radiation	Radiation of various frequencies and energies, from gamma rays, X-rays, ultraviolet, visible and infrared light, to radio waves. It consists of waves with both an electrical and a magnetic component that can travel through matter.
Electron-withdrawing groups	Chemical functionalities that are conjugated with carbon–carbon double or triple bonds and by resonance withdraw electrons from the unsaturated carbon furthest away, also referred to as EWGs.
Electrophile	Any molecule or ion that can accept a pair of electrons from a nucleophile and form a new covalent bond.
Electrophilicity	The efficiency with which an electrophile reacts with various nucleophiles, compared to other electrophiles.
Elemental composition	The number and nature of the various atoms that constitute a molecule. The elemental composition of water is H_2O, and that of ethanol is C_2H_6O.
β-Elimination	An elimination of two parts from adjacent atoms in a molecule, which results in an unsaturation. The reverse of an addition to a double or triple bond.
Embryo	The first stage of the development of an organism: for a human, this occupies the first 8 weeks of intrauterine life.

Emulsion	A suspension of small (approximately 1 μm in diameter) lipid droplets in a water solution.
Enantiomers	A pair of achiral molecules that are mirror images, having identical chemical properties except for their interaction with other achiral molecules, for example, biomolecules such as proteins and DNA.
Endocrine disruptors	Compounds that have an effect on the endocrine system without being hormones themselves.
Endocytosis	A transport mechanism by which cells absorb solid or liquid material. The plasma membrane invaginates and is pinched off to form a small intracellular vesicle which is dissolved.
Endogenous	Having an origin from within the body.
Endonuclease	An enzyme that hydrolyzes internal phophodiester bonds in, for example, DNA.
Endoplasmic reticulum	A cell organelle located in the cytoplasm consisting of a network of membrane tubules, vesicles, and sacs interconnected with the nuclear envelope, where, for example, protein synthesis and metabolism of exogenous compounds take place.
Enterohepatic circulation	The recycling of chemicals that are excreted by the liver with the bile by their reabsorption (in most cases after conversion by the intestinal bacteria) in the intestines and transport with the blood back to the liver.
Epidemiological investigation	The comparison of the frequency of an illness, for example, the number of cases of cancer in an group of individuals exposed to, for example, a chemical at the workplace with the frequency in a nonexposed but otherwise similar control group.
Essential chemicals	Chemicals that are required for normal and optimal body functions but are not produced in adequate amounts by the body itself and have to be supplied in the food.
Esterases	Enzymes that hydrolyze carboxylic acid esters to the corresponding carboxylic acids and alcohols. Some are able to hydrolyze thioesters to carboxylic acids and thiols, and

even amides, ureas, phosphoric esters, and thiophosphoric esters.

Eukaryote
A cell, or an organism with cells, that has a nucleus.

Eutrophication
The enrichment of organisms with nutrients, or growth-limiting factors, something that may disturb ecological equilibria and thereby affect ecosystems.

Excretion
The transport of chemicals present inside the body to the outside via, for example, the lungs (in the expired air), the liver (in the feces), or the kidneys (in the urine).

Exocytosis
A transport mechanism by which cells can discharge material to the extracellular fluid by fusing the membrane of intracellular vesicles with the plasma membrane.

Exogenous
Having an origin from outside the body.

Exonuclease
An enzyme that hydrolyzes terminal phosphodiester bonds in, for example, DNA.

Extracellular fluid
The water solution that surrounds our cells, comprising the interstitial fluid and the plasma and accounting for 30% of the water in the body.

Facilitated diffusion
Passive protein-assisted transport by proteins that recognize certain chemicals and help them to diffuse through a membrane.

FAD/FADH$_2$
Flavin adenine dinucleotide, an agent that takes part in biological oxidations and reductions. During oxidations FAD is reduced to FADH$_2$, and *vice versa*.

Ferric ion
Fe^{3+}.

Ferrous ion
Fe^{2+}.

Filtration
Passive protein-assisted transport through channel proteins that form transmembrane pores in the cell membranes.

First-pass effect
The effect whereby nutrients absorbed in the intestines, the organ intended for the uptake of all exogenous chemicals except molecular oxygen, pass through the liver, where they are modified before they are distributed to other organs by the blood.

Fetus
The later stage of the development of an organism, after the embryo: for a human

	this is the period between 8 weeks and birth.
Frameshift mutation	A point mutation caused by the deletion or insertion of
	one or several base pairs in DNA. It changes the reading frame for protein synthesis.
Freons	Chemically extremely stable derivatives of the smallest hydrocarbons such as methane and ethane in which one or several of the hydrogen atoms have been exchanged for chlorine and/or fluorine. Freons lacking hydrogens are called chlorofluorocarbons or CFCs
Fumigant	An airborne biocide in general, but often refers to pesticides which will disinfect the soil from various parasites, for example, nematodes.
Gamete	A germ cell that has matured to a reproductive cell capable of fusing with a similar cell of the opposite sex during fertilization, also called a sex cell.
Gastrointestinal tract	The mouth, esophagus, stomach, and small and large intestines.
Gene	A sequence of nucleotides in the genome that has a specific function, for example, coding for a protein or regulating the transcription of another gene.
Genome	The genetic content of a cell.
Genotoxicity	The potency of something, for example, a chemical, to damage the genetic material and thereby cause a toxic effect.
Germ cell	A cell line in animals from which the next generation of gametes will be derived.
Glomerular filtration	The transport of the essentially protein-free plasma from the blood to Bowman's capsule in the glomerulus.
Glomerulus	The unit where the blood is filtered in the nephron, intimately associated with Bowman's capsule.
Glycolipid	A plasma membrane component with an attached carbohydrate, situated so that the carbohydrate is at the extracellular surface.
Glycoprotein	A plasma protein with one or several carbohydrates attached, situated so that the

	carbohydrate(s) are at the extracellular surface.
Glycoside	A carbohydrate derivative in which the hemiacetal hydroxyl group has been replaced by another group, the so called aglycone, attached by a heteroatom. If the carbohydrate is glucose, glycosides are called glucosides.
Glycoside bond	A covalent bond between a carbohydrate (the acetal carbon) to the heteroatom in alcohols (forming *O*-glycosides), thiols (forming *S*-glycosides) and amines (forming *N*-glycosides).
Golgi apparatus	A membrane organelle that processes and secretes the newly synthesized proteins in the cell.
Greenhouse effect	The warming of the atmosphere due to the reflection of infrared radiation from the earth by gases such as water vapor, carbon dioxide, and CFCs. In a greenhouse, heat retention results from the corresponding effect of the glass panes reflecting infrared radiation, but also from the conservation of warm air.
GTP	Similar to ATP, but containing the base guanine instead of adenine.
Heme	An iron-containing group that is essential for the function of, for example, hemoglobin and the cytochromes.
Hemoglobin	A protein composed of four polypeptide chains with a heme group, located in the red blood cells and responsible for transporting molecular oxygen.
Halons	Chemically stable halogenated derivatives of the smallest hydrocarbons, which may contain all four halogens but are always brominated.
Heat capacity	The amount of energy required to increase the temperature of a given amount of a substance by 1 °C.
Heat of evaporation	The amount of energy required to convert a given amount of a liquid into its gaseous state.
Helper T cell	A type of T cell that aids and regulates B cells so that the production of antibodies is enhanced.

Hepatotoxic	Toxic to the liver.
Herbicide	A chemical that is toxic to plants and may have found a use as a weed-killer.
Histone proteins	Proteins with many basic amino acids that are complexed with DNA in the chromosomes of eukaryotes.
Hormone	An endogenous compound that is synthesized and secreted into the blood by special cells, and that has a strong effect on a physiological function in another part of the body.
Humification	The processes in which the products of the degradation of dead organisms in the soil react with each other and with products of microorganisms to form complex polymers that are of vital importance for the organisms living in the soil.
Hybrid orbital	An orbital formed by the combination of two or several atomic orbitals.
Hydrocarbon	A compound that may be saturated or unsaturated but is composed of only hydrogen and carbon atoms.
Hydrogen bond	A relatively strong dipole–dipole attractive force between hydrogen atoms attached to strongly electronegative atoms and other strongly electronegative atoms.
Hydrophilic	Loving water, from Greek.
Hydrophobic	Fearing water, from Greek.
Hydrophobic effect	The tendency of nonpolar molecules and groups to stay together in a separate phase and not be dissolved by water solutions.
Hydrosphere	The parts of the world that contain water: oceans, seas, lakes, rivers, floods, streams, and rills.
Hypersensitivity	An acquired sensitivity to an antigen that can result in a serious reaction if the same antigen is encountered again.
Inducer	An effector molecule that is responsible for the induction of enzyme synthesis.
Induction	The initiation or increase of enzyme synthesis in response to an environmental stimulus.
Inductive effect	The polarization of the electron density of a covalent bond because one end has more electronegative substituents. The inductive

effect will stabilize any negative charge present in the positive end of the dipole and any positive charge present in the negative end.

Inflammation
: A nonspecific response of the immune system to an injury or the presence of an antigen characterized by swelling, pain, and increased blood flow in the affected region.

Intermediate
: An intermediary product formed between successive reaction steps.

Interstitial fluid
: The extracellular fluid surrounding the cells of a tissue, excluding the blood plasma.

Inversion (mutation)
: A chromosome aberration that is characterized by the reversal of a chromosome segment.

Inversion (weather)
: A weather condition characterized by lower temperature of the air that is close to the ground compared to the air at higher altitudes. The cooler air will remain stagnant, and the inversion over a city will create problems as the concentrations of air pollutants increase.

Irreversible
: Proceeding essentially in one direction only.

Isomers
: Compounds with identical molecular formulae but with different chemical structures and therefore different properties.

Ketene group
: A carbon–carbon double bond in which one carbon forms an additional double bond, with an oxygen, and the other carbon binds two groups (e.g., hydrogen and/or alkyl groups), that is, $R_1R_2C=C=O$.

Larynx
: The part of the respiratory system that connects the pharynx and trachea and contains the vocal cords.

Leucocyte
: A white blood cell.

Ligand
: A chemical that by noncovalent forces will bind to a molecule, normally a protein.

Ligases
: A class of enzymes that catalyze reactions that combine two molecules.

Lipophilic
: Loving fat, from Greek.

Lipophobic
: Fearing fat, from Greek.

Lipoprotein
: A complex between fat and protein which may be rich in the former (low-density

	plasma protein) or the latter (high-density plasma protein).
Lithosphere	Stone and rock material of the earth's crust extending to approximately 100 km below the surface of the earth.
Lymph nodes	Small organs that are situated at the lymph vessels, where lymphocytes are formed and stored.
Lymphocyte	A cell type which mainly is responsible for the specific defense of the immune system against foreign matter, a type of leucocyte.
Lysosome	Cell organelle surrounded by a single membrane, which contains digestive enzymes that are used to break down materials (e.g., macromolecules and even bacteria) that have been absorbed by endocytosis, as well as damaged components of the cell itself.
Macrophage	A cell type that is part of the immune system and whose function is to take care of foreign matter by phagocytosis.
Malignant tumor	A mass of cancer cells that divide rapidly and invade surrounding tissue and, if not treated, will normally kill an organism.
Membrane	A structural barrier that surrounds cells and organelles composed of phospholipids and other amphipathic compounds as well as proteins. It regulates the flow of many chemicals through the cell and provides a framework for many enzymes and receptors.
Micro-, μ	A millionth part. One microliter (μL) is 0.000 001 liter and one micrometer (μm) is 0.000 001 m.
Microbes	Minute organisms, for example bacteria, fungi, and protozoa.
Microsomal	Associated with the smooth endoplasmic reticulum, especially of the liver cells.
Microtubules	Filaments that provide internal support for the cell, maintain and change the shape of cells, and participate in the movements of organelles and cell components.
Microvilli	Fingerlike projections which increase the absorptive surface of the epithelial cells of the small intestine and the kidney nephron.

Milli-, m	A thousandth part; one milliliter (mL) is 0.001 liter and one millimeter (mm) is 0.001 m.
Mineralized	Completely degraded to inorganic materials, for example, carbon dioxide, nitrate, and water, leaving no organic materials behind.
Mitochondria	The cell organelle that is responsible for the production of energy in the form of ATP.
Molecular weight	The weight (in grams) of one mole (6.02×10^{23} molecules) of a compound.
Mutation	A hereditary alteration in the nucleotide sequence of the DNA of a cell.
NAD/NADH	A coenzyme that facilitates the transfer of hydrogens in various reactions.
Nano-, n	A billionth part; one nanolitre (nL) is 0.000 000 001 liter and one nanometer (nm) is 0.000 000 001 m.
Nematicide	An agent that will kill nematodes, small worms (<1 mm long) that are parasites in many economically important plants.
Nephron	The functional unit of the kidney, composed of the glomerulus, Bowman's capsule, proximal tubule, Henle's loop, distal tubule, and collecting duct.
Nephrotoxic	Toxic to the kidneys.
Neuron	A nerve cell.
Neurotoxic	Toxic to the nervous system.
Neurotransmitter	A compound that is released by a nerve cell as a response to stimulation and that will act on other excitable cells.
N-Nitrosamide group	An amide with a nitroso group connected to the amide nitrogen by a nitrogen–nitrogen bond.
N-Nitrosamine group	An amine with a nitroso group connected to the amine nitrogen by a nitrogen–nitrogen bond.
Nonbonding electrons	See unshared electrons.
Nuclear envelope	A double membrane that surrounds the nucleus of a cell.
Nuclear pores	Openings in the nuclear envelope that facilitate the passing of RNA from the nucleus to the endoplasmic reticulum.
Nucleophile	A molecule or an ion that is able to donate an unshared electron pair to an electrophile

	in such a way that a new covalent bond is formed.
Nucleophilic addition	A reaction between a nucleophile and an electrophile that in some way can accommodate the additional electron pair, for example, resulting in the addition of the nucleophile to a double or triple bond in the electrophile.
Nucleophilic substitution	A reaction between a nucleophile and an electrophile that has a leaving group that is forced to leave the electrophile together with an electron pair.
Nucleophilicity	The efficiency with which a nucleophile reacts with various electrophiles, compared to that of other nucleophiles.
Nucleoside	A building block of nucleic acids, composed of a sugar (ribose or 2-deoxyribose) bonded to a purine or pyrimidine base by a β-N-glycoside bond.
Nucleotide	An ester between phosphoric acid and one of the sugar hydroxyl groups (position 3 or 5) of a nucleoside.
Nucleus	A membrane-enclosed organelle of eukaryotic cells that contains the chromosomes.
Octane number	The amount (in %) of isooctane (2,2,4-trimethylpentane) in a isooctane/heptane mixture that has the same knock properties as the petrol being evaluated.
Octet rule	The tendency of the atoms in the first rows of the periodic system to react in ways to achieve a filled outer shell of eight valence electrons.
Edema	The accumulation of excess fluid in the interstitial space.
Operator gene	A nucleotide sequence that is recognized by a specific repressor protein, which by binding to the operator inhibits the transcription of the associated genes.
Orbital	A region in space where an electron or a pair of electrons spends most (90–95%) of its time.
Organ	A collection of tissues that form a structural unit and collaborate to serve a common function.
Organ system	A collection of organs and tissues that collaborate to serve an overall function.

Organelle	Macroscopic cellular components that are separated from the rest of the cytoplasm and perform specialized functions.
Organogenesis	The period during gestation when the major organs of the body form.
Osmotic pressure	An important driving force for the transport of a chemical between solutions of different composition that are separated by, for example, a membrane.
Ovum	The female gamete.
Oxidative phosphorylation	A process that takes place in the mitochondria and that makes use of the energy released by the oxidation of nutrients to produce ATP from ADP and inorganic phosphate.
Oxyhemoglobin	Hemoglobin that is combined with molecular oxygen.
Pancreas	A gland, connected by a duct to the small intestine, which secretes hormones (e.g., insulin) as well as digestive enzymes and bicarbonate into the intestines.
Pedosphere	The top soil layer of the lithosphere, formed by the weathering of rocks and stone and the decay of dead organisms.
Peptide bond	A covalent bond between the amino group of one amino acid and the carboxylic group of another, with the elimination of water, to form an amide.
PNS	Peripheral nervous system: those parts of the nervous system that do not include the CNS.
Persistent	Long-lived: a compound that is very difficult for various organisms to degrade and that withstands sunlight and other chemicals.
pH	The measure of the acidity of a water solution. At pH 7 a solution is neutral, while it is acidic at lower values and basic at higher. The pH value corresponds to the negative logarithm (base 10) of the hydronium ion concentration.
Phagocyte	A leucocyte that attacks and destroys any foreign cell to which antibodies have bound to surface antigens.
Phagocytosis	A form of endocytosis in which larger particles (e.g., cells or parts of cells) are

	absorbed by a cell by the invagination of the plasma membrane.
Pharynx	The throat, the common passage for both air and food.
Phospholipid	A lipid with a diacylglycerol backbone to which a phosphate group plus a strongly polar or ionic group is attached – an important component of cell membranes.
Phosphorylation	The addition of a phosphate group to an organic molecule.
Pinocytosis	A form of endocytosis in which liquid is absorbed by a cell by the invagination of the plasma membrane.
Placenta	The border between fetal and maternal tissues in the uterus, where the exchange of nutrients and excretion products between the two circulations takes place.
Plasma	The blood minus cells.
Plasma cell	Active B cell secreting antibodies.
Plasma membrane	The outer barrier of cells, which separates the cytoplasm from the extracellular fluid.
Polycyclic aromatic hydrocarbons	Hydrocarbons composed of three or more fused aromatic rings formed as a by-product during the combustion of organic material – also called PAHs.
Polymer	A big molecule that consists of many single units (monomers) that have been linked together.
Polymerase	An enzyme that assembles identical or similar subunits into a large molecule or a polymer, for example, DNA.
Polypeptide	A chain of at least 20 amino acids linked by peptide bonds.
ppb	Parts per billion, for example, $1 \mu l$ per m^3 or $1 mg$ per tonne.
ppm	Parts per million, for example $1 mL$ per m^3 or $1 g$ per tonne.
Pre-electrophilic	An organic compound that is converted to a significantly electrophilic compound by the mammalian metabolism or by some other biological or chemical transformation.
Prokaryote	A cell, or an organism with cells, that lacks a nucleus.
Promoter gene	A nucleotide sequence adjacent to the operator gene at a structural gene, to which

	the decoding enzymes bind and initiate the transcription if the operator is not blocked by a repressor.
Protease	An enzyme that breaks the peptide bonds between amino acids in peptides.
Protein	A macromolecule consisting of one or several polypeptide chains.
Protein kinase	A class of enzymes that catalyze the addition of a phosphate group to certain amino acids of proteins.
Radical	Any chemical with one or several unpaired electrons.
Reactive oxygen species	Collective term used for the reactive species formed when molecular oxygen is reduced: superoxide, hydrogen peroxide, and the hydroxyl radical.
Receptors	Proteins normally situated in the plasma membranes to which ligands bind and exert their biological effect.
Recessive	A gene, or its corresponding trait, is recessive if it is only manifest in organisms that have identical genes for the same trait (a homozygote). A mutation is called recessive if the potential effects are not directly observable.
Repressor	A protein that binds to an operator gene and prevents the transcription of the associated genes by inhibiting the decoding enzymes from binding to the promotor gene.
Resonance stabilization	The lowering of the energy and consequent stabilization of a molecule or an ion due to resonance.
Resonance structure	An electronically alternative but still reasonable structure that can be assumed to make a significant contribution to the resonance hybrid.
Reversible	Proceeding in either direction, permitting the establishment of an equilibrium.
Ribosome	An organelle situated in the endoplasmic reticulum consisting of RNA and proteins, whose function is to synthesize proteins with the amino acid sequence specified by a gene.
RNA	Ribonucleic acid, a polynucleotides in which the sugar is ribose and the bases adenine,

cytosine, guanine or uracil. The transmitter of information from DNA to the proteins.

Rodenticide — Toxic to rodents.

Scavengers — Antioxidants such as carotene, vitamin C, and vitamin E, that react with and inactivate radicals.

Solute — The component in a solution that is dissolved by the solvent.

Somatic cells — All body cells except the germ cells and the gametes.

Sorbate — The chemical that is being sorbed during adsorption and absorption.

Sorbent — The organic or inorganic surface (two dimensions) or material (three dimensions) that sorbs a sorbate.

Sorption — The adsorption of a chemical onto a surface or the absorption of it into a material.

Sperm — The male gamete.

Structural gene — A gene that codes for a polypeptide.

Sulfate group — Present in organic sulfates – esters of sulfuric acid containing a completely oxidized sulfur atom, $-O-S(=O)_2-O-$.

Sulfide group — As an ether group but with S instead of O, $-S-$.

Sulfonate group — Present in sulfonic acids, in which the sulfur atom is oxidized so that it binds two oxygen atoms with double bonds and one with a single bond, $-S(=O)_2-O-$.

Sulfone group — A sulfide group in which the sulfur atom is oxidized so that it binds two oxygen atoms with double bonds, $-S(=O)_2-$.

Sulfoxide group — A sulfide group in which the sulfur atom is oxidized so that it binds one oxygen atom with a double bond, $-S(=O)-$.

Suppressor T cell — A type of T cell that regulates B cells so that the production of antibodies is reduced.

Surface tension — The effect whereby molecules at the surface of a liquid have higher energy than those in the bulk, so that, as all systems try to decrease their energy as much as possible, a liquid will minimize its surface. This gives rise to the surface tension, which, for example, permits some insects to walk on water.

Synapse — A space into which a chemical neurotransmitter is released as a response to an action

	potential in a nerve cell and diffuses through in order to influence the activity of an adjacent nerve cell.
T cells	Lymphocytes capable of becoming the effectors of the cell-mediated response of the immune system.
Tautomers	Isomers which differ in the location of a hydrogen atom and a double bond relative to a heteroatom, in equilibrium with each other.
Teratogenicity	The potency of something, for example, a chemical, to increase the number of teratogenic effects or to make them appear at an earlier stage than would be the case with spontaneous effects.
Thiol group	As a hydroxyl group but with S instead of O, –SH.
Threshold level	The minimal dose of a toxic chemical that will cause a harmful effect.
Trachea	The airway that connects the larynx with the brochi.
Transcription	The transfer of genetic information from DNA (a gene) to a RNA molecule, by enzymes called transcriptases.
Transition state	An unstable and transient species formed during the course of a chemical reaction, it has the highest energy and cannot be isolated.
Translocation	A chromosome aberration that is character-ized by the change in position of a chromo-some segment.
Triglyceride	An ester of glycerol (1,2,3-propanetriol) and three fatty acids.
Tubular reabsorption	The transfer of compounds from the kidney tubule in the nephron to the blood in the capillaries surrounding it.
Tubular secretion	The transfer of compounds from the blood in the capillaries surrounding the kidney tubuli to the primary urine.
Umbilical cord	The cord between the placenta and the fetus, containing the umbilical arteries and vein.
Unshared electron pair	Valence electrons that are not involved in covalent bonds but that may participate in the reactions of the chemical, also called nonbonding electrons.

Valence electrons	Electrons in the outermost (valence) electron shell of an atom, which may participate in bonds to other atoms.
van der Waals forces	Weak intermolecular attractive forces caused by the induction of weak dipoles in nonpolar bonds by neighboring molecules.
Vein	A larger, thin-walled vessel that carries blood back to the heart.
Vesicle	A small intracellular body surrounded by a membrane.
Villi	Projections from the surface of the small intestine that increase its overall surface and are covered by a single layer of epithelial cells.
Vitamin	Essential compound that must be present in small amounts to maintain normal health and growth.
Wavelength	The distance between two wave peaks in electromagnetic radiation, directly correlated to the energy of the radiation (radiation with shorter wavelength contains more energy and is more hazardous).
Zwitterion	An electrically neutral molecule that has both a positive and a negative charge in different parts, for example, an amino acid.

Index

Chemistry, Health, and Environment. Olov Sterner
© 2010 WILEY-VCH Verlag GmbH & Co. KGaA, Weinheim
ISBN: 978-3-527-32582-5